Proceedings of the First Interdisciplinary CHESS Interactions Conference

Proceedings of the First Interdisciplinary CHESS Interactions Conference

Saskatoon, Saskatchewan, Canada   17 – 20 August 2009

editors

Chary Rangacharyulu
University of Saskatchewan, Canada

Emmanuel Haven
University of Leicester, UK

World Scientific

NEW JERSEY • LONDON • SINGAPORE • BEIJING • SHANGHAI • HONG KONG • TAIPEI • CHENNAI

Published by

World Scientific Publishing Co. Pte. Ltd.
5 Toh Tuck Link, Singapore 596224
USA office: 27 Warren Street, Suite 401-402, Hackensack, NJ 07601
UK office: 57 Shelton Street, Covent Garden, London WC2H 9HE

British Library Cataloguing-in-Publication Data
A catalogue record for this book is available from the British Library.

PROCEEDINGS OF THE FIRST INTERDISCIPLINARY CHESS INTERACTIONS CONFERENCE

ISBN-13 978-981-4295-88-8
ISBN-10 981-4295-88-4

Printed in Singapore by B & Jo Enterprise Pte Ltd

FOREWORD

As the pursuits of the sciences, arts, literature or engineering or any conceivable discipline, are all anthropocentric endeavors which are solely depending on human logic, creativity and intuition, it is unlikely that the methodologies of them all are totally disconnected. Even amongst two disciplines, which are seemingly quite farther apart and completely orthogonal, there may be some common threads in modes of thought, lines of discourse, methods of analysis and interpretation of data.

Thus, a conference exploring the methodologies of diverse disciplines which brings together active inter-disciplinarians from various fields of study may serve the purpose of finding commonalities, identifying distinct features and initiating discussions which can, with some concerted efforts, foster inter-disciplinary dialog among practicing researchers and young minds. At such conference, ideally, people will emphasize how they do things, how they reason, how they interpret (but less so on what they do). It is expected that these dialogs would, while sharing this information, lead to cross-fertilization and even to some collaborative research ventures. This was the aim of the first international interdisciplinary conference on Computer science, Humanities, Engineering/ Education/Economics and ScienceS (natural and social sciences) interactions, also known as the First Interdisciplinary CHESS Interactions Conference.

The seeds for this conference were sown in the Summer of 2006, as we got together for the Third Feynman Festival conference at the University of Maryland, USA. We picked Saskatoon in August 2009 as the venue for the first gathering. From the very early planning stages of the conference, we were well aware, that a conference of this type can easily get out of hand, get diluted and become a forgotten episode. In order to avoid such cataclysm, we assigned ourselves a few tasks. We set ourselves as a first task to invite serious researchers whose work influenced more than just their main discipline. Secondly, we reminded our speakers, regardless of their international status, that the two main goals of this conference were i) to share their methodologies with people from diverse areas of research and ii) to draw parallels between their field of study and other disciplines as much as possible. Our third task consisted in imploring them to submit papers for inclusion in the proceedings.

We believe that this proceedings volume shows that we were quite successful in attracting serious scholars of international renown from a diversity of disciplines.

Given that this is a first attempt at organizing such a unique conference, we were not disappointed to observe that some contributors decided not to leave their comfort zones. As we have indicated above, we actively asked contributors to explore the common and distinct features of their field(s) of study with those of other disciplines. We also reminded them to carefully assess constraints and be aware of misconceptions which may be generated when one borrows concepts and techniques from other fields.

Almost all contributions in this proceedings volume fall into either one or more categories of generation, practice and communication of knowledge across disciplines. Many authors spoke about applications of a set of techniques to more than one area of study.

Not all presentations and discussions of the conference are included in this proceedings volume. The reader will find an extensive discussion of statistical physics approaches as they apply to cognitive psychology, economics and market analysis and phase transitions as they pertain to religion and languages (Tsallis, Stanley, Jackson, Switzer and Ausloos). Knowledge acquisition and epistemology are the subjects of more than one contribution (Juurlink, Gupta, Pollack and Roychoudhury). Methods of quantum mechanics and its applications were the subject of the papers by Goyal, Busemeyer and Khrennikov. Some of them ventured into pointing out the pitfalls and mishaps that might befall if one imports metaphors of one discipline to another one oblivious to the subtle distinctions amongst them (Jackson, Porter, Rangacharyulu and Haven). Csapó elucidated symmetry and asymmetry as underlying currents in music composition, while Augsburg focused on integrating different modes of inquiry for school teachers.

We have many people and organizations to thank for the success of this conference. The Honorable Rob Norris, Minister of Advanced Education, Government of Saskatchewan (Canada) was a supporter and provided funds. We have also received financial support from the Universities of Saskatchewan and Leicester (UK). Janusz Kozinski, Dean of Engineering at the University of Saskatchewan, not only sponsored this conference, but he also lent his staff to actively help us in organizing the event. Brij Verma of the Division of Science gave a helping hand in organizing the local events. As always, Marj Granrude and Angelika Ortlepp of the Physics and Engineering Physics department provided unstinting help. M.P.M. Nair donated his time, energy and technical prowess in videotaping the full conference. Karen Tanino and Anne-Marie Cey were indispensable in pulling the conference together and getting the word out. We also thank Peter Jackson (Dean of Social Science at the University of Leicester) and Alan Bryman (Head of Department of the School of Management at the University of Leicester) for their ongoing support.

Thanks also to Stephen Garrett (from the Mathematics Department at the University of Leicester) for suggesting Paul Towers (also in the Mathematics Department) as 'the' gentleman who would take care of all our formatting problems. Stephen was indeed correct!

We must also thank Philippe Haven for having supplied one of his drawings to grace the cover of this proceedings volume.

Preparations for a sequel to this conference, which will be held at the University of Leicester in the Summer of 2011, are underway. We look forward to a continuation of engaging intellectual discourses of diverse disciplines enriching each other.

Chary Rangacharyulu (Dept. of Physics and Engineering Physics; University of Saskatchewan (Canada))

Emmanuel Haven (School of Management; University of Leicester (UK))

Congress Participants

Conference Programme

Monday August 17 - 2009

1:00 – 4:00 Tutorial/Workshop by Dr. M. Gupta
Room 103 Physics Building
Fuzzy Neural Computing Systems: Theory and Applications
4:00 – 5:00 Tour of Canadian Light Source
Meet outside Room 103 Physics Building
5:00 – 7:00 Wine & Cheese Reception
Registration – *Geology Atrium*

Tuesday August 18 – 2009

8:00 – 8:30 Opening Ceremony
8:30 – 9:30 Prof. E. Stanley
Boston University, USA
Liquid Water, the "Most Complex" Fluid
9:30 – 10:15 Prof. L. Switzer
Concordia University, Canada
Market Efficiency and the Risks and Returns of Dynamic Trading Strategies with Commodity Futures
10:15 – 10:30 COFFEE BREAK
10:30 – 11:15 Prof. M. Gupta
University of Saskatchewan, Canada
On the Morphology of Uncertainty in Human Perception and Cognition
11:20 Transportation to Dinner
Bus departs from U of S for the Radisson Hotel
11:45 – 1:15 LUNCH – *Radisson Hotel*
Guest Speaker: Dr. G. Küchler
Lufthansa Systems, Germany
One Day in the Airline Industry or What it Takes to Run a Highly Chaotic System at the Highest Standard of Safety and Security
2:15 – 3:00 Prof. B. De Baets
University of Gent, Belgium
Rock-Paper-Scissors Revisited
3:00 – 3:45 Prof. C. Rangacharyulu
University of Saskatchewan, Canada
Structures and Symmetries in Physics
3:45 – 4:00 COFFEE BREAK
4:00 – 4:45 Prof. C. Roychoudhuri
University of Connecticut, USA
Consilience: Unity of Knowledge by Synthesizing an Evolving Epistemology
4:45 – 5:30 Prof. T. Augsburg
San Francisco State University and Stanford University, USA
Integrating Different Modes of Inquiry for Pre-service Teachers

Wednesday August 19 – 2009

8:30 – 9:30 Prof. C. Tsallis
CBPF, Brazil & Santa Fe Inst, USA
Entropy, a Unifying Concept
9:30 - 10:15 Dr. T. Porter
University of Saskatchewan, Canada
Impossibility of a Theory of Everything
10:15 – 10:30 COFFEE BREAK
10:30 – 11:30 Prof. G. Pollack
University of Washington, USA
Water, Energy, and Life: Fresh Views from the Water's Edge
11:30 – 12:00 Prof. J. Vassileva
University of Saskatchewan, Canada
Social Computing – A New Interdisciplinary Study
12:00 – 1:00 LUNCH – *Marquis Hall*
1:00 – 1:45 Prof. B. Juurlink
Alfaisal University, Saudi Arabia
Fumbling Towards Ecstasy
1:45 – 2:30 Dr. P. Goyal
Perimeter Institute, Canada
What is Quantum Theory Telling Us About How Nature Works?
2:30 – 3:00 Prof. K. Tanino
University of Saskatchewan, Canada
CHESS Interactions – Common Research Themes from a Plant Agriculturist's Perspective
3:00 – 4:30 Group Photo of Conference Participants
Graduate Student Poster Session and Coffee (*Foyer, Arts 241*)
5:30 Transportation to Banquet
Bus departs from U of S for ***Dakota Dunes***. There will be a half hour stop at the ***Radisson Hotel*** before the banquet.
7:00 – 9:30 BANQUET – *Dakota Dunes*
Guest Speaker/Performer: Prof. Gyula Csapó
University of Saskatchewan, Canada

Thursday August 20 – 2009

8:30 – 9:30 Prof. J. Busemeyer
Indiana University, USA
Comparison of Quantum and Bayesian Inference Models
9:30 – 10:15 Prof. G. Rodgers
Brunel University, United Kingdom
Statistical Physics and Healthcare Systems
10:15 – 10:30 COFFEE BREAK
10:30 – 11:15 Prof. E. Haven
University of Leicester, United Kingdom
Title of Talk TBA
11:15 – 12:00 Prof. M Ausloos
University of Liege, Belgium
On Religion and Language Evolutions Seen Through Agent Based Models

12:00 – 12:30 Conclusion
12:30 – 1:30 LUNCH – *Marquis Hall*
Guest Speaker: Prof. Richard Florizone
University of Saskatchewan, Canada
Title of Talk TBA
2:00 – 6:00 Optional Tours of Batoche or the Western Development Museum
Advanced sign-up at the registration table

CONTENTS

SWITCHING PHENOMENA

H. E. STANLEY,[1,*] S. V. BULDYREV,[1,2] G. FRANZESE,[3] S. HAVLIN,[4]
F. MALLAMACE,[5] M. G. MAZZA,[1] P. KUMAR,[6] V. PLEROU,[1] T. PREIS,[7]
K. STOKELY,[1] and L. XU[1]

[1] *Center for Polymer Studies and Departments of Physics,*
Chemistry, Biomedical Engineering, and Physiology
Boston University, Boston, MA 02215 USA
[*] *E-mail: hes@bu.edu*
[2] *Department of Physics, Yeshiva University*
500 West 185th Street, New York, NY 10033 USA
[3] *Departament de Física Fondamental, Universidad de Barcelona*
Diagonal 647,08028 Barcelona, Spain
[4] *Minerva Center and Department of Physics,*
Bar-Ilan University, 52900 Ramat-Gan, Israel
[5] *Dipartimento di Fisica, Univ. Messina*
Vill. S. Agata, C.P. 55, I-98166 Messina, ITALY
[6] *Center for Studies in Physics and Biology, Rockefeller University,*
New York, NY 10021 USA
[7] *Institute of Physics, Johannes Gutenberg University Mainz,*
Staudingerweg 7, D-55128 Mainz, Germany

One challenge of biology, medicine, and economics is that the systems treated by these serious scientific disciplines can suddenly "switch" from one behavior to another, even though they possess no perfect metronome in time. As if by magic, out of nothing but *randomness* one finds remarkably fine-tuned processes in time. The past century has, philosophically, been concerned with placing aside the human tendency to see the universe as a fine-tuned machine. Here we will address the challenge of uncovering how, through randomness (albeit, as we shall see, strongly correlated randomness), one can arrive at some of the many temporal patterns in physics, economics, and medicine and even begin to characterize the switching phenomena that enable a system to pass from one state to another. We discuss some applications of correlated randomness to understanding switching phenomena in various fields. Specifically, we present evidence from experiments and from computer simulations supporting the hypothesis that water's anomalies are related to a switching point (which is not unlike the "tipping point" immortalized by Malcolm Gladwell), and that the bubbles in economic phenomena that occur on all scales are not "outliers" (another Gladwell immortalization).

1. Switching Phenomena: Introduction

The title I have given to this talk I owe in part to interactions with biological and medical researchers. Many of them believe that *randomness* means *uncorrelated randomness.* They learn that statistical physics deals with random phenomena, so they assume that our field cannot possibly yield any insights into the real world as they correctly know that no system in which they are interested corresponds to simple *uncorrelated randomness.* Hence, we found using the adjective "correlated" helped persuade our collaborators that what we do may possibly be applicable to systems in which they are interested.

Indeed, 100 years ago most of our understanding of systems was based on the assumption that the constituents were basically uncorrelated, and so we understood well ideal gases, and ideal paramagnets. When correlations were introduced, gases developed a completely new phenomenon, the liquid state. And paramagnets developed a completely new magnetic phenomenon, the ferromagnetic state.

In this talk, I will discuss recent applications of correlated randomness to three areas of science for which statistical physics is becoming useful: liquid water, economics, and physiology/medicine.

2. Switching Phenomena in Water

It is widely acknowledged that if we do not understand water we cannot understand biology. Scientifically, water is the prototype complex fluid. It is not a simple, "bag-of-marbles" liquid, but a "bag of tetrahedra"–charged tetrahedra interacting with long-range Coulomb forces.

Our approach is based on the fact that water has a tetrahedral local geometry. In this sense water shares features with other liquids such as silicon, silica, and carbon. Because water is both tetrahedral and charged it means that a simple Lennard-Jones potential is not sufficient to describe its complexity. One way to modify the Lennard-Jones potential to provide at least a simplified description is to bifurcate the single minimum into two minima. The first minimum, at a closer distance, corresponds to two pentamers (a water molecule and its four neighbors) of water interacting with each other in a rotated configuration. The second minimum, at a greater distance, occurs in the unrotated position (Fig. 1). This second position is a deeper minimum because although the pentamers are farther apart there is the potential for hydrogen bonding between the molecules and we can see the beginnings of an ice-like hexagonal structure.[1–3]

The important point is that there are two minima with the outer one corresponding to a larger specific volume—because the distance is larger—and a lower entropy. The possibility is that liquid water could at low temperatures condense not into a single phase—as we anticipate when a gas with a simple interaction like a Lennard-Jones potential condenses into a fluid—but into two different phases. This possibility was first raised by Takahashi 60 years ago and various elaborations of this model have been made by a number of people.[4–6] Hence at low temperatures there will be enhanced entropy and volume fluctuations, and enhanced anticorrelations between entropy and volume (since the deeper outer well corresponds to lower entropy but higher specific volume). The consequences of this fact qualitatively explain the phenomenon we were describing at the beginning—volume fluctuations are increased, entropy fluctuations are increased, and cross-fluctuations of volume and entropy are decreased. This singularity-free picture has been actively studied for the past 31 years.[7–11] Additional scenarios are discussed in[12] (see Fig. 2).

It has been speculated that at low temperatures there will be a genuine phase transition in which the single component liquid actually separates into two different phases. The implications of this when applied to real water molecules produce a phase diagram of liquid water (Fig. 3). This liquid-liquid phase transition hypothesis was first proposed on the basis of computer simulations.[13] A first test of this hypothesis is that at very low temperature we find a glassy phase, not unlike that of any other liquid except that at high pressure this glassy phase suddenly switches from a low-density form to a high-density form. These two forms correspond to the two different local arrangements characteristic of water tetrahedra. The order parameter jump between these two phases is not a trivial amount, but on the order of 30 percent.[14]

Between the liquid and glassy phases of water we have a region in which water does not exist as a liquid—a "No Man's Land." The hypothesis that follows from the reasoning we have just described is that this first order phase transition line known to separate the two amorphous forms of solid water extends into this No Man's Land and ultimately terminates at a critical point. Just as the glassy water first-order transition line separates a low-density amorphous phase from a high-density amorphous phase of water, so also this extension of the line into the liquid region separates a low-density liquid from a high-density liquid. The power-law behavior uncovered over the years by Angell, Anisimov and collaborators corresponds to the fact that the extension of this first-order line beyond the critical point—the

"Widom line," defined to be the locus of maximum correlation length.[15–18] Experiments on a path approaching the Widom line display phenomena that initially look as though there will be a divergence, as the correlations increase. However since the correlation length itself is not infinite, there will ultimately be rounding.

The degree of ordering and disordering of local tetrahedrality of water upon changing thermodynamic variables, such as pressure or temperature, offers a simple measure of order and disorder in the case of liquid water. Kumar and his collaborators derive a relation for the subset of the structural entropy which is associated with the degree of local tetrahedral ordering.[16] They find that the most relevant part of the total entropy fluctuations (or the specific heat) of water comes from the tetrahedral fluctuations – suggesting that the fluctuations of local tetrahedrality of water contributes the most to the total specific heat.[16] Moreover, the specific heat associated with the tetrahedral ordering C_P^Q behaves identically, within error bars, to the total specific heat C_P^{total} (Fig. 4), consistent with the qualitative idea that the principal contributor to temperature dependence of the entropy are the angular variables, not the translational variables. In particular, both functions display clear maxima at the Widom line, which is equivalent to saying that at the Widom line the entropy fluctuations are a maximum (and the dependence of entropy on temperature is a maximum).

This phase diagram is hypothesized, but it has not been proved. What has been proved is that computer simulations using tried and tested models of liquid water confirm the broad features of this phase diagram (see, e.g.,[19–22] and refs. therein). But computer models of water (like computer models of anything) are subject to the charge "garbage-in, garbage-out"—you get out what you put in. All computer models of complex systems such as liquid water are of necessity simplifications.

Here we are guided by exact solutions of simplified models. Simplified models are designed less for the purpose of matching experimental detail, but rather to capture the essential physical features of a real system to explore the effect of these features. A family of models introduced first by Jagla seems to reveal the fact that in order for a system to display a liquid-liquid phase transition, the essential feature of an interaction potential must be that it has not one but rather two characteristic length scales (see Ref.[23] and references therein).

Current experiments on this problem are of two sorts. The first is a set of experiments inspired by Mishima that involves probing the No Man's Land by studying the metastable extensions of the melting lines of the various high-pressure polymorphs of ice: ice III, ice V, ice IV, and ice XII.[24,25] Two of these lines clearly display "kinks." Since the slope of any melting line is the difference of the volume change divided by the entropy change of the two phases that coexist at that line, if there is a change in slope there must be a change in these quantities. Since there is no change in the crystal part, there must be a change in the liquid part. This means the liquid must undergo a jump in either its volume or its entropy or both. That is the definition of a first-order phase transition.

The second sort of experiments avoid the existence of a no-man's land by using water confined either in nanotubes or near the surface of a macromolecule. These experiments, carried out largely in the research groups of S.-H. Chen and F. Mallamace, are consistent with the possibility of a liquid-liquid phase transition.[26–41] Other novel collective behavior of water is under active investigation, e.g., S. Han and collaborators have found evidence for a solid-liquid critical point in confined water films.[42]

3. Switching Phenomena in Economics

One quarter of any newspaper with a financial section is filled with economic fluctuation data. Most economic graphs look approximately like the one we get when we plot the S&P 500 stock index as a function of time over 40 years (Fig. 5). We can compare this empirical data with that generated by a simple uncorrelated biased random walk, a model first used over 100 years ago by Bachelier. At first it seems that there is little difference, but looking more closely we see events in the real data that do not have counterparts in the random walk. Black Monday in October 1987 is reflected in the real data, which shows a loss of about 30 percent of the total value of the market in just one day. In the random walk we do not see fluctuations anywhere near this magnitude because the probability of taking n steps in the same direction of a random walk is $(1/2)^n$—it decreases exponentially with n.

Economists nevertheless have traditionally used this uncorrelated biased Gaussian random walk to describe real economic data, relegating events such as Black Monday to the dustbin category of "outliers." We physicists do not like to do things this way. We do not take Newton's law seriously part of the time, and then—if we suddenly see an example of what appears to be levitation—simply call it an "outlier." We like to find laws that describe all examples of a phenomenon. This is a strong motivation for we physicists to step in and try our hand—we sense a delicious scientific challenge.

We return to our two graphs, the S&P 500 stock index as a function of time over 40 years and the simple uncorrelated biased random walk, and plot not the absolute value of the index but instead the *change* in the index (the numerical derivative, the "return"). We normalize that by the standard deviation. We look in Fig. 6 over a 13-year period rather than our original 40-year period of Fig. 5 and see, e.g., that on Black Monday the fluctuations were more than 30 standard deviations (both positive and negative) for the day. We also note obvious time correlations in the volatility. Perhaps more striking is to look at the other curve, the uncorrelated random walk, and see the Gaussian distribution for the fluctuations—with no fluctuations greater than five standard deviations. The "outliers" that the economists are content to live with are any fluctuations of the actual data that are greater than five standard deviations. In this 13-year period there are exactly 64, i.e., 2^6. If we count only those fluctuations of the actual data that are greater than ten standard deviations, we get exactly 8, i.e., 2^3. If we count only those that are greater than 20, we get one, i.e., 2^0: Black Monday. Each time we double the x-axis we change the y-axis by a power of 2^3. This behavior is consistent with a functional equation of the form[43]

$$f(\lambda x) \sim \lambda^p f(x) \tag{1}$$

Functional equations have as their solutions functional forms, and the solution to the homogeneous functional equation is a power-law form. From visual inspection of the raw data in Fig. 6, we see a power law of the "inverse cubic" form

$$f(x) = x^{-3} \tag{2}$$

which corresponds to a functional equation, a scaling equation, with $p = -3$.

If we replace our visual examination of these two graphs with a close computer analysis of not just the S&P 500 stock index but every stock transaction over an extended time period (approximately one GB of data), we find[44,45] that the actual graph giving the number of times a fluctuation exceeds a given amount as a function of that amount is perfectly straight on log-log paper *out to 100 standard deviations* (Fig. 7). The slope of the line, α, is indistinguishable from the value $\alpha = 3$ that we guessed from visual inspection. Note also that our slope is outside the Lévy stable regime.[46]

This is how we find laws in statistical physics, but finding new empirical laws in economics is useful. It is useful for practical reasons, one can quantify risk—produce an actual number for the chance some fluctuating quantity changes by a given amount. It is also useful for conceptual reasons, since knowing the empirical laws places constraints on the range of acceptable theoretical idea.

When critical phenomena was studied 40 years ago, the empirical part was a very important contributor toward our ultimate understanding of phase transitions and critical phenomena. Uncovering empirical facts led to the recognition of regularities to which certain approaches could be applied, e.g., the Widom scaling hypothesis and the Wilson renormalization group.[43] So also in economics we can perhaps first discover empirical regularities—e.g., the inverse cubic law—that will prove useful in ultimately understanding the economy. We have the beginnings of an explanation, but it is only the beginning since the current theory explains the inverse cubic law of price changes, as well as the "half cubic law" of trade volume[47] but does not explain the strange nature of the temporal correlations. The autocorrelation function of price changes decays exponentially in time so rapidly that after 20 min it is in the level of "noise." However the autocorrelation function of changes in the *absolute value* of the price (called the "volatility") decays with a power law of exponent approximately 0.3.[48]

However, how can we characterize trend switching processes in financial markets? At first glance, price fluctuations on financial markets appear to be completely random. In order to uncover the patterns formed by correlated randomness, one has to find an appropriate "metronome" that is able to "synchronize" these sudden trend reversals occurring on time scales from very large down to very small. One leading candidate as a metronome for financial markets is the local extreme value in the price time series.

Such an analysis can provide insight into switching processes in complex systems in general and financial systems in particular. The study of dramatic crash events is limited by the fortunately rare number of such events. Increasingly, one seeks to understand the current financial crisis by comparisons with the depression of the 1930's. Here we ask if the smaller financial crises—trend switching processes on all time scales—also provide information of relevance for large crises. If this is so, then the large abundance of data on smaller crises should provide quantifiable statistical laws for *bubbles on all scales*.

To analyze switching processes of financial fluctuations, we first propose how a switching process can be quantitatively analyzed. Let $p(t)$ be the transaction price of trade t, which is a discrete variable $t = 1, \ldots, T$. A transaction price $p(t)$ is defined to be a *local maximum* of order Δt if there is no higher transaction price in the interval $t - \Delta t \leq t \leq t + \Delta t$, and is defined to be a *local minimum* of order Δt if there is no lower transaction price in this interval (Fig. 8).[49]

However, the time series of transaction prices provides only the metronome, with which we analyze volume fluctuations $v(t)$ from one beat of the metronome to the next. We use 500 price time series of all 500 S&P500 constituents (2.6 million data points) and a price time series of the German DAX Future (14 million data points).[50] The volume is the number of contracts traded in each individual transaction in case of the German market and the number of traded stocks per day in case of the US market. For the analysis, we introduce a renormalized time scale ε between successive extrema—successive beats of the metronome. Thus, $\varepsilon = 0$ corresponds to the beginning of a trend and $\varepsilon = 1$ indicates the end of a trend (Fig. 8a). We analyze a range of ε for the interval $0 \leq \varepsilon \leq 2$, so we can analyze trend switching processes both before as well as after the critical value $\varepsilon = 1$. We find that new price extrema in the price are linked with peaks in the volume time series (Figs. 8b and 8c) and, more surprisingly, that the volume obeys power law scaling both before as well as after the critical point $\varepsilon = 1$ (Figs. 8d and 8e).

This switching mechanism has no scale, for time scales varying over 9 orders of magnitude down to the smallest possible time scale—the scale of single transactions measured in units of 10 ms. Thus, the well-known catastrophic bubbles occurring on large time scales may not be outliers but in fact single dramatic events caused by the inherent, scale-free behavior related to the formation of increasing and decreasing trends on time scales from very large down to very small.

One reason the economy is of interest to statistical physicists is that, like an Ising model, it is a system made up of many subunits. The subunits in an Ising model are called spins, and the subunits in the economy are buyers and sellers. During any unit of time these subunits of the economy may be either positive or negative as regards perceived market opportunities. People interact with each other, and this fact often produces what economists call "the herd effect." The orientation of whether we buy or sell is influenced not only by our neighbors but also by news. If we hear bad news, we may be tempted to sell. So the state of any subunit is a function of the states of all the other subunits and of a field parameter.

4. Switching Phenomena in Physiology & Medicine

The most magical switching phenomena come to light when one examines biology and medicine. A developing embryo suddenly switches from one state to another, almost as if there were a master hand guiding it. A healthy individual suddenly develops a systemic and fatal disease, almost as if "fate" has decreed that the time is up for living. Or genetic diseases like cancer–

the gene is present yet the patient remains healthy until—suddenly—tumor growth switches on. And, tragically, molecules suddenly begin to aggregate in the brains when one contracts Alzheimer disease. Why? On the microscopic level, chain molecules switch from an unfolded state to a reproducible folded state for reasons not at all understood. Even everyday phenomena involve switching—from wake to sleep, and then from sleep to wake. Actually there are countless sleep-related switchings occurring "while sleeping"—from one sleep stage to another, and then to a brief wake state, followed by more sleep.[51]

Not only switchings occur, but fractal forms abound. We now know, e.g., that nerve cells adopt patterns with well-defined fractal dimensions[52] but we do not know why. We know that the noncoding DNA has long-range power law (fractal) correlations among its constituent base pairs.[53,54] We are only beginning to understand the relation between the various fractal forms that abound and the various forms of switchings. We do know that the two phenomena appear sometimes to be related; e.g., we now know that sleep-wake switching interacts with the fractal scaling behavior of heartbeat intervals.[55]

Diseases appear to fall into different "universality classes." For example, Alzheimer's disease is only one form of aggregation disease. Others may include ALS (amyotrophic lateral sclerosis) and Parkinson's disease, and many think they all involve amyloid mediated aggregation.

Since we do not really know for sure what AD is, our group's approach is to try to discover what it is.[56–60] What initially triggers AD? One of the current hypotheses is that it is triggered by a phase transition. In a liquid-to-crystal phase transition, the liquid nucleates to a lower free-energy state called a crystal. In AD, it looks as if the analogous event is a two-step process—first of protein folding and then of protein aggregation. So one starts with a protein, specifically a peptide, in a correlated random configuration, which first folds itself and then aggregates into a form called an amyloid fibril. These amyloid fibrils aggregate and form microscopic objects in the brain.

The protein that aggregates in Alzheimer disease is actually a peptide, a fragment of a protein, and it comes in two forms: one with 40 amino acids and the other with 42 amino acids. The extra two amino acids are hydrophobic. The "first three minutes" of Alzheimer's disease involves the aggregation of the entire peptide. The first step seems to be the formation of what are called paranuclei, the joining together of a small number—e.g., 6 or 12—peptides, and then the aggregation of the paranuclei into larger objects. The time scale here is slow, so the study of this phenomenon by molecular dynamics is problematic. A typical timescale for a molecular dynamics simulation is on the order of nanoseconds, and here we are talking about minutes. We need to do something to speed up the simulation, and thus we draw on the concept of universality.

One thing we learn from universality is that completely different fluids behave the same way near phase transitions. They even have identical critical exponents, regardless of the details. The critical factor does not seem to reside in the details of the molecule but in the fact that they have an attractive part. Similarly, in the spirit of universality, one finds the same generic collective behavior when one coarse grains the actual peptide by replacing each amino acid group by only four balls: three in the peptide backbone and one representing the side group.

A typical simulation might start with 28 of these coarse-grained peptides at time zero.[61] Then using the algorithm[62,63] that speeds up the simulation by ten orders of magnitude, one can achieve aggregation of these peptides in a reasonable amount of computer time (on the order of days of computer time). The structure of what we found is of considerable interest to those studying Alzheimer's disease. In particular, we find aggregation, with remarkably reproducible microscopic detail concerning exactly what sticks to what. This is terribly important because if we know what sticks to what, we can imagine covering the sticky spots and hence offering some potentially useful hints concerning possible therapies for Alzheimer's disease. That is, if we know what is sticking to what we can begin to think about how to block that sticking process.

5. Outlook

The realization that a host of geometric phenomena share analogous features of scaling, universality, and renormalization was not fully appreciated in 1969.[64] Later still, the interplay between geometric and thermal phenomena revealed new complexity such as higher order critical points.[65] Today we are still struggling to understand the basis of seemingly ubiquitous phenomena, such as the Zipf law, which appear to work not only for word usage but for describing the size distribution of firms.[66] Finally, the *dynamics* of firm growth (and firm shrinkage) is a problem with little understanding.[67–70] Also, the *cross-correlations* between price changes of different stocks can be analyzed using the techniques of random matrix theory.[71]

Price growth (and price collapse) has traditionally been viewed as a complex form of random walk, while now we are beginning to understand this ubiquitous economic phenomenon as a sequence of switches (or "bubbles") of all sizes ranging in time scales over 10 orders of magnitude from msec to days.[72,73] Perhaps most dramatic are the various novel forms of switching phenomena (a catastrophic cascade of failures) that can occur when one damages a single node of one of several coupled networks.[74]

6. Acknowledgements

In addition to those joining as co-authors of this article, there are many other collaborators, a complete list of whom appears on the webpage http://polymer.bu.edu/hes. Our work could not have been performed without the generous support of many agencies, in the USA primarily NSF, NIH, ONR, and DTRA and the Bechtel Foundation.

References

1. M. Canpolat, F. W. Starr, M. R. Sadr-Lahijany, A. Scala, O. Mishima, S. Havlin, and H. E. Stanley, "Local Structural Heterogeneities in Liquid Water under Pressure," *Chem. Phys. Lett.* **294**, 9–12 (1998).
2. F. W. Starr, J. K. Nielsen, and H. E. Stanley, "Fast and Slow Dynamics of Hydrogen Bonds in Liquid Water," *Phys. Rev. Lett.* **82**, 2294–2297 (1999).
3. P. Kumar, G. Franzese, and H. E. Stanley, "Dynamics and Thermodynamics of Water," *J. Phys.: Condens. Matt.* **20**, 244114 (2008).
4. P. C. Hemmer and G. Stell, "Fluids with Several Phase Transitions," *Phys. Rev. Lett.* **24**, 1284–1287 (1970).
5. M. R. Sadr-Lahijany, A. Scala, S. V. Buldyrev and H. E. Stanley, "Liquid State Anomalies for the Stell-Hemmer Core-Softened Potential," *Phys. Rev. Lett.* **81**, 4895–4898 (1998).
6. G. Franzese, G. Malescio, A. Skibinsky, S. V. Buldyrev, and H. E. Stanley, "Generic Mechanism for Generating a Liquid-Liquid Phase Transition," *Nature* **409**, 692–695 (2001).
7. H. E. Stanley, "A Polychromatic Correlated-Site Percolation Problem with Possible Relevance to the Unusual behavior of Supercooled H_2O and D_2O," *J. Phys. A* **12**, L329–L337 (1979).
8. H. E. Stanley and J. Teixeira, "Interpretation of the Unusual Behavior of H_2O and D_2O at Low Temperatures: Tests of a Percolation Model," *J. Chem. Phys.* **73**, 3404–3422 (1980).
9. A. Geiger and H. E. Stanley, "Low-density patches in the hydrogen-bonded network of liquid water: Evidence from molecular dynamics computer simulations," *Phys. Rev. Lett.* **49**, 1749–1752 (1982).
10. S. Sastry, P. Debenedetti, F. Sciortino, and H. E. Stanley, "Singularity-Free Interpretation of the Thermodynamics of Supercooled Water," *Phys. Rev. E* **53**, 6144–6154 (1996).
11. L. Bosio, J. Teixeira, and H. E. Stanley, "Enhanced Density Fluctuations in

Supercooled H_2O, D_2O, and Ethanol-Water Solutions: Evidence from Small-Angle X-Ray Scattering," *Phys. Rev. Lett.* **46**, 597–600 (1981).
12. K. Stokely, M. G. Mazza, H. E. Stanley, and G. Franzese, "Effect of Hydrogen Bond Cooperativity on the Behavior of Water," *Proc. Natl. Acad. Sci. USA* **107**, 1301–1306 (2010).
13. P. H. Poole, F. Sciortino, U. Essmann, and H. E. Stanley, "Phase Behavior of Metastable Water," *Nature* **360**, 324–328 (1992).
14. O. Mishima, "Reversible First-Order Transition between Two H_2O Amorphs at -0.2 GPa and 135 K," *J. Chem. Phys.* **100**, 5910–5912 (1994).
15. L. Xu, P. Kumar, S. V. Buldyrev, S.-H. Chen, P. H. Poole, F. Sciortino, and H. E. Stanley, "Relation between the Widom Line and the Dynamic Crossover in Systems with a Liquid-Liquid Critical Point," *Proc. Natl. Acad. Sci. USA* **102**, 16558–16562 (2005).
16. P. Kumar, S. V. Buldyrev, H. E. Stanley, "A Tetrahedral Entropy for Water," *Proc. Natl. Acad. Sci. USA* **106**, 22130–22134 (2009).
17. G. Franzese, and H. E. Stanley, "The Widom Line of Supercooled Water," *J. Phys.: Condens. Mat.* **19**, 205126 (2007).
18. P. Kumar, G. Franzese, and H. E. Stanley, "Predictions of Dynamic Behavior under Pressure for Two Scenarios to Explain Water Anomalies," *Phys. Rev. Lett.* **100**, 105701 (2008).
19. M. Yamada, S. Mossa, H. E. Stanley, F. Sciortino, "Interplay Between Time-Temperature-Transformation and the Liquid-Liquid Phase Transition in Water," *Phys. Rev. Lett.* **88**, 195701 (2002).
20. S. Harrington, P. H. Poole, F. Sciortino, and H. E. Stanley, "Equation of State of Supercooled SPC/E Water," *J. Chem. Phys.* **107**, 7443–7450 (1997).
21. G. Franzese, M. I. Marqués, and H. E. Stanley, "Intra-Molecular Coupling as a Mechanism for a Liquid-Liquid Phase Transition," *Phys. Rev. E* **67**, 011103 (2003).
22. M. G. Mazza, K. Stokely, E. G. Strekalova, H. E. Stanley, and G. Franzese, "Cluster Monte Carlo and Numerical Mean Field Analysis for the Water Liquid-Liquid Phase Transition," *Comp. Phys. Comm.* **180**, 497–502 (2009).
23. Z. Yan, S. V. Buldyrev, N. Giovambattista, and H. E. Stanley "Structural Order for One-Scale and Two-Scale Potentials" Phys. Rev. Lett. **95**, 130604 (2005).
24. O. Mishima and H. E. Stanley, "Decompression-Induced Melting of Ice IV and the Liquid-Liquid Transition in Water," *Nature* **392**, 164–168 (1998).
25. O. Mishima and H. E. Stanley, "The Relationship between Liquid, Supercooled and Glassy Water," *Nature* **396**, 329–335 (1998).
26. F. Mallamace, "The Liquid Water Polymorphism," *Proc. Natl. Acad. Sci. USA* **106**, 15097–15098 (2009).
27. J.-M. Zanotti, M.-C. Bellissent-Funel, and S.-H. Chen, "Experimental Evidence of a Liquid-Liquid Transition in Interfacial Water," *Europhys. Lett.* **71**, 91–97 (2005).
28. L. Liu, S.-H. Chen, A. Faraone, C.-W. Yen, and C. Y. Mou, "Pressure Dependence of Fragile-to-Strong Transition and a Possible Second Critical Point in Supercooled Confined Water," *Phys. Rev. Lett.* **95**, 117802 (2005).

29. F. Mallamace, C. Branca, M. Broccio, C. Corsaro, C. Y. Mou, and S. H. Chen, "The Anomalous Behavior of the Density of Water in the Range 30K $< T <$ 373K," *Proc. Natl. Acad. Sci. USA* **104**, 18387–18392 (2007).
30. F. Mallamace, C. Corsaro, M. Broccio, C. Branca, N. González-Segredo, J. Spooren, S.-H. Chen, and H. E. Stanley, NMR " Evidence of a Sharp Change in a Measure of Local Order in DEEPLY Supercooled Confined Water," *Proc. Natl. Acad. Sci. USA* **105**, 12725–12729 (2008).
31. C. Corsaro, J. Spooren, C. Branca, N. Leone, M. Broccio, C. Kim. S.-H. Chen, H. E. Stanley, and F. Mallamace, "Clustering Dynamics in Water/Methanol Mixtures: A Nuclear Magnetic Resonance Study at 205K $< T <$ 295K," *J. Phys. Chem. B* **112**, 10449–10454 (2008).
32. S.-H. Chen, F. Mallamace, C. Y. Mou, M. Broccio, C. Corsaro, A. Faraone, and L. Liu, "The Violation of the Stokes-Einstein Relation in Supercooled Water," *Proc. Natl. Acad. Sci. USA* **103**, 12974–12978 (2006).
33. P. Kumar, S. V. Buldyrev, S. L. Becker, P. H. Poole, F. W. Starr, and H. E. Stanley, "Relation between the Widom line and the Breakdown of the Stokes–Einstein Relation in Supercooled Water," *Proc. Natl. Acad. Sci. USA***104**, 9575–9579 (2007).
34. F. Mallamace, M. Broccio, C. Corsaro, A. Faraone, D. Majolino, V. Venuti, L. Liu, C. Y. Mou, and S.-H. Chen, "Evidence of the Existence of the Low-Density Liquid Phase in Supercooled, Confined Water," *Proc. Natl. Acad. Sci. USA* **104**, 424–428 (2007).
35. P. Kumar, Z. Yan, L.Xu, M. Mazza, S. Buldyrev, S.-H. Chen, S. Sastry, and H. E. Stanley, "Glass Transition in Biomolecules and the Liquid-Liquid Critical Point of Water," *Phys. Rev. Lett.* **97**, 177802 (2006).
36. M. Lagi, X. Chu, C. Kim, F. Mallamace, P. Baglioni, and S.-H. Chen, "The Low-Temperature Dynamic Crossover Phenomenon in Protein Hydration Water: Simulations vs Experiments," *J. Phys. Chem. B* **112**, 1571–1575 (2008).
37. L. Xu, F. Mallamace, Z. Yan, F. W. Starr, S. V. Buldyrev, and H. E. Stanley, "Appearance of a Fractional Stokes-Einstein Relation in Water and a Structural Interpretation of Its Onset," *Nature Physics* **5**, 565–569 (2009).
38. F. Mallamace, C. Branca, C. Corsaro, N. Leone, J. Spooren, H. E. Stanley, and S.-H. Chen, "Dynamical Crossover and Breakdown of the Stokes-Einstein Relation in Confined Water and in Methanol-Diluted Bulk Water," *J. Phys. Chem. B* **114**, 1870–1878 (2010).
39. Y. Zhang, A. Faraone, W. A. Kamitakahara, K.-H. Liu, C.-Y. Mou, and S.-H. Chen, "Experimental Evidence of the Existence of a Liquid-Liquid Critical Point in Supercooled Confined Water," *Proc. Natl. Acad. Sci. USA* (to be published).
40. P. Kumar, G. Franzese, S. V. Buldyrev, and H. E. Stanley, "Dynamics of Water at Low Temperatures and Implications for Biomolecules," in *Aspects of Physical Biology: Biological Water, Protein Solutions, Transport and Replication*, Series: Lecture Notes in Physics, Vol. 752, edited by G. Franzese and M. Rubi (Springer Berlin, 2008), p. 3–22.
41. G. Franzese, K. Stokely, X.-Q. Chu, P. Kumar, M. G. Mazza, S.-H. Chen, and

H. E. Stanley, "Pressure Effects in Supercooled Water: Comparison between a 2D Model of Water and Experiments for Surface Water on a Protein," *J. Phys: Condens. Matt.* **20**, 494210 (2008).
42. S. Han, M. Y. Choi, P. Kumar, and H. E. Stanley, "Solid-Liquid Critical Point in Confined Water Films" Nature Physics (in publication stage).
43. H. E. Stanley, "Scaling, Universality, and Renormalization: Three Pillars of Modern Critical Phenomena," *Rev. Mod. Phys.* **71**, S358–S366 (1999).
44. V. Plerou, P. Gopikrishnan, L. A. N. Amaral, M. Meyer, and H. E. Stanley, "Scaling of the Distribution of Price Fluctuations of Individual Companies," *Phys. Rev. E* **60**, 6519–6529 (1999).
45. P. Gopikrishnan, V. Plerou, L. A. N. Amaral, M. Meyer, and H. E. Stanley, "Scaling of the Distributions of Fluctuations of Financial Market Indices," *Phys. Rev. E* **60**, 5305–5316 (1999).
46. K. Matia, Y. Ashkenazy, and H. E. Stanley, "Multifractal Properties of Price Fluctuations of Stocks and Commodities," *Europhys. Letters* **61**, 422–428 (2003).
47. X. Gabaix, P. Gopikrishnan, V. Plerou, and H. E. Stanley, "Institutional Investors and Stock Market Volatility," *Quarterly Journal of Economics* **121**, 461–504 (2006).
48. P. Cizeau, Y. Liu, M. Meyer, C.-K. Peng, and H. E. Stanley, "Volatility Distribution in the S&P500 Stock Index" *Physica A* **245**, 441–445 (1997).
49. T. Preis, J. J. Schneider, and H. E. Stanley, "Switching Processes in Financial Markets," Science (submitted, 2009).
50. T. Preis and H. E. Stanley, "Switching Phenomena in a System with No Switches," J. Stat. Phys. (in press), DOI: 10.1007/s10955-009-9914-y (2010).
51. C.-C. Lo, L. A. N. Amaral, S. Havlin, P. Ch. Ivanov, J.-H. Peter, T. Penzel, and H. E. Stanley, "Dynamics of Sleep-Wake Transitions during Sleep," *Europhys. Lett.* **57**, 625–631 (2002).
52. F. Caserta, H. E. Stanley, W. Eldred, G. Daccord, R. Hausman, and J. Nittmann, "Physical Mechanisms Underlying Neurite Outgrowth: A Quantitative Analysis of Neuronal Shape," *Phys. Rev. Lett.* **64**, 95–98 (1990).
53. S. V. Buldyrev, A. L. Goldberger, S. Havlin, C.-K. Peng, M. Simons, and H. E. Stanley, "Generalized Lévy Walk Model for DNA Nucleotide Sequences," *Phys. Rev. E* **47**, 4514–4523 (1993).
54. S. V. Buldyrev, A. L. Goldberger, S. Havlin, C.-K. Peng, H. E. Stanley, M. H. R. Stanley, and M. Simons, "Fractal Landscapes and Molecular Evolution: Analysis of Myosin Heavy Chain Genes," *Biophys. J.* **65**, 2673–2679 (1993).
55. P. Ch. Ivanov, A. Bunde, L. A. N. Amaral, S. Havlin, J. Fritsch-Yelle, R. M. Baevsky, H. E. Stanley, and A. L. Goldberger, "Sleep-Wake Differences in Scaling Behavior of the Human Heartbeat: Analysis of Terrestrial and Long-Term Space Flight Data," *Europhys. Lett.* **48**, 594–600 (1999).
56. R. B. Knowles, C. Wyart, S. V. Buldyrev, L. Cruz, B. Urbanc, M.E. Hasselmo, S. Havlin, H. E. Stanley and B. T. Hyman, "Plaque-Induced Neural Network Disruption in Alzheimer's Disease," *Proc. Nat. Acad. Sci. USA* **96**, 5274–5279 (1999).
57. B. T. Hyman, H. L. West, G. W. Rebeck, S. V. Buldyrev, R. N. Mantegna, M.

Ukleja, S. Havlin, and H. E. Stanley, "Quantitative analysis of senile plaques in Alzheimer disease: Observation of log-normal size distribution and of differences associated with apolipoprotein E genotype and trisomy 21 (Down syndrome)," *Proc. Nat. Acad. Sci. USA* **92**, 3586–3590 (1995).
58. L. Cruz, B. Kutnjac-Urbanc, S. V. Buldyrev, R. Christie, T. Gómez-Isla, S. Havlin, M. McNamara, H. E. Stanley, and B. T. Hyman, "Aggregation and Disaggregation of Senile Plaques in Alzheimer Disease," *Proc. Nat. Acad. Sci. USA* **94**, 7612–7616 (1997).
59. H. E. Stanley, S. V. Buldyrev, L. Cruz, T. Gomez-Isla, S. Havlin, B.T.Hyman, R. Knowles, B. Urbanc, and C. Wyart, "Statistical physics nd Alzheimer's disease" [Proc. Bar-Ilan Conf], *Physica A* **249**, 460–471 (1998).
60. S. V. Buldyrev, L. Cruz, T. Gomez-Isla, S. Havlin, H. E. Stanley, B. Urbanc and B. T. Hyman "Description of Microcolumnar Ensembles in Association Cortex and their Disruption in Alzheimer and Lewy Body Dementias," *Proc. Nat. Acad. Sci. USA* **97**, 5039–5043 (2000); accompanied on page 5019–5021 by a 3-page invited commentary E. G. Jones, "Microcolumns in the cerebral cortex," which places this paper into a broader context.
61. B. Urbanc, L. Cruz, S. Yun, S. V. Buldyrev G. Bitan, D. B. Teplow, and H. E. Stanley "*In Silico* Study of Amyloid β Protein Folding and Oligomerization," *Proc. Natl. Acad. Sci. USA* **101**, 17345–17350 (2004).
62. N. Dokholyan, S. V. Buldyrev, H. E. Stanley and E. I. Shakhnovich, "Discrete Molecular Dynamics Studies of the Folding of a Protein-Like Model," *Folding & Design* **3**, 577–587 (1998).
63. N. V. Dokholyan, S. V. Buldyrev, H. E. Stanley, and E. I. Shakhnovich, "Identifying the Protein Folding Nucleus Using Molecular Dynamics," *J. Mol. Biol.* **296**, 1183–1188 (2000)
64. A. Coniglio and H. E. Stanley, "Screening of Deeply Invaginated Clusters and the Critical Behavior of the Random Superconducting Network," *Phys. Rev. Lett.* **52**, 1068–1071 (1984).
65. H. E. Stanley, R. J. Birgeneau, P.J. Reynolds, and J. F. Nicoll, "Thermally Driven Phase Transitions near the Percolation Threshold in Two Dimensions," *J. Phys. C: Solid State Phys.* **9**, L553–560 (1976).
66. M. H. R. Stanley, S. V. Buldyrev, S. Havlin, R. Mantegna, M.A. Salinger, and H. E. Stanley, "Zipf Plots and the Size Distribution of Firms," *Economics Lett.* **49**, 453–457 (1995).
67. D. Fu, F. Pammolli, S. V. Buldyrev, M. Riccaboni, K. Matia, K. Yamasaki, and H. E. Stanley, "The Growth of Business Firms: Theoretical Framework and Empirical Evidence," *Proc. Natl. Acad. Sci. USA* **102**, 18801–18806 (2005).
68. M. Riccaboni, F. Pammolli, S. V. Buldyrev, L. Ponta, and H. E. Stanley, "The Size Variance Relationship of Business Firm Growth Rates," *Proc. Natl. Acad. Sci. USA* **105**, 19595–19600 (2008).
69. L. A. N. Amaral, S. V. Buldyrev, S. Havlin, H. Leschhorn, P. Maass, M. A. Salinger, H. E. Stanley, and M. H. R. Stanley, "Scaling Behavior in Economics: I. Empirical Results for Company Growth," *J. Phys. I France* **7**, 621–633 (1997).

70. Y. Lee, L. A. N. Amaral, D. Canning, M. Meyer, and H. E. Stanley, "Universal Features in the Growth Dynamics of Complex Organizations," *Phys. Rev. Lett.* **81**, 3275–3278 (1998).
71. V. Plerou, P. Gopikrishnan, B. Rosenow, L.A.N. Amaral, T. Guhr, and H. E. Stanley, "A Random Matrix approach to Financial Cross-Correlations," *Phys. Rev. E* **66**, 066126 (2002).
72. T. Preis, J. J. Schneider, and H. E. Stanley, "New Laws Describing Trend Switching Processes in Financial Markets," *Proc. APFA7*, Tokyo (2009).
73. T. Preis and H. E. Stanley, "Trend Switching Processes in Financial Markets," *Econophysics Approaches to Large-Scale Business Data and Financial Crisis*, edited by M. Takayasu et al. DOI 10.1007/978-4-431-53853-0 (Springer, Tokyo, 2010).
74. S. V. Buldyrev, R. Parshani, G. Paul, H. E. Stanley, and S. Havlin, "Catastrophic Cascade of Failures in Interdependent Networks," *Nature* (in press).
75. T. Preis, W. Paul, and J. J. Schneider, "Fluctuation Patterns in High-Frequency Financial Asset Returns," *Europhys. Lett.* **82**, 68005 (2008).
76. A. Shleifer, *Inefficient Markets: An Introduction to Behavioral Finance* (Oxford University Press, Oxford, 2000).
77. D. Helbing, I. Farkas, and T. Vicsek, "Simulating Dynamical Features of Escape Panic," *Nature* **407**, 487–490 (2000).

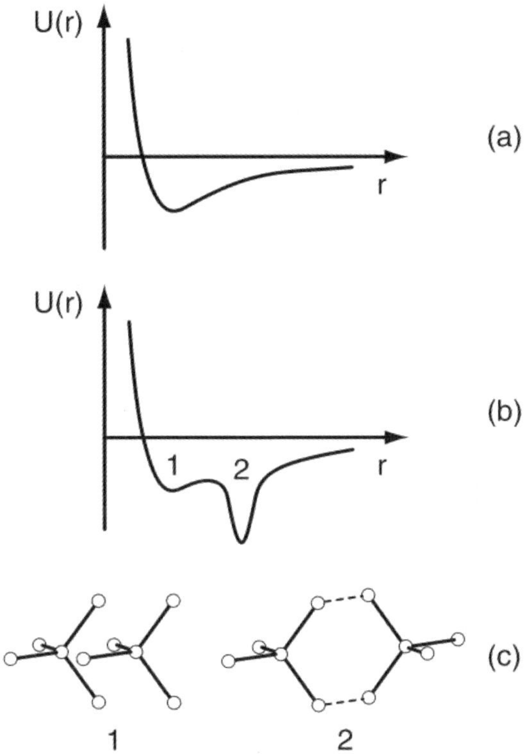

Fig. 1. Physical arguments relating to the plausibility of the existence of the known liquid-gas critical point C and the hypothesized LDL-HDL critical point C′. (a) Idealized system characterized by a pair interaction potential with a single attractive well. At low enough T ($T < T_c$) and high enough P ($P > P_c$), the system condenses into the "liquid" well shown. (b) Idealized system characterized by a pair interaction potential whose attractive well has two sub-wells, the outer of which is deeper and narrower. For low enough T ($T < T_{c'}$) and low enough P ($P < P_{c'}$), the one-phase liquid can "condense" into the narrow outer "LDL" sub-well, thereby giving rise to a LDL phase, and leaving behind the high-density liquid phase occupying predominantly the inner subwell. (c) Two idealized interaction clusters of water molecules ("Walrafen pentamers") in configurations that may correspond to the two sub-wells of (b). This figure is courtesy of Dr. Osamu Mishima.

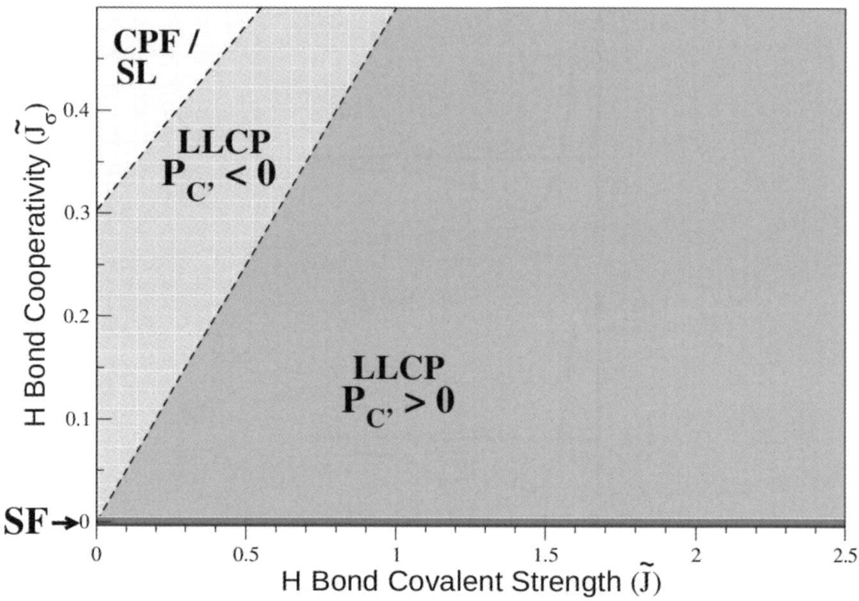

Fig. 2. Possible scenarios for water for different values of H bond energies $\tilde{J}$ of the covalent (or directional) component, and $\tilde{J}_\sigma$ of the cooperative (or three-body) component, obtained from MF calculations.[12] (i) If $\tilde{J}_\sigma = 0$ (red line along x–axis), the singularity free (SF) scenario is realized, independent of $\tilde{J}$. (ii) For large enough $\tilde{J}_\sigma$, water would possess a first–order liquid–liquid phase transition line terminating at the liquid–gas spinodal—the critical point free (CPF) scenario; the liquid spinodal would retrace at negative pressure, as in the stability limit (SL) scenario (yellow region in top left). (iii) For other combinations of $\tilde{J}$ and $\tilde{J}_\sigma$, water would be described by the liquid–liquid critical point (LLCP) scenario. For larger $\tilde{J}_\sigma$, the LLCP is at negative pressure (ochre region between dashed lines). For smaller $\tilde{J}_\sigma$, the LLCP is at positive pressure (orange region in bottom right). Dashed lines separating the three different regions correspond to MF results of the microscopic cell model. The $P - T$ phase diagram evolves continuously as $\tilde{J}$ and $\tilde{J}_\sigma$ change. Courtesy of Kevin Stokely and Marco G. Mazza.

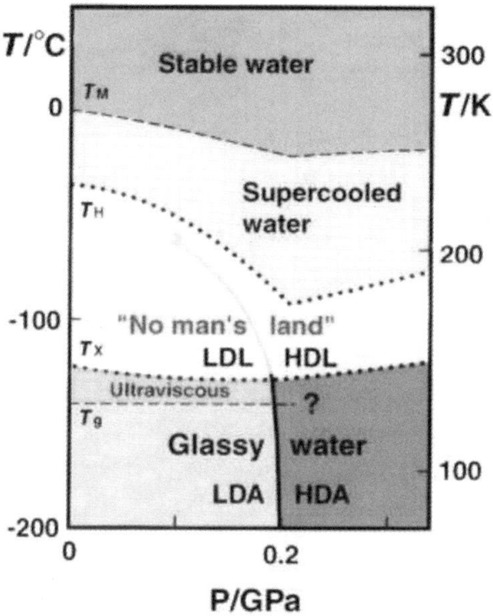

Fig. 3. Schematic illustration indicating the various phases of liquid water (color-coded). This figure is courtesy of Dr. Osamu Mishima.

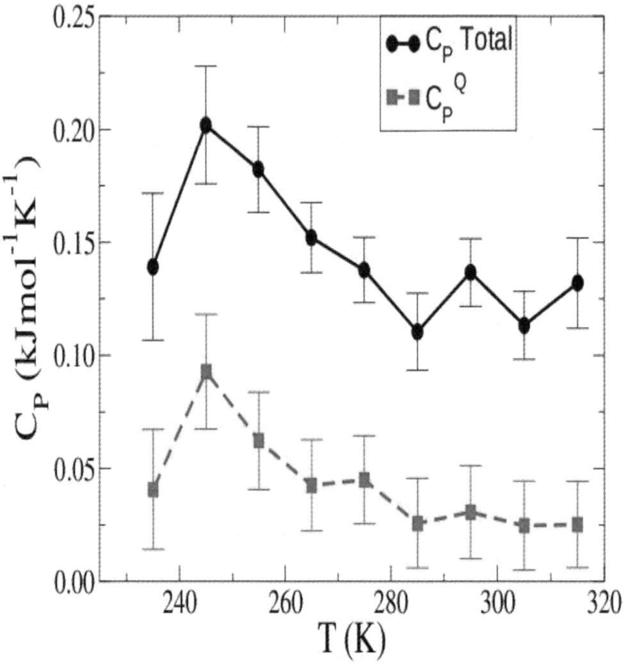

Fig. 4. Constant pressure specific heat C_P and specific heat associated with tetrahedral entropy C_P^Q of water at atmospheric pressure for TIP5P model of water. Courtesy of Pradeep Kumar.

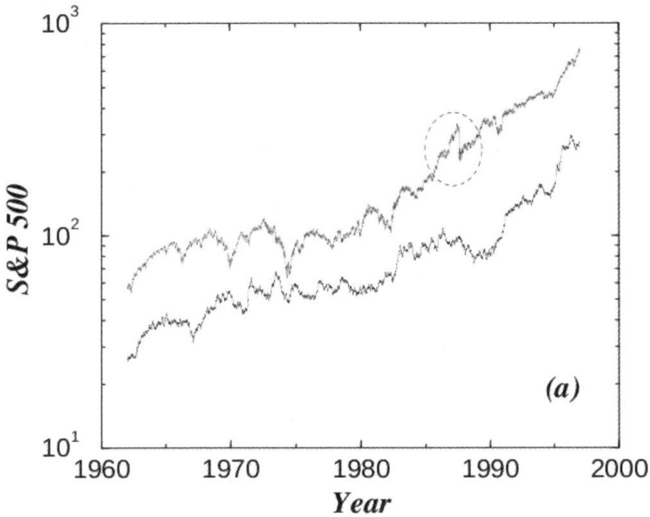

Fig. 5. The S&P 500 index is the sum of the market capitalizations of 500 companies. The sharp jump seen in 1987 is the market crash of October 19. Comparison of the time evolution of the S&P 500 for the 35-year period 1962–96 (top line) and a biased Gaussian random walk (bottom line). The random walk has the same bias as the S&P 500 —approximately 7% per year for the period considered. This figure is courtesy of Vasiliki Plerou and Parameswaran Gopikrishnan.

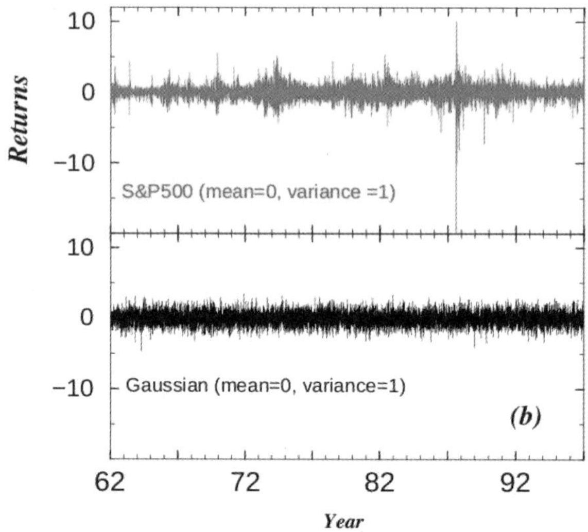

Fig. 6. Sequence of 10-min returns for the S&P 500, normalized to unit variance, compared with sequence of i.i.d. Gaussian random variables with unit variance, which was proposed by Bachelier as a model for stock returns. Note that, in contrast to the top curve, there are no "extreme" events in the bottom curve. This figure is courtesy of Vasiliki Plerou and Parameswaran Gopikrishnan.

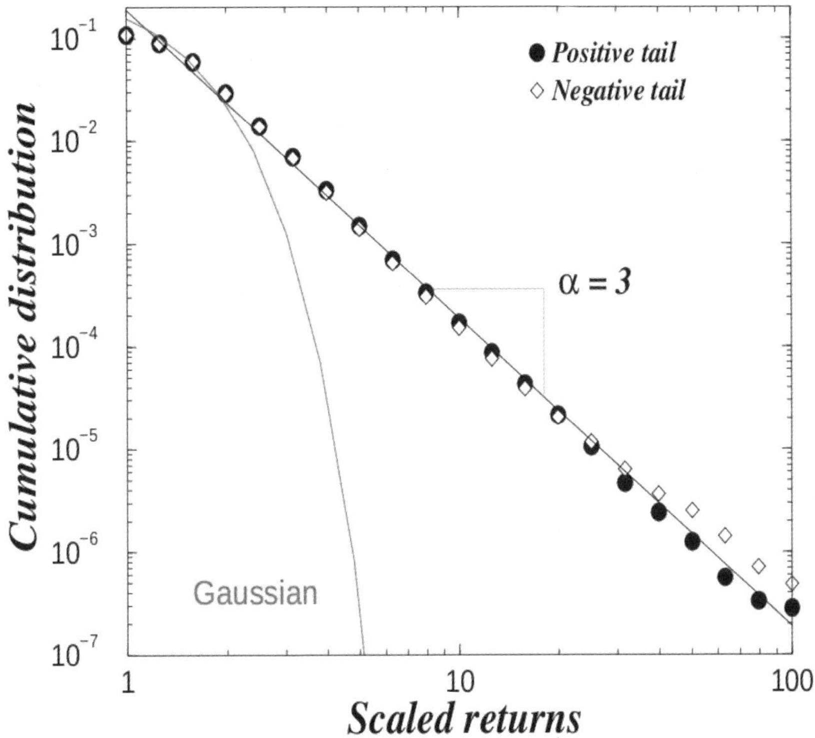

Fig. 7. Cumulative distributions of the positive and negative tails of the normalized returns of the 1000 largest companies in the TAQ database for the 2-year period 1994–1995. The solid line is a power-law regression fit in the region $2 \leq x \leq 80$. Shown for comparison is the behavior of a Gaussian function with the same value of the standard deviation. Note the striking fact that traditional theories based upon assumed Gaussian functional forms for this distribution predict that the probability of a fluctuation larger than 5 standard deviations is approximately 8 orders of magnitude smaller than the probability of an everyday event. In contrast, the empirical data demonstrate that the probability of a fluctuation larger than 100 standard deviations is approximately 8 orders of magnitude smaller than the probability of an everyday event. This figure is courtesy of Vasiliki Plerou and Parameswaran Gopikrishnan.

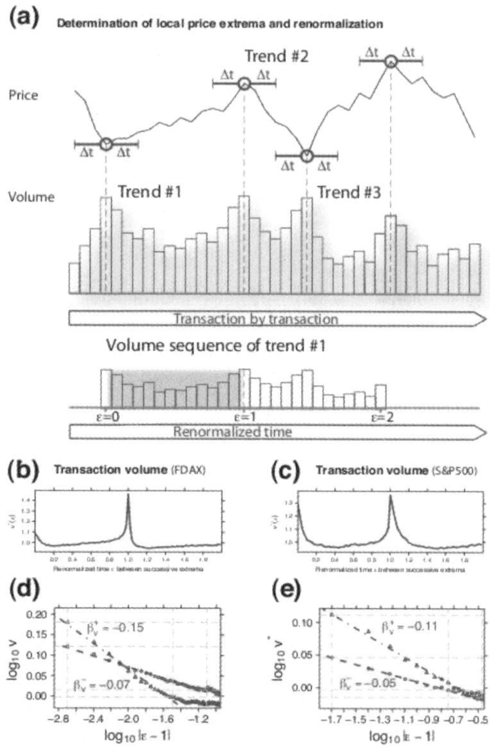

Fig. 8. Schematic visualization of trend segregation. Positive trends start at local price minima (red circles) and end at local maxima (blue circle)—and vice versa. A transaction price $p(t)$ is a local maximum if there is no higher transaction price in the interval $t-\Delta t \leq t \leq t+\Delta t$. Analogously, $p(t)$ is a local minimum if there is no lower transaction price in the interval $t-\Delta t \leq t \leq t+\Delta t$. We assign $\varepsilon = 0$ to the start of each trend, and $\varepsilon = 1$ to the end of each trend as shown for the first trend. In order to study trend switching processes—both before as well as after the end of a trend—we consider additionally the subsequent sequence of identical length. In the region around $\varepsilon = 1$, we find scale-free behavior of related quantities—volume and inter-trade time.[49,50,75] This behavior is consistent with "self-organized" macroscopic interactions among many traders,[76] not unlike "tension" in a pedestrian crowd.[77] (b) Averaged volume sequence $v^*(\varepsilon)$ of the German DAX Future time series. Δt ranges from 50 to 100 transactions (ticks). Extreme values of the price coincide with sharp peaks in the volume time series. (c) A very similar behavior is obtained for the averaged volume sequence $v^*(\varepsilon)$ of S&P500 stocks. Here, Δt ranges from 10 days to 100 days. (d) Log–log plot of the FDAX transaction volumes (50 ticks $\leq \Delta t \leq$ 1000 ticks) before reaching an extreme price value ($\varepsilon < 1$, circles) and after reaching an extreme price value ($\varepsilon > 1$, triangles). The straight lines correspond to power-law scaling with exponents $\beta_{\bar{v}}^{+} = -0.155 \pm 0.004$ and $\beta_{\bar{v}}^{-} = -0.068 \pm 0.001$. (e) Log–log plot of the transaction volumes shown in (c) indicates a power-law behavior with exponents $\beta_{\bar{v}}^{+} = -0.109 \pm 0.003$ and $\beta_{\bar{v}}^{-} = -0.052 \pm 0.001$ which are similar to our results on short time scales. This figure is courtesy of T. Preis.

ENTROPY, A UNIFYING CONCEPT: FROM PHYSICS TO COGNITIVE PSYCHOLOGY

CONSTANTINO TSALLIS[1,2] and ALEXANDRA C. TSALLIS[3]

[1] *Centro Brasileiro de Pesquisas Fisicas and National Institute of Science and Technology for Complex Systems, Rua Xavier Sigaud 150, 22290-180 Rio de Janeiro-RJ, Brazil*
E-mail: tsallis@cbpf.br

[2] *Santa Fe Institute, 1399 Hyde Park Road, Santa Fe, NM 87501, USA*

[3] *Instituto de Psicologia, Universidade Federal do Rio de Janeiro, Avenida Pasteur 250, Pavilhão Nilton Campos, 22290-240 Rio de Janeiro-RJ, Brazil*

Together with classical, relativistic and quantum mechanics, as well as Maxwell electromagnetism, Boltzmann-Gibbs (BG) statistical mechanics constitutes one of the main theories of contemporary physics. This theory primarily concerns *inanimate matter*, and at its generic foundation we find nonlinear dynamical systems satisfying the ergodic hypothesis. This hypothesis is typically guaranteed for systems whose maximal Lyapunov exponent is *positive*. What happens when this crucial quantity is *zero* instead? We suggest here that, in what concerns thermostatistical properties, we typically enter what in some sense may be considered as a new world — the world of *living systems* — . The need emerges, at least for many systems, for generalizing the basis of BG statistical mechanics, namely the Boltzmann-Gibbs-von Neumann-Shannon entropic functional form, which connects the macroscopic, thermodynamic quantity, with the occurrence probabilities of microscopic configurations. This unifying approach is briefly reviewed here, and its widespread applications — from physics to cognitive psychology — are overviewed. Special attention is dedicated to the learning/memorizing process in humans and computers. The present observations might be related to the gestalt theory of visual perceptions and the actor-network theory.

1. Introduction

It is the fate of all successful scientific theories to eventually become generalized. We have several such examples in physics. The superb classical, Newtonian mechanics starts sensibly failing when the velocities that are involved approach that of light in the vacuum. Special, and then general, relativity, appropriately generalize it, basically replacing the Galilean invariance by the Lorentz one. On a different front, Newtonian mechanics also starts failing when the masses that are involved are comparable to that of an electron. As well known, quantum mechanics appropriately generalize it in that sense. At a later stage, both generalizations of classical mechanics were unified, Dirac equation incorporating, as particular instances, Newton's dynamical law — the golden relation of classical mechanics —, Einstein's dynamical law — the golden relation of relativity —, and Schroedinger's equation — the golden relation in quantum mechanics —. This powerful theory is sometimes referred to as relativistic quantum mechanics. Similarly, Maxwell equations have been conveniently generalized through classical and quantum field theories.

It is our viewpoint that an analogous path is to be followed for Boltzmann-Gibbs (BG) statistical mechanics *. Indeed, as we shall see later on, the BG entropy, hence the BG statistical mechanics, find their notorious utility for systems whose N elements are essentially *weakly* correlated. The precise meaning of "weakly" will be addressed later on, in connection with both the thermodynamic extensivity of the entropy and the Central Limit Theorem (CLT) of theory of probabilities. When these correlations are *strong* enough, the central BG concepts either become mathematically ill-defined, or loose their primary utility.

The above generalizations have crucial relations (see, for instance,[1]) with the independent *physical universal constants* G, c (or more conveniently c^{-1}), h, and k_B (or more conveniently k_B^{-1}) †. All physical quantities whose units are based on *space*, *mass*, *time* and *temperature* (or equivalently *space*, *mass*, *time* and *electric current*) can be expressed as combinations of powers of these four universal constants, as noticed long ago by Planck.[2,4]

---

*It is to be noted that already this simple expression is considered provocative by some. Indeed, some physicists tend to just use the expression *statistical mechanics*, considering that *Boltzmann-Gibbs statistical mechanics* is an unnecessary pleonasm. It is nevertheless unavoidable the thought that, at the end of the XIXth century, physicists tended to simply use *mechanics*, since Newtonian mechanics was the only one physically acceptable. Today, when both relativistic and quantum mechanics are well established in the scientific community, the use of expressions such as *classical mechanics*, or *Newtonian mechanics*, has nowadays evolved into a costumary, harmless, and useful tradition. In fact, "provocative" expressions are very frequent in science. For example, the expression *kinetic theory of gases*, profusely used by Boltzmann and some physicists at his time, was considered provocative by others. Indeed, the word *kinetic* evoques something which is in motion. We all know today that this "something" are molecules, atoms. But in Boltzmann time, many physicists did not even accept the mere existence of atoms. As late as in the 1890s, Mach, influential on relativity and quantum mechanics, was still saying "I do not believe in atoms".

†The electron charge e does not need to be included here. Indeed, through the (pure number) *fine-structure constant* $\alpha \equiv 2\pi e^2/hc = 1/137.035999...$, e is related to c and h.

Newtonian mechanics corresponds to $G = c^{-1} = h = k_B^{-1} = 0$. Newtonian gravitation generalizes it in the sense that it allows for $G \neq 0$. Special relativity is a different generalization in the sense that it allows for $c^{-1} \neq 0$ (c had already been introduced in theoretical physics, since it was historically englobed within Maxwell electromagnetism). Quantum mechanics is still another generalization in the sense that it allows for $h \neq 0$. Maxwell-Boltzmann statistical mechanics (i.e., BG statistical mechanics for classical systems) is still another generalization in the sense that it allows for incomplete information about (the state of) the system, and introduces $k_B^{-1} \neq 0$. Examples of more complex generalizations are general relativity (which simultaneously allows for $(G, c^{-1}) \neq (0,0)$), relativistic quantum mechanics (which simultaneously allows for $(h, c^{-1}) \neq (0,0)$), Fermi-Dirac and Bose-Einstein statistics (which simultaneously allow for $(h, k_B^{-1}) \neq (0,0)$), the still elusive quantum gravity (in principle simultaneously allowing for $(G, h, c^{-1}) \neq (0,0,0)$). Let us finally mention statistical quantum gravity, at present time a definitively quite distant hypothetical physical theory, which would in principle allow for $(G, c^{-1}, h, k_B^{-1}) \neq (0,0,0,0)$.

1.1. *Entropy* ‡

The ubiquitous concept of *energy* is associated with the *possibilities* for the configurations of a mechanical system (e.g., the eigenvalues of the Hamiltonian of a quantum system defined in a specific Hilbert space). The concept of *entropy*[6] emerges in an even larger domain, since it can be defined for any system, mechanical or not, which admits a set of *probabilities* for its possible configurations. For instance, if we are dealing with a quantum mechanical system, the set of probabilities typically is that corresponding to the eigenvectors of the Hilbert space. Epistemologically speaking, entropy is one of the most subtle concepts in physics. Entropy and energy together constitute the basis on which statistical mechanics — one of the pillars of contemporary physics — is constructed.

The entropy, initially defined by Clausius for thermodynamical purposes, connects the macroscopic and microscopic worlds. Its most elementary form is the logarithmic one, first introduced by Boltzmann and refined by Gibbs, von Neumann, Shannon, Jaynes and others. For a finite discrete

‡For the present text to be conveniently self-contained, several parts of[5] are reproduced here.

set of probabilities $\{p_i\}$ is given by

$$S_{BG} = -k\sum_{i=1}^{W} p_i \ln p_i \qquad \left(\sum_{i=1}^{W} p_i = 1\right), \tag{1}$$

where BG stands for *Boltzmann-Gibbs.* The conventional constant k is typically taken to be the Boltzmann universal constant k_B for thermostatistical systems, or taken to be unity in information theory. For the particular case of equal probabilities, i.e., $p_i = 1/W$, $\forall i$, we have

$$S_{BG} = k \ln W\,, \tag{2}$$

carved on stone in Boltzmann grave in Vienna. Expression (1) enables the construction of the remarkably useful physical theory frequently referred to as *Boltzmann-Gibbs statistical mechanics.*

Since Boltzmann, many entropic forms have been introduced, and reintroduced again and again (see[1,6] and references therein for details). In 1988,[7] a generalized form, namely

$$S_q = k\frac{1-\sum_{i=1}^{W} p_i^q}{q-1} \qquad \left(\sum_{i=1}^{W} p_i = 1;\; q \in \mathcal{R};\; S_1 = S_{BG}\right), \tag{3}$$

was proposed as the basis for generalizing BG statistical mechanics (into a theory now known as *nonextensive statistical mechanics* [§]). For the particular case of equal probabilities, expression (3) becomes

$$S_q = k\frac{W^{1-q}-1}{1-q}\,. \tag{4}$$

By introducing the $q-$*logarithmic* function

$$\ln_q x \equiv \frac{x^{1-q}-1}{1-q} \qquad (x>0;\; q\in\mathcal{R};\; \ln_1 x = \ln x)\,, \tag{5}$$

expressions (3) and (4) can be respectively rewritten as follows:

$$S_q = k\sum_{i=1}^{W} p_i \ln_q \frac{1}{p_i} = -k\sum_{i=1}^{W} p_i^q \ln_q p_i = -k\sum_{i=1}^{W} p_i \ln_{2-q} p_i\,, \tag{6}$$

[§]The word *nonextensive* is to be associated here with the fact that the total energy of mechanical long-range-interacting many-body systems is nonextensive, in contrast with the case of short-range-interacting many-body systems, whose total energy is *extensive* in the thermodynamical sense.

and

$$S_q = k \ln_q W\,. \tag{7}$$

The axiomatics associated with this entropy have been quite explored and interesting characterizations have emerged. For details we may address the reader to,[8–12] among others.

1.2. *Additivity versus extensivity*

We adopt for *entropy additivity* the definition given in Penrose's classical book,[13] namely that an entropy S is said *additive* if, for any two *probabilistically independent* systems A and B, i.e., for $p_{i,j}^{A+B} = p_i^A p_j^B$, $\forall (i,j)$, we have that

$$S(A+B) = S(A) + S(B)\,, \tag{8}$$

where $S(A+B) \equiv S(\{p_{i,j}^{A+B}\})$, $S(A) \equiv S(\{p_i^A\})$, and $S(B) \equiv S(\{p_i^B\})$.

From definition (3) it is straightforward to prove that, for any two probabilistically independent systems A and B,

$$\frac{S_q(A+B)}{k} = \frac{S_q(A)}{k} + \frac{S_q(B)}{k} + (1-q)\frac{S_q(A)}{k}\frac{S_q(B)}{k}\,. \tag{9}$$

Therefore, S_{BG} is additive, whereas S_q $(q \neq 1)$ is nonadditive ¶.

Entropic extensivity is a concept in some sense more subtle than additivity. An entropy S of a given system constituted by N elements is said to be *extensive* if

$$0 < \lim_{N\to\infty} \frac{S(N)}{N} < \infty\,, \tag{10}$$

i.e., if $S(N) \propto N$ for $N >> 1$. We see therefore that additivity only depends on the specific mathematical connection between the macroscopic entropy functional and the probabilities of the configurations of the system. Extensivity depends on this but *also* on the specific system, more precisely on the nature of the correlations of its elements, and therefore of its collective configurations.

¶We remark that additivity is obtained whenever $(1-q)/k \to 0$. We see that this can occur in two different manners: $q \to 1$ for fixed k, and $k \to \infty$ for fixed q. The latter corresponds to the infinite temperature T limit, since in all thermostatistical systems T always appears in the form kT, thus having the dimension of an energy.

The distinction between additivity and extensivity has already been illustrated in simple probabilistic systems.[14] It has also been shown for the so-called *block entropy* (entropy of a subsystem of the entire system) of strongly quantum entangled fermionic and bosonic systems.[15,16]

For the probabilistic system it has been shown[14] that

$$S_{q_{ent}}(N) \propto N \quad (N >> 1)\,, \tag{11}$$

where *ent* stands for *entropy*, N is the number of (strongly correlated) binary random variables, and

$$q_{ent} = 1 - \frac{1}{d}\,, \tag{12}$$

$d = 1, 2, 3, \ldots$ characterizing the width of an infinitely long strip of nonvanishing probabilities of a probability triangle asymptotically satisfying the Leibnitz rule (i.e., asymptotically scale-invariant).

For a (one-dimensional) fermionic system it has been shown that, at criticality at vanishing temperature, we analytically have[15]

$$\lim_{N\to\infty} S_{q_{ent}}(N, L) \propto L \quad (L >> 1)\,, \tag{13}$$

where L is the number of first-neighboring spins (or analogous elements) within an infinitely long ($N \to \infty$) chain, and

$$q_{ent} = \frac{\sqrt{9 + c^2} - 3}{c}\,, \tag{14}$$

where c is the *central charge* ($c = 1/2$ for the Ising ferromagnet, and $c = 1$ for the isotropic XY ferromagnet, in the presence of a critical transverse magnetic field in both cases). We verify that q_{ent} monotonically increases from zero to one when c increases from zero to infinity.

For a (one-dimensional) spin-S random Heisenberg-like ferromagnet it has been shown that, at criticality at vanishing temperature, we numerically have[16]

$$q_{ent} = 1 - \frac{1.67}{c_{eff}}\,, \tag{15}$$

where the central charge c_{eff} is given by

$$c_{eff} = \ln(2S + 1)\,. \tag{16}$$

We verify that when S increases from zero to infinity, c_{eff} increases from zero to infinity, and q_{ent} increases from minus infinity to one. We see therefore that, while Eq. (14) imposes $q_{ent} \geq 0$, Eq. (15) allows for $q_{ent} < 0$.

For the (two-dimensional) bosonic system the results are qualitatively the same ($q_{ent} \le 1$).[15] However they have been established only numerically, not analytically.

The generic scenario which emerges is that, for a vast class of systems (but certainly not all), a value q_{ent} exists such that $S_{q_{ent}}(N) \propto N \ \ (N \to \infty)$, where N is the number of elements of the system under consideration (which might be the entire system, or only a large part of it). For standard systems, we have that $q_{ent} = 1$; for various classes of anomalous systems, we have $q_{ent} \neq 1$. The situation is depicted in Fig. 1.

SYSTEMS	ENTROPY S_{BG} (additive)	ENTROPY S_q (q<1) (nonadditive)
Short-range interactions, weakly entangled blocks, etc	EXTENSIVE	NONEXTENSIVE
Long-range interactions (QSS), strongly entangled blocks, etc	NONEXTENSIVE	EXTENSIVE

Fig. 1. Entropic additivity and entropic extensivity are different concepts. The BG entropy is additive, whereas S_q (for $q \neq 1$) is nonadditive. Extensivity depends on the system: for standard systems, S_{BG} is extensive, whereas it can be nonextensive for anomalous systems; S_q is the other way around.

1.3. *Central limit theorems*

The *Central Limit Theorem* within theory of probabilities basically states that the *sum* of a large number N of *independent* (or quasi-independent in some specific sense) random variables whose variance is *finite* converge, after appropriate centering and rescaling, to a Gaussian (i.e., $p(x) \propto e^{-\beta x^2}$). This distribution constitutes an attractor in the space of distributions, and is therefore thought to be the reason for the ubiquity of Gaussian distributions in nature. If the single distribution has a *divergent* variance instead (and also satisfies some supplementary mathematical conditions), the attractors are the celebrated Lévy distributions. The situation changes drastically if strong correlations exist among the N random variables. Depending on the nature of the correlations very many types of attractors might emerge. There is however a special class of strong correlations, referred to as *q-independence*,[17] for which the attractors are q-Gaussians (i.e., $p(x) \propto e_q^{-\beta x^2}$, where the q-exponential function is defined as the inverse of the q-logarithmic one defined in Eq. (5)), if a specific generalized variance is *finite*. If this variance *diverges* instead, the attractors are the so-called (q, α)-stable distributions: see[18] for full details. The schematic description of these four theorems is presented in Fig. 2.

The physical-mathematical interpretation of the class of strong correlations named as q-independence is not yet fully elucidated. However, it might well be that q-independence between N random variables implies (strict or asymptotic) probabilistic scale-invariance in the sense that

$$\int dx_N \, h_N(x_1, x_2, ..., x_N) \sim h_{N-1}(x_1, x_2, ..., x_{N-1}) \quad (N \to \infty) \,. \qquad (17)$$

Although probably necessary, this property is surely not sufficient. Indeed, (strictly or asymptotically) scale-invariant probabilistic models (with finite values for the appropriately generalized variance) have been analytically solved, some of them yielding, in the $N \to \infty$ limit, q-Gaussians, whereas other models yield distributions which numerically are amazingly close to q-Gaussians, but which definitively are *not* exactly q-Gaussians. Models that yield q-Gaussians are available in;[20,21] models that have been proved[22] to be not exactly q-Gaussians are presented in.[23,24]

Like Gaussians, q-Gaussians also are ubiquitous [‖] in natural, artificial and social systems. What could be the cause of such fact? It could very well be precisely the theorem appearing in Fig. 2 which corresponds to q-Gaussian attractors. In what follows we shall exhibit various (exact or nearly exact) q-Gaussian distributions: in Section 2 for dissipative one-dimensional dissipative maps, in Section 3 for long-range-interacting many-body classical Hamiltonians. These two systems share a crucial property, namely that they have a maximal Lyapunov exponent which approaches zero, thus excluding strong chaos. In Section 4 we make some connections of the present entropic ideas with cognitive psychology. Finally we conclude in

---

[‖]The word *ubiquitous* is here used not in the strict sense of being everywhere, but only in the loose sense of being found very frequently.

	$q=1$ [*independent*]	$q \neq 1$ (*i.e.*, $Q \equiv 2q-1 \neq 1$) [*globally correlated*]
$\sigma_Q < \infty$ $(\alpha = 2)$	$\mathbb{F}(x) =$ *Gaussian* $G(x)$, *with same* σ_1 *of* $f(x)$	$\mathbb{F}(x) = G_q(x) \equiv G_{(3q_1-1)/(1+q_1)}(x)$, *with same* σ_Q *of* $f(x)$ $G_q(x) \sim \begin{cases} G(x) & \textit{if } \lvert x \rvert \ll x_c(q,2) \\ f(x) \sim C_q / \lvert x \rvert^{2/(q-1)} & \textit{if } \lvert x \rvert \gg x_c(q,2) \end{cases}$, *with* $\lim_{q \to 1} x_c(q,2) = \infty$
$\sigma_Q \to \infty$ $(0 < \alpha < 2)$	$\mathbb{F}(x) =$ *Levy distribution* $L_\alpha(x)$, *with same* $\lvert x \rvert \to \infty$ *behavior* $L_\alpha(x) \sim \begin{cases} G(x) & \textit{if } \lvert x \rvert \ll x_c(1,\alpha) \\ f(x) \sim C_\alpha / \lvert x \rvert^{1+\alpha} & \textit{if } \lvert x \rvert \gg x_c(1,\alpha) \end{cases}$ *with* $\lim_{\alpha \to 2} x_c(1,\alpha) = \infty$	$\mathbb{F}(x) = L_{q,\alpha}$, *with same* $\lvert x \rvert \to \infty$ *asymptotic behavior* $L_{q,\alpha} \sim \begin{cases} G_{\frac{2(1-q)-\alpha(1+q)}{2(1-q)-\alpha(3-q)},\,\alpha}(x) \sim C^*_{q,\alpha} / \lvert x \rvert^{\frac{2(1-q)-\alpha(3-q)}{2(1-q)}} & \textit{(intermediate regime)} \\ G_{\frac{2\alpha q-\alpha+3}{\alpha+1},\,2}(x) \sim C^L_{q,\alpha} / \lvert x \rvert^{(1+\alpha)/(1+\alpha q-\alpha)} & \textit{(distant regime)} \end{cases}$

Fig. 2. $N^{1/[\alpha(2-q)]}$-scaled attractors $\mathcal{F}(x)$ when summing $N \to \infty$ q-independent identical random variables with symmetric distribution $f(x)$ with Q-variance $\sigma_Q \equiv \int_{-\infty}^{\infty} dx\, x^2\, [f(x)]^Q / \int_{-\infty}^{\infty} dx\, [f(x)]^Q$ ($Q \equiv 2q-1$; $q_1 = (1+q)/(3-q)$; $q \geq 1$). *Top left:* The attractor is the Gaussian sharing with $f(x)$ the same variance σ_1 (standard CLT). *Bottom left:* The attractor is the α-stable Lévy distribution which shares with $f(x)$ the same asymptotic behavior, i.e., the coefficient C_α (Lévy-Gnedenko CLT, or α-generalization of the standard CLT). *Top right:* The attractor is the q-Gaussian which shares with $f(x)$ the same $(2q-1)$-variance, i.e., the coefficient C_q (q-generalization of the standard CLT, or q-CLT). *Bottom right:* The attractor is the (q,α)-stable distribution which shares with $f(x)$ the same asymptotic behavior, i.e., the coefficient $C^L_{q,\alpha}$ (q-generalization of the Lévy-Gnedenko CLT and α-generalization of the q-CLT). The case $\alpha < 2$, for both $q = 1$ and $q \neq 1$ (more precisely $q > 1$), further demands specific asymptotics for the attractors to be those indicated; essentially the divergent q-variance must be due to fat tails of the power-law class, excepting for possible logarithmic corrections (for the $q = 1$ case see, for instance,[19] and references therein).

Section 5 by mentioning various systems presented in the literature which, like those described in detail in the present Sections 2 and 3, exhibit q-Gaussians.

2. Unimodal One-dimensional Dissipative Maps

Let us here concentrate on a paradigmatic one-dimensional dissipative dynamical system, namely the logistic map, defined as $x_{t+1} = 1 - a x_t^2$, where a is the map parameter ($0 \leq a \leq 2$), $-1 \leq x_t \leq 1$, and $t = 0, 1, 2, \ldots$. Our object of interest is the sum

$$y = \sum_{i=N_0+1}^{N_0+N} (x_i - \langle x \rangle) \tag{18}$$

in the vicinity of chaos threshold $a_c = 1.4011..$, where N_0 is the number of transient steps (typically $N_0 >> 1$) and

$$\langle x \rangle = \frac{1}{n_{ini}} \frac{1}{N} \sum_{j=1}^{n_{ini}} \sum_{i=1}^{N} x_i^{(j)} \tag{19}$$

is the average over a large number of N iterates as well as a large number n_{ini} of randomly chosen initial values $x_1^{(j)}$ of iterates of the map. This problem has been addressed firstly in[25] and a closer look has been given very recently in.[26] Here, we try to further clarify the study and analyse it in a more compact manner.

Generically, the problem at hand is the form of the probability distribution of the random variable given in Eq. (18). Indeed, the ordinary Central Limit Theorem (CLT) (yielding the Gaussian form) is applicable when the Lyapunov exponent is positive, but, when approaching the edge of chaos from above (i.e., for $a > a_c$), an infinite number of values of a accumulate which violate this condition. The situation becomes then quite subtle, as we shall review here. Essentially, strong correlations between the iterates of the map emerge. In[26] it is shown that the problem is much more complex than the ordinary CLT, which is the case when the map is at the chaotic regime (e.g., for $a = 2$). One needs now to be careful on how we approach the chaos threshold point (a_c) and how the number of iterations (N) to be used increases to infinity. In mathematical language, this means that two limits are to be performed simultaneously, namely, $(a - a_c) \to 0$ and $1/N \to 0$. It is argued in[26] that the limit distributions appear to be of q-Gaussian type if these two limits are performed simultaneously in the following special way (which satisfies an appropriate scaling relation). We first choose a value of a above and close to a_c. We then calculate the quantity n defined as follows:

$$n = -\frac{\log |a - a_c|}{\log \delta}, \tag{20}$$

where $\delta = 4.6692011...$ is the Feigenbaum constant. We then denote by k the nearest integer value of $2n$, and define N^* through

$$N^* = 2^k. \tag{21}$$

The q-Gaussians numerically appear to gradually emerge when we choose $N = N^*$ and keep making $(a - a_c) \to 0$ (hence $N = N^* \to \infty$). Let us refer to this region as the q-Gaussian probability distribution functions (PDFs) one and analyse its borders. To do this, typical values for a have been used: they are given in Table 1, as well as their related parameters; a values can easily be taken so that $2n$ values would be obtained with the same precision. Each group with the same precision enables the construction of a linear curve in the space $1/N$ vs $(a - a_c)^s$, where $s \equiv \ln 4/\ln\delta \simeq 0.9$: see Fig. 3. Numerical inspection has shown that no other q-Gaussian linear curves occur at the left of the largest slope and at the right of the smallest

Table 1. The values of map parameter a and N^* used in this work. The values of $2n$ (obtained from the scaling relation), and the values of q and β (estimated from simulations) are also listed. In this work, the critical value a_c is approximated as $a_c = 1.4011551890920505$.

a	$a - a_c$	$2n$	$N^* = 2^k$	q	β
1.40159888	0.00044369	10.02	2^{10}		
1.40125021721	0.00009503	12.02	2^{12}	1.70	6.6
1.40117554121	0.00002035	14.02	2^{14}		
1.40115954790	0.00000436	16.02	2^{16}		
1.40152683	0.00037164	10.25	2^{10}		
1.40123478	0.00007959	12.25	2^{12}	1.70	6.5
1.401172235	0.00001705	14.25	2^{14}		
1.4011588398	0.00000365	16.25	2^{16}		
1.4014862	0.00033101	10.40	2^{10}		
1.401226075	0.00007088	12.40	2^{12}	1.70	6.8
1.401170372	0.00001518	14.40	2^{14}		
1.401158441	0.00000325	16.40	2^{16}		
1.401464065	0.00030888	10.49	2^{10}		
1.401221341	0.00006615	12.49	2^{12}	1.68	6.7
1.4011693567	0.00001417	14.49	2^{14}		
1.4011582234	0.00003034	16.49	2^{16}		
1.40145934	0.00030415	10.51	2^{11}		
1.40122033	0.00006514	12.51	2^{13}	1.61	6.8
1.40116914	0.00001395	14.51	2^{15}		
1.401158177	0.00000299	16.51	2^{17}		
1.40138924	0.00023405	10.85	2^{11}		
1.401205317	0.00005013	12.85	2^{13}	1.65	6.5
1.401165925	0.00001074	14.85	2^{15}		
1.4011574883	0.00000230	16.85	2^{17}		
1.40136531	0.00021012	10.99	2^{11}		
1.40120019	0.00004500	12.99	2^{13}	1.63	6.5
1.401164827	0.00000964	14.99	2^{15}		
1.4011572532	0.00000206	16.99	2^{17}		

slope in Fig. 3. All q-Gaussian lines appear to exist only between these two extremes; see examples in Fig. 4 as well as Fig. 5 where (q, β) pairs of all studied cases are plotted.

The q-Gaussian-like PDFs are lost as soon as the scaling relation (Eq. (20)) is ignored, i.e., if values for N larger or smaller that N^* are used. Two new regions emerge. If we use values for N that are sensibly smaller than N^*, or, in other words, the value of a that is being used is too close to the critical point, peaked PDF's are observed (see also[27,28]). On the other extreme, if we use values for N that are sensibly larger than N^*, we observe in most of the cases Gaussian PDF's, i.e., $P(y) = e^{-y^2/(2\sigma^2)}/\sqrt{2\pi\sigma^2})$. Two representative examples for the two regions outside the q-Gaussian one are indicated with dashed lines (magenta) in Fig. 3 and illustrated in Fig. 6.

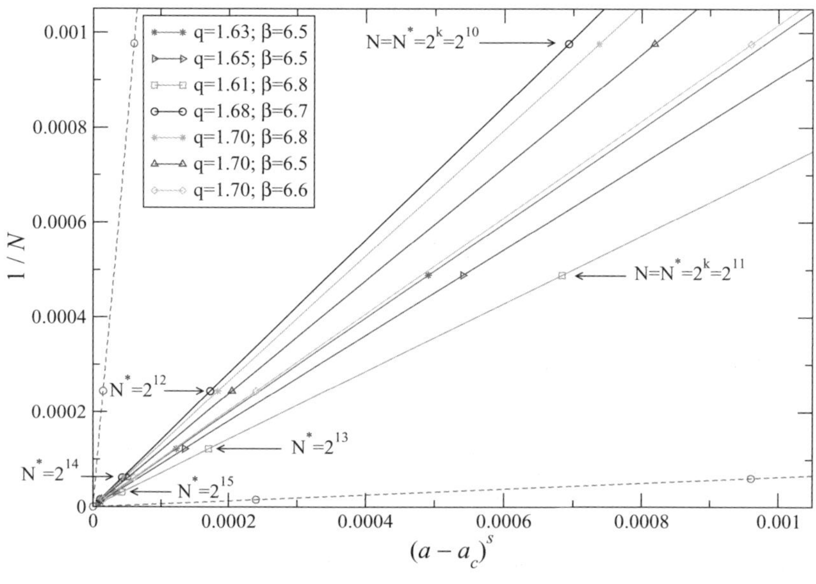

Fig. 3. View of the parameter region $1/N$ vs $(a - a_c)^s$ with $s = \ln 4/\ln \delta \simeq 0.9$ coming directly from the scaling relation (20). Typical values of a are shown, and their correspondiong values of N^* are determined from the scaling relation (20). Any other possible choices for N^* yield lines that remain between the lines with the largest and the smallest slopes shown in the figure. If one approaches the critical point (origin) along any of these lines in this region, the probability distribution function (PDF) appears to gradually approach, excepting for a small oscillating contribution, a q-Gaussian. The regions at the left of the largest slope and at the right of the smallest slope are not accessible as far as N^* values are concerned. Four of the seven q-Gaussian examples are presented in Fig. 4; the peaked (almost vertical line) and the Gaussian (almost horizontal line) examples are presented in Fig. 6.

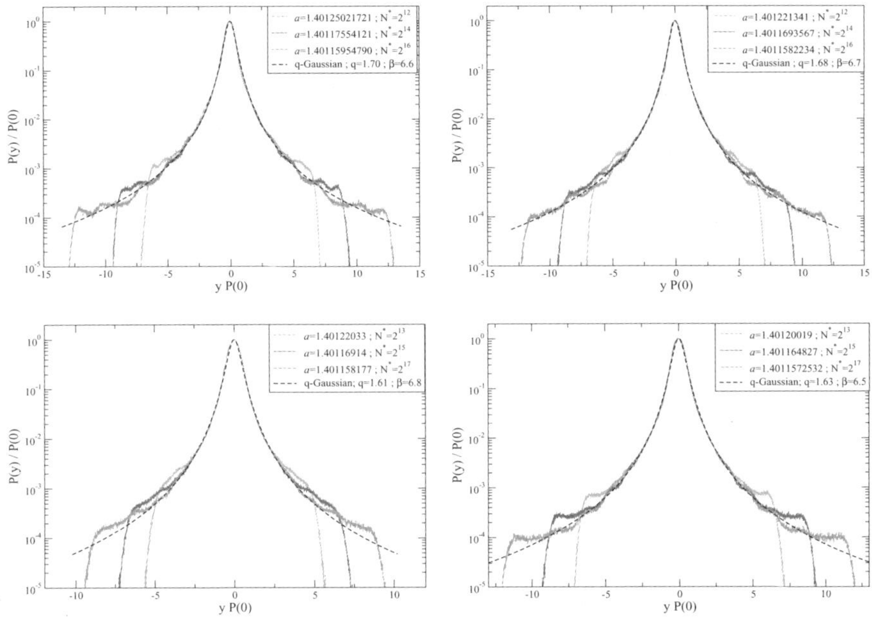

Fig. 4. Data collapse of PDF's for the cases $N = N^* = 2^k$ for four of the seven representative examples given in Table 1.

Very recently, a quite different one-dimensional dissipative map has been studied as well.[29] It is the sine-circle map at its edge of chaos, whose road to chaos, namely the so-called quasi-periodic one, is of a very different nature than that of the logistic map edge of chaos. Once again, q-Gaussian attractors numerically appear to emerge (see Figs. 7 and 8). With one interesting difference though: the values of q are now *below* unity, in contrast with the logistic ones, which are *above* unity. Like for the logistic map, when the Lyapunov exponent is not zero but positive, Gaussians emerge instead (see Fig. 9), consistently with the standard Central Limit Theorem.

3. Long-range-interacting Many-body Classical Hamiltonian Systems

Let us consider the following classical Hamiltonian of N interacting planar classical rotators on a d-dimensional (simple hypercubic) lattice:[30]

$$\mathcal{H} = \sum_{i=1}^{N} \frac{p_i^2}{2} + \frac{1}{2} \sum_{i,j} \frac{1 - cos(\theta_i - \theta_j)}{r_{ij}^{\alpha}} \quad (\alpha \geq 0)\,, \tag{22}$$

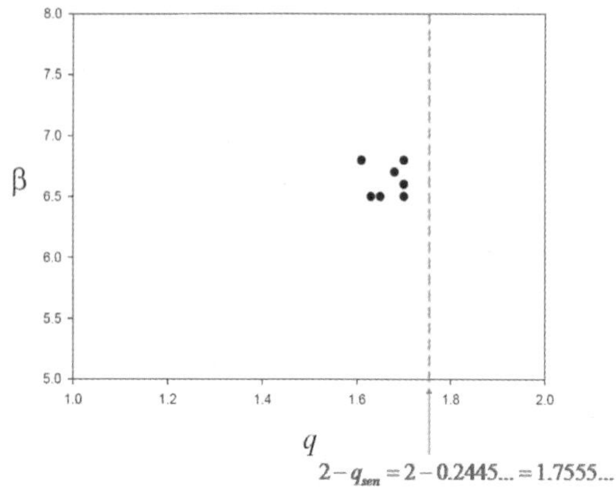

Fig. 5. The parameters (q, β) corresponding to the seven q-Gaussians indicated in Fig. 3. These specific seven examples appear to exclude the value $2 - q_{sen} = 1.7555...$, which could have been a plausible result. At the present numerical precision, even if quite high, it is not possible to infer whether the analytical result corresponding to the present observations would be only one or a set of q-Gaussians, assuming that exact q-Gaussians are involved, on top of which a small oscillating component possibly exists.

where r_{ij} runs over all possible distances within the d-dimensional lattice. The particular case $\alpha = 0$ is referred to in the literature as the HMF model,[31] and has been intensively studied in the last decade (in its standard representation the coupling constant is divided by N, which artificially makes the total energy extensive in the thermodynamical sense). Its dynamical molecular approach has exhibited a variety of interesting phenomena: see, for instance,[1] and references therein. In particular, for the isolated system (microcanonical ensemble) at energy per particle equal to 0.69, long-standing quasi-stationary states (QSS) emerge when certain classes of initial conditions (usually called *water-bag* initial conditions) are used. Within the water-bag conditions, one may consider initial magnetization equal to zero (usually referred to as $M = 0$), or equal to its maximal value (usually referred to as $M = 1$), or values in between. Such choices influence the specific trajectory of the full system within its $2N$-dimensional phase space (Gibbs Γ space). It has been shown that, for $M = 1$ initial conditions (possibly for virtually all values of initial M), *ergodicity is broken*. Indeed the *summed* (over n equidistant instants) one-velocity marginal PDF differs when we take ensemble-average or time-average: see[32] and references therein. Many of the time-averaged PDF's numerically approach a q-Gaussian (see Fig. 10 for one such example). Although no analytical proof is available at the present time, this might be a consequence of the

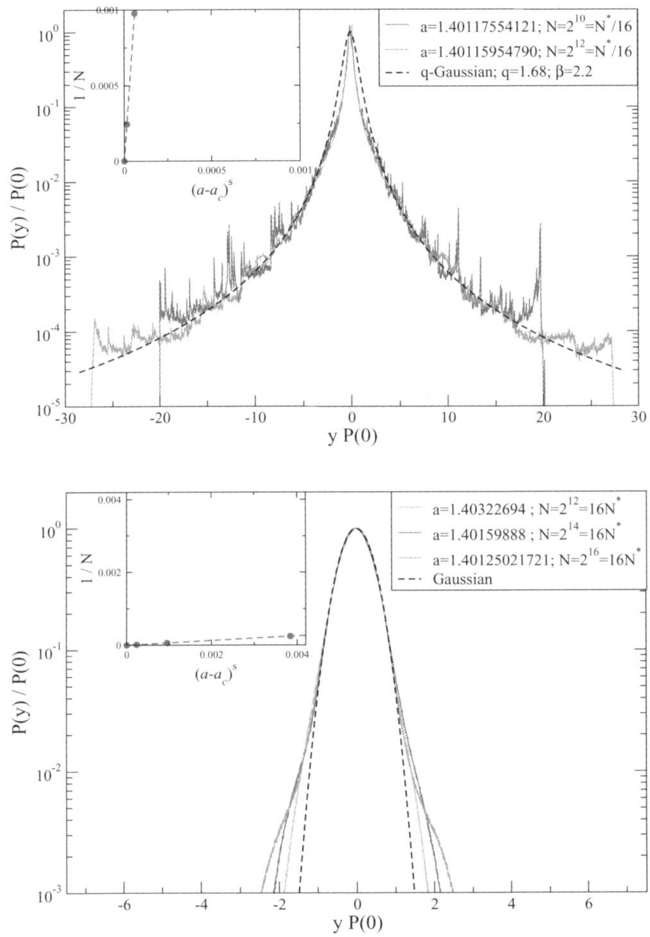

Fig. 6. Representative examples from the peaked region (left) and the Gaussian region (right).

q-Central Limit Theorem,[17] within which q-Gaussians are the attractors in the space of PDF's. The finite-size effects are illustrated in Fig. 10.

4. Cognitive Psychology - Learning/memorizing Process

Learning and memorization are basic functions of cognitive abilities of humans and others. A study,[34] briefly reviewed here, has been carried on some time ago along such lines in connection with the nonadditive entropy S_q

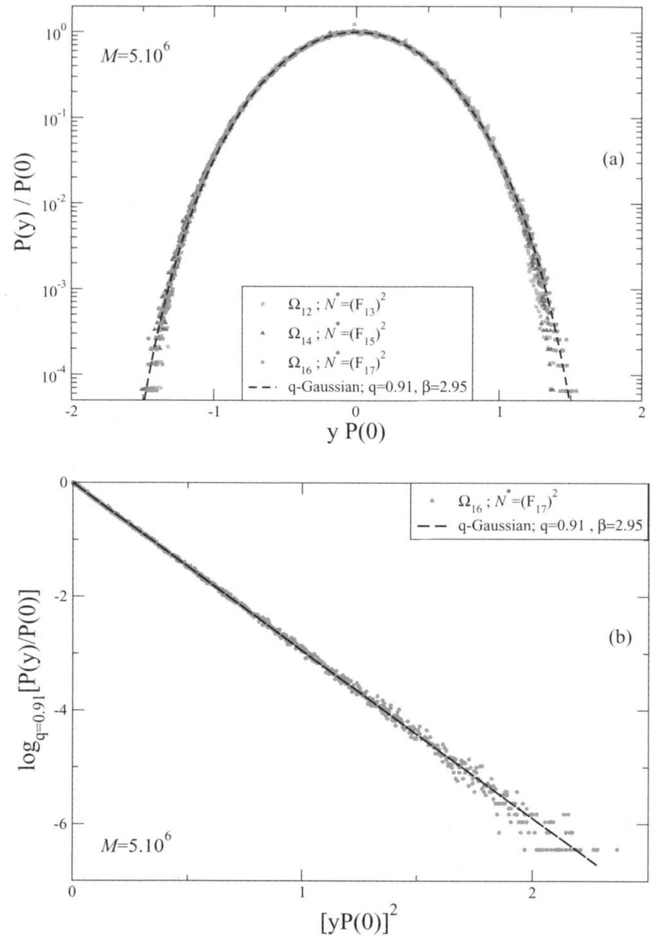

Fig. 7. q-Gaussian emerging near the edge of chaos of the sine-circle map. $K = 1$ and the value of Ω is gradually approaching the critical value at which the Lyapunov exponent vanishes. Values of Ω corresponding to *even* exponents for N [(a) and (b) correspond to two different representations of the same numerical data]. The black dashed line corresponds to the q-Gaussian $P(y)/P(0) = e_q^{-\beta[yP(0)]^2}$ with $(q, \beta) = (0.91, 2.95)$. From,[29] where all details can be seen.

and its associated nonextensive statistical mechanics. The task consisted in memorizing a specific 5×5 matrix, as well as 7×7 matrices, chosen once for ever. The population to which this task was applied was 120 university students. The experimental results, as well as its comparison with a percep-

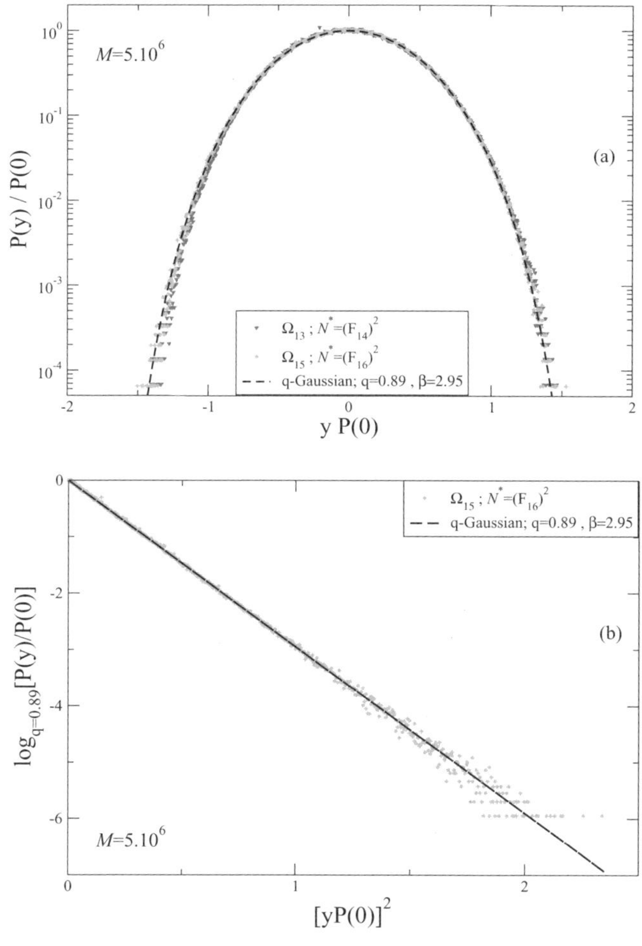

Fig. 8. q-Gaussian emerging near the edge of chaos of the sine-circle map. $K = 1$ and the value of Ω is gradually approaching the critical value at which the Lyapunov exponent vanishes. Values of Ω corresponding to *odd* exponents for N [(a) and (b) correspond to two different representations of the same numerical data]. The black dashed line correspond to the q-Gaussian $P(y)/P(0) = e_q^{-\beta[yP(0)]^2}$ with $(q, \beta) = (0.89, 2.95)$. From,[29] where all details can be seen.

tron learning with a q-generalized Langevin equation are presented in Fig. 11. We verify that this specific learning appears to occur in a nonextensive (or *global*) manner, i.e., with $q \neq 1$. This fact confirms our initial hypothesis and gives substance to our proposal of the notion of *homo metaphoricus* (humans being animals capable of doing metaphors, precisely due to the

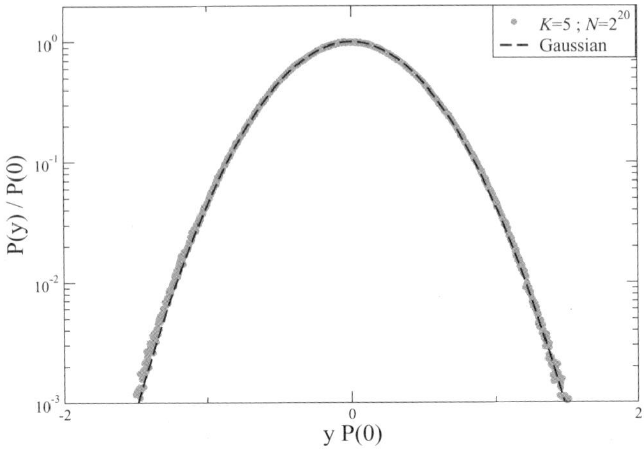

Fig. 9. Gaussian emerging at $(K,\Omega) = (5, 0.606661063469)$, where the Lyapunov exponent is positive. The black dashed line corresponds to the Gaussian $P(y)/P(0) = e^{-\beta[yP(0)]^2}$. From.[29]

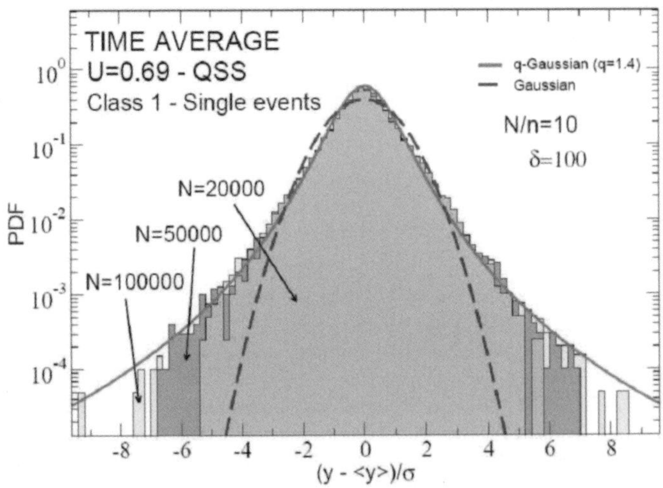

Fig. 10. Velocity distribution for the HMF model at its quasi-stationary state (QSS) [Data from Fig. 2 of[32]]. The finite-N effects are visible, gradually approaching a (possible) q-Gaussian when both N and n are diverging. For comparison, a Gaussian is shown as well.

nonextensive nature of their learning/memorizing). In fact, today we may revisit the various aspects involved in the experiments. Some of them can be thought within the frame of the *actor-network-theory*.[35–38] More precisely, we did follow the production of the effects of the connected actors during the experiment. For instance, a relevant one was that, when the participants answered our final questions about how they did to memorize/learn the matrices, they systematically expressed that, at the early stages of the execution of the task, they looked for *establishing a strategy*. The nonextensivity (or globality) of the activity has therefore to be consistently thought in connection to the choice of a specific strategy by the participants.

The above experiment has been recently redone for a quite more complex task, namely the learning of languages.[39,40] Typical results are reproduced in Figs. 12 and 13 for two different experiments. The results are intriguingly similar to those obtained for the learning of matrices. Further approaches representing cognitive tasks with computational models can be seen in.[41–43]

Finally we may remark that the nonextensive (global) nature (i.e., $q \neq 1$) of the above cognitive phenomena certainly suggests strong correlations between the elements playing the relevant roles. This fact unavoidably makes us to think that something similar might be at the basis of the gestalt theory of visual perceptions.[44,45]

5. Final Remarks

Many q-Gaussian-like distributions have been observed in recent years in natural, artificial and social systems. Obviously, only for a mathematically formulated model, the hope exists to possibly prove analytically that the relevant distribution precisely, and not only approximately, is a q-Gaussian. In all other cases, we can only expect for increasingly high-precision indications from real experiments or observations. Computational evidence can and does provide important hints, however never a proof.

This said, let us mention in what follows some of the many other systems where q-Gaussians have been used to approach the observed PDF's: (i) The velocity distribution of (cells of) *Hydra viridissima* follows a $q = 3/2$ PDF;[46] (ii) The velocity distribution of (cells of) *Dictyostelium discoideum* follows a $q = 5/3$ PDF in the vegetative state and a $q = 2$ PDF in the starved state;[47] (iii) The velocity distribution in defect turbulence;[48] (iv) The velocity distribution of cold atoms in a dissipative optical lattice;[51] (v) Velocity distribution during silo drainage;[49,50] (vi) The velocity distribution in a driven-dissipative 2D dusty plasma, with $q = 1.08 \pm 0.01$ and $q = 1.05 \pm 0.01$ at temperatures of $30000\,K$ and $61000\,K$ respectively;[52] (vii)

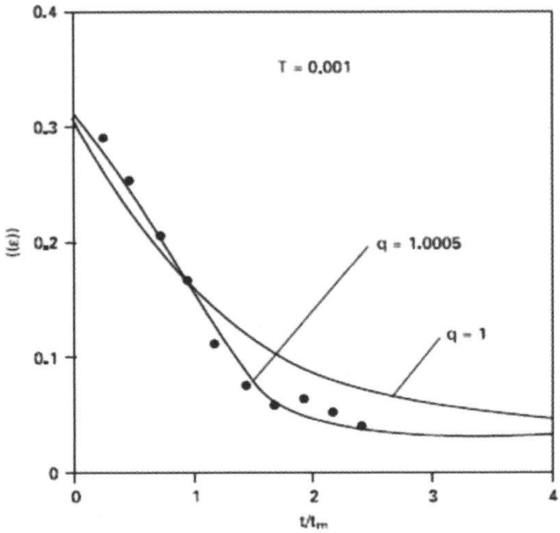

Fig. 11. Time-dependence of the average error during the learning/memorization of the 5×5 matrix for humans (full dots) and for q-generalized perceptrons (solid lines). From.[34]

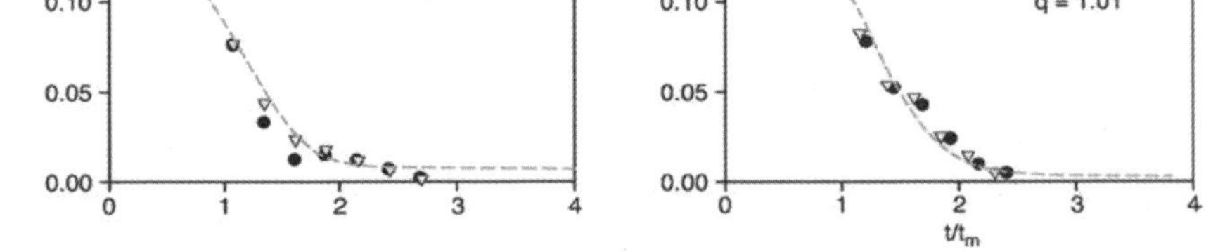

Fig. 12. Time-dependence of the average error during the learning/memorization of words by humans (full dots) and by q-generalized perceptrons (dashed lines). Shallow (a) and deep (b) orthography data in experiment 1. See details in.[39]

The spatial (Monte Carlo) distributions of a trapped $^{136}Ba^{+}$ ion cooled by various classical buffer gases at $300\,K$;[53] (viii) The distributions of price returns at the stock exchange;[54–56] (ix) The distributions of returns of magnetic field fluctuations in the solar wind plasma as observed in data from

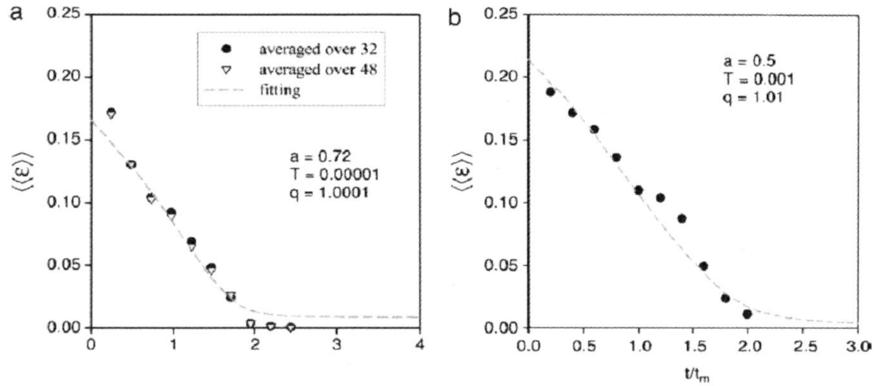

Fig. 13. Time-dependence of the average error during the learning/memorization of words by humans (full dots) and by q-generalized perceptrons (dashed lines). Shallow (a) and deep (b) orthography data in experiment 2. See details in.[39]

Voyager 1[57] and from Voyager 2;[58] (x) The distributions of returns of the avalanche sizes in the Ehrenfest's dog-flea model;[59] (xi) The distributions of returns of the avalanche sizes in the self-organized critical Olami-Feder-Christensen model, as well as in real earthquakes;[60] (xii) The distributions of angles in the HMF model;[61] (xiii) The distribution of stellar rotational velocities in the Pleiades;[62] (xiv) The distribution of intensities of X-ray emission from black hole candidates, low mass neutron stars and high mass neutron stars.[63] Some indirect evidence is available as well: although no q-Gaussian distribution has been directly observed in some relevant physical quantity, a q-exponential relaxation has been seen in various paradigmatic spin-glass substances through neutron spin echo experiments.[64]

Clearly, the simplest hypothesis which would explain the ubiquity of q-Gaussians is the applicability of the q-Central Limit Theorem to these complex systems. The involved random variables would, in such case, be expected to be q-independent.[17] The present belief is that (probabilistic) scale-invariance is necessary but not sufficient for q-independence. Further studies are needed to clarify the applicability of such ideas to the systems mentioned above, as well as possibly to others.

Acknowledgments

We are grateful to O. Afsar and U. Tirnakli for authorizing us to reproduce their recent results on the sine-circle map; analogously to T. Hadzibeganovic and S.A. Cannas concerning their results in.[39] One of us (CT) thanks C.

Rangacharyulu and E. Haven for very warm hospitality in Saskatoon. This work has been partially supported by CNPq and Faperj (Brazilian agencies).

References

1. C. Tsallis, *Introduction to Nonextensive Statistical Mechanics - Approaching a Complex World* (Springer, New York, 2009).
2. M. Planck, *Uber irreversible Strahlugsforgange*, Ann d. Phys. (4) (1900) 1, S. 69-122 (see also[3]).
3. M.J. Duff, *Comment on time-variation of fundamental constants*, hep-th/0208093 (2004).
4. M. Planck, *Verhandlungen der Deutschen Physikalischen Gessellschaft* **2**, 202 and 237 (1900) (English translation: D. ter Haar, S. G. Brush, Planck's Original Papers in Quantum Physics (Taylor and Francis, London, 1972)].
5. C. Tsallis and U. Tirnakli, *Nonadditive entropy and nonextensive statistical mechanics - Some central concepts and recent applications*, J. Phys. (2010), in press, 0911.1263 [cond-mat.stat-mech].
6. C. Tsallis, *Entropy*, in *Encyclopedia of Complexity and Systems Science*, ed. R.A. Meyers (Springer, Berlin, 2009), 11 volumes [ISBN: 978-0-387-75888-6].
7. C. Tsallis, *Possible generalization of Boltzmann-Gibbs statistics*, J. Stat. Phys. **52**, 479-487 (1988).
8. R.J.V. dos Santos, *Generalization of Shannon's theorem for Tsallis entropy*, J. Math. Phys. **38**, 4104-4107 (1997).
9. S. Abe, *Axioms and uniqueness theorem for Tsallis entropy*, Phys. Lett. A **271**, 74 (2000).
10. F. Topsoe, *Factorization and escorting in the game-theoretical approach to non-extensive entropy measures*, Physica A **365**, 91-95 (2006).
11. F. Topsoe, *Towards operational interpretations of generalized entropies*, J. Phys. (2010), in press.
12. A. Ohara, H. Matsuzoe and S. Amari, *A dually flat structure on the space of escort distributions*, J. Phys. (2010), in press.
13. O. Penrose, *Foundations of Statistical Mechanics: A Deductive Treatment* (Pergamon, Oxford, 1970), page 167.
14. C. Tsallis, M. Gell-Mann and Y. Sato, *Asymptotically scale-invariant occupancy of phase space makes the entropy S_q extensive*, Proc. Natl. Acad. Sc. USA **102**, 15377-15382 (2005).
15. F. Caruso and C. Tsallis, *Nonadditive entropy reconciles the area law in quantum systems with classical thermodynamics*, Phys. Rev. E **78**, 021101 (2008).
16. A. Saguia and M.S. Sarandy, *Nonadditive entropy for random quantum spin-S chains*, 0912.3718 [cond-mat.stat-mech].
17. S. Umarov, C. Tsallis and S. Steinberg, *On a q-central limit theorem consistent with nonextensive statistical mechanics*, Milan J. Math. **76**, 307-328 (2008).
18. S. Umarov, C. Tsallis, M. Gell-Mann and S. Steinberg, *Generalization of*

symmetric α-stable Lévy distributions for $q \neq 1$, J. Math. Phys. (2010), in press.
19. J.P Bouchaud and A. Georges, *Anomalous diffusion in disordered media: Statistical mechanisms, models and physical applications*, Phys. Rep. **195**, 127 (1990).
20. A. Rodriguez, V. Schwammle and C. Tsallis, *Strictly and asymptotically scale-invariant probabilistic models of N correlated binary random variables having* q–*Gaussians as* $N \to \infty$ *limiting distributions*, JSTAT P09006 (2008).
21. R. Hanel, S. Thurner and C. Tsallis, *Limit distributions of scale-invariant probabilistic models of correlated random variables with the q-Gaussian as an explicit example*, Eur. Phys. J. B (2009), DOI: 10.1140/epjb/e2009-00330.
22. H.J. Hilhorst and G. Schehr, *A note on q-Gaussians and non-Gaussians in statistical mechanics*, J. Stat. Mech. (2007) P06003.
23. L.G. Moyano, C. Tsallis and M. Gell-Mann, *Numerical indications of a q-generalised central limit theorem*, Europhys. Lett. **73**, 813-819 (2006).
24. W.J. Thistleton, J.A. Marsh, K.P. Nelson and C. Tsallis, *q-Gaussian approximants mimic non-extensive statistical-mechanical expectation for many-body probabilistic model with long-range correlations*, Cent. Eur. J. Phys. **7**, 387-394 (2009).
25. U. Tirnakli, C. Beck and C. Tsallis, *Central limit behavior of deterministic dynamical systems*, Phys. Rev. E **75**, 040106 (2007).
26. U. Tirnakli, C. Tsallis and C. Beck, *A closer look at time averages of the logistic map at the edge of chaos*, Phys. Rev. E **79**, 056209 (2009).
27. M.A. Fuentes and A. Robledo, *Stationary distributions of sums of marginally chaotic variables as renormalization group fixed points*, 0912.2942 [cond-mat.stat-mech].
28. M.A. Fuentes and A. Robledo, *Renormalization group structure for sums of variables generated by incipiently chaotic maps*, 0912.2930 [cond-mat.stat-mech].
29. O. Afsar and U. Tirnakli, *Central limit behavior of the sine-circle map at the quasi-periodic edge of chaos*, 1001.2689 [cond-mat.stat-mech] (2010).
30. C. Anteneodo and C. Tsallis, *Breakdown of the exponential sensitivity to the initial conditions: Role of the range of the interaction*, Phys. Rev. Lett. **80**, 5313 (1998).
31. M. Antoni and S. Ruffo, Phys. Rev. E **52**, 2361 (1995).
32. A. Pluchino, A. Rapisarda and C. Tsallis, *Comment on "Ergodicity and central limit theorem in systems with long-range interactions" by Figueiredo A. et al*, Europhys. Lett. **85**, 60006 (2009).
33. C. Tsallis, G. Bemski and R.S. Mendes, *Is re-association in folded proteins a case of nonextensivity?*, Phys. Lett. A **257**, 93 (1999).
34. A.C. Tsallis, C. Tsallis, A.C.N. Magalhaes and F.A. Tamarit, *Human and computer learning: An experimental study*, Complexus **1**, 181 (2003).
35. B. Latour and S. Woolgar, *Laboratory Life: The Social Construction of Scientific Facts* (Sage, Los Angeles, 1979).
36. B. Latour, *Reassembling the Social: An Introduction to Actor-Network-Theory* (Oxford University Press, Oxford, 2005).

37. J. Law and J. Hassard, eds., *Actor Network Theory and After*, Sociological Review Monographs (Blackwell, Oxford, 1999).
38. V. Despret, *Hans, le cheval qui savait compter* (Empêcheurs de penser en rond, Paris, 2003).
39. T. Hadzibeganovic and S.A. Cannas, *A Tsallis' statistics based neural network model for novel word learning*, Physica A **388**, 732-746 (2009).
40. T. Takahashi, T. Hadzibeganovic, S.A. Cannas, T. Makino, H. Fukui and S. Kitayama, *Cultural neuroeconomics of intertemporal choice*, Neuroendocrinology Lett. **30**, 185-191 (2009).
41. L.A.V. Carvalho, D.Q. Mendes and R.S. Wedemann, *Creativity and delusions: The dopaminergic modulation of cortical maps*, Proceedings of the 2003 International Conference on Computational Science (ICCS 2003), eds. P.M.A. Sloot et al, *Lecture Notes in Computer Science* **2657**, 511-520, 2003 (Springer-Verlag, Berlin, 2003).
42. R.S. Wedemann, R. Donangelo and L.A.V. de Carvalho, *Nonextensivity in a memory network access mechanism*, Braz. J. Phys. **395**, 495-499 (2009).
43. R.S. Wedemann, R. Donangelo and L.A.V. de Carvalho, *Generalized memory associativity in a network model for the neuroses*, Chaos **19**, 015116 (2009).
44. M. Wertheimer, *Gestalt Theory*, in *A Source Book of Gestalt Psychology*, ed. Ellis W.D. (Routledge and Kegan Paul, London, 1938), pages 1-11.
45. W. Kohler, *Gestalt psychology today*, American Psychologist **14**, 727-734 (1959).
46. A. Upadhyaya, J.-P. Rieu, J.A. Glazier and Y. Sawada, *Anomalous diffusion and non-Gaussian velocity distribution of Hydra cells in cellular aggregates*, Physica A **293**, 549 (2001).
47. A.M. Reynolds, *Can spontaneous cell movements be modelled as Lévy walks?*, Physica A **389**, 273-277 (2010).
48. K.E. Daniels, C. Beck and E. Bodenschatz, *Defect turbulence and generalized statistical mechanics*, Physica D **193**, 208 (2004).
49. R. Arevalo, A. Garcimartin and D. Maza, *Anomalous diffusion in silo drainage*, Eur. Phys. J. E **23**, 191-198 (2007) [DOI10.1140/epje/i2006-10174-1].
50. R. Arevalo, A. Garcimartin and D. Maza, *A non-standard statistical approach to the silo discharge*, in *Complex Systems - New Trends and Expectations*, eds. H.S. Wio, M.A. Rodriguez and L. Pesquera, Eur. Phys. J.-Special Topics **143** (2007) [DOI: 10.1140/epjst/e2007-00087-9].
51. P. Douglas, S. Bergamini and F. Renzoni, *Tunable Tsallis distributions in dissipative optical lattices*, Phys. Rev. Lett. **96**, 110601 (2006); G.B. Bagci and U. Tirnakli, *Self-organization in dissipative optical lattices*, Chaos **19**, 033113 (2009).
52. B. Liu and J. Goree, *Superdiffusion and non-Gaussian statistics in a driven-dissipative 2D dusty plasma*, Phys. Rev. Lett. **100**, 055003 (2008).
53. R.G. DeVoe, *Power-law distributions for a trapped ion interacting with a classical buffer gas*, Phys. Rev. Lett. **102**, 063001 (2009).
54. L. Borland, *Closed form option pricing formulas based on a non-Gaussian stock price model with statistical feedback*, Phys. Rev. Lett. **89**, 098701 (2002).

55. L. Borland, *A theory of non-gaussian option pricing*, Quantitative Finance **2**, 415 (2002).
56. S.M.D. Queiros, *On non-Gaussianity and dependence in financial in time series: A nonextensive approach*, Quant. Finance **5**, 475-487 (2005).
57. L.F. Burlaga and A.F.-Vinas, *Triangle for the entropic index q of non-extensive statistical mechanics observed by Voyager 1 in the distant heliosphere*, Physica A **356**, 375 (2005).
58. L.F. Burlaga and N.F. Ness, *Compressible "turbulence" observed in the heliosheath by Voyager 2*, Astrophys. J. **703**, 311-324 (2009).
59. B. Bakar and U. Tirnakli, *Analysis of self-organized criticality in Ehrenfest's dog-flea model*, Phys. Rev. E **79**, 040103(R) (2009).
60. F. Caruso, A. Pluchino, V. Latora, S. Vinciguerra and A. Rapisarda, *Analysis of self-organized criticality in the Olami-Feder-Christensen model and in real earthquakes*, Phys. Rev. E **75**, 055101(R)(2007).
61. L.G. Moyano and C. Anteneodo, *Diffusive anomalies in a long-range Hamiltonian system*, Phys. Rev. E **74**, 021118 (2006).
62. J.C. Carvalho, R. Silva, J.D. do Nascimento and J.R. de Medeiros, *Power law statistics and stellar rotational velocities in the Pleiades*, Europhys. Lett. **84**, 59001 (2008).
63. M.A. Moret, V. de Senna, G.F. Zebende and P. Vaveliuk, *X-ray binary systems and nonextensivity*, Physica A **389**, 854-858 (2010).
64. R.M. Pickup, R. Cywinski, C. Pappas, B. Farago and P. Fouquet, *Generalized spin glass relaxation*, Phys. Rev. Lett. **102**, 097202 (2009).

FUMBLING TOWARDS ECSTASY*: A JOURNEY TO UNDERSTAND A SMALL CORNER OF THE UNIVERSE

BERNHARD H.J. JUURLINK

College of Medicine, Alfaisal University,
Riyadh, Kingdom of Saudi Arabia
E-mail: bjuurlink@gmail.com

What is common amongst the sciences, humanities and fine arts is creativity. A scientist is simply an artist with his/her head in the clouds BUT feet anchored firmly in terra firma – a scientist's creativity must be reflective of the world in which he/she lives to a much greater extent than that of the artist. Creativity is the ability to see connections others have trouble seeing, to see the unfamiliar in the familiar. Although often considered to be a rare gift, creativity is a fundamental property of living things. How are things connected? This question is fundamental and necessary for life; hence, it is hardwired into all living things, especially in animal and animal-like beings. Is this a nutrient or poison? Is the oxygen tension appropriate for my survival? Is that predator or prey? Is this the direction to water? to minerals? In animals the hard-wiring is over-written by soft-wired programming as childhood turns to youth and then adulthood - animal adults are less curious than the young. This is, of course necessary, since the brain is slowly establishing some sort of map of the external world in relationship to that of the body, i.e., the brain is establishing a terra cognita; life is not possible in a perpetual terra informa. One cannot be overburdened with investigating all possible, and seemingly impossible, connections since this leads to insanity. Curiosity must have some boundaries, it must have some reasonably firm base off which to spring. Creativity requires some, usually inchoate, knowledge about which boundaries to destroy and which ones to erect. It requires some anchor to which possible connections can be tied. But it also must allow the existence of apparent contradictions and terra cognita ideally should be terra cognita infecta. It must allow for accidental findings that may resolve the apparent contradictions. This paper attempts to delineate some of the creative aspects that are held in common between the poet and the scientist.

1. Introduction

This essay is my perspective on what is knowledge, how we obtain knowledge, what is creativity and what is the essential difference between the poet and the

* taken from the title of a song by Sarah McLachlin

scientist. It is also an essay on some of my experiences that led to new research directions. I have read widely over the years and must have encountered most of the ideas presented in this essay somewhere in my readings; however, the source of these ideas are forgotten. This forgetting reminds one of a quotation by Albert Einstein: *"Education is what remains after one has forgotten what one has learned in school"*.

2. What is Knowledge?

2.1 The Search for Patterns of Familiarity

Our brains are organized to search for patterns that are familiar, that have some meaning to our survival. This is illustrated in Figure 1. We clearly see patterns, but what are we seeing? Is it abstracted from some entity that we 'know'? For those that appreciate abstract art it may be sufficient to appreciate the geometrical and colour patterns seen in the original figure. But for many of us there is an unease that can only be ameliorated by recognizing something familiar or by identifying some meaning that can be attributed to the figure or identifying some relevant context.

Figure 1. Abstracted patterns

Figure 2 illustrates the source of the abstracted patterns in Figure 1. Figure 2 is full of meaning for some. In it we recognize a variation of the unique dress of the First Nations dancer of the great plains and parklands of North America. For those of us who have seen First Nations dancers it reminds us of the unique dissonant harmony of the dance that reflects some aspect of the traditional life from a century or more ago. The dancer may also make us reflect on some of the cultural traditions of the Lakota or other First Nations people.

So what is knowledge? Figures 1 and 2 suggest that knowledge in some sense is the recognition of patterns that are not only related to the world around us but that that have some meaning in the context of our survival, either our physical survival or mental survival.

2.2 How To Construct A Knowledge Framework?

Knowledge is more than just a recognition of patterns. For us humans it is a mental reconstruction on how we 'imagine' the world to be. Even simple organisms such as a bacterium must have "knowledge' of how the world around them is organized. Bacteria have means to detect features of the world around them that affect their survival. Unlike with us, the bacterial detection and response systems are intimately related to their genomes. The bacterial perspective on the world around them is far simpler and may give us some insight into what is "knowledge". Let us consider the "lac operon" present in certain gut bacteria.

Figure 2. First Nations Dancer (with face obscured) from the great plains and parkland regions of North America

The sugar glucose is a major energy source for gut bacteria as it is for us. Some gut bacteria such as *Escherichia coli* (*E. coli*) can utilize lactose, the milk sugar, as an energy source. Lactose is a sugar dimer comprised of glucose and galactose. In order for *E. coli* to utilize lactose it must first express the gene that encodes β-galactosidase that converts lactose into glucose and galactose as well as lactose permease, a protein that promotes lactose entry into the cell. Since transcription of genes into messenger RNA (mRNA) and the subsequent translation of the mRNA into protein requires energy as well as amino acid building blocks it is more efficient for the bacterium not to express β-galactosidase and lactose permease when there is an absence of lactose, for example, if the organism inhabited by the bacterium is not drinking milk. How *E. coli* does this was initially delineated by two French researchers, Francois Jacob and Jacques Monod in the 1960's. For their work in understanding the genetic mechanisms behind regulation of lactose metabolism in bacteria, Jacob and Monod received the Nobel Prize in Physiology/Medicine in 1965.

DNA or deoxyribonucleic acid as most everyone now knows is a polymer comprised of four nucleic acid bases (adenine {A}, thymine ({T}, cytidine {C} and guanine {G}). Furthermore, DNA is organized as a double stranded polymer with the bases hydrogen-bonding to each other in the following pairing: A:T and G:C. Information is encoded in the specific sequence of bases in one of the strands. From a simplistic point of view a typical gene is comprised of a promoter DNA base sequence and a protein-encoding DNA base sequence (Fig. 3). The promoter region, which can bind to specific proteins, regulates how readily the protein-encoding region is transcribed by RNA polymerase into a messenger RNA (mRNA). In bacteria the promoter region can be bound by repressor proteins that decrease the rate of transcription or can be naked and, thus, allow transcription of the protein-encoding region into an mRNA.

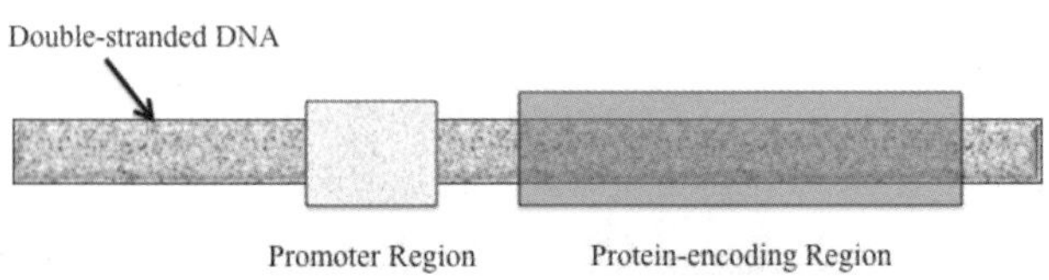

Figure 3. Organization of the gene into promoter and protein-encoding regions.

The information encoded in mRNA then is used by transfer RNAs (tRNAs) and the ribosomes to synthesize a specific sequence of amino acids forming an amino acid polymer, i.e., protein. Such proteins have specific functions, for example, β-galactosidase converts lactose into the sugars glucose and galactose.

2.2.1 The lac operon

The lac operon is comprised of three genes (conventionally described as LacZ, LacY and LacA) involved in lactose metabolism as well as an upstream promoter region. Normally the promoter region of the lac operon genes is occupied by lac repressor protein inhibiting the expression of the lac operon genes (Fig. 4A). The lac repressor gene is constitutively expressed, i.e., it is always expressed. In the absence of lactose the promoter of the lac operon is occupied by the lac repressor protein: in such a condition little transcription occurs and only low levels of the operon genes are expressed. What are the protein-encoding genes of the operon? There are three, only two of which are critical in allowing the metabolism of lactose: 1) β-galactosidase (LacZ), the enzyme that breaks the bonds between the two sugars (galactose and glucose) that comprise lactose. 2) lactose permease gene (LacY) whose protein product increases the probability of lactose entering the bacterium. 3) β-galactoside transacetylase (LacA), this latter inactivates the potentially toxic galactose. The low levels of β-galactosidase that is constitutively expressed will convert some of the lactose to allolactose, an isomer (alternative form) of lactose. Allolactose binds to the lac repressor causing the repressor to change conformation such that it no longer binds to the lac promoter (Fig. 4B), thus allowing expression of β-galactosidase, lactose permease and β-galactoside transacetylase – this then enables the bacterium to obtain glucose from lactose and use the glucose thus obtained as a fuel.

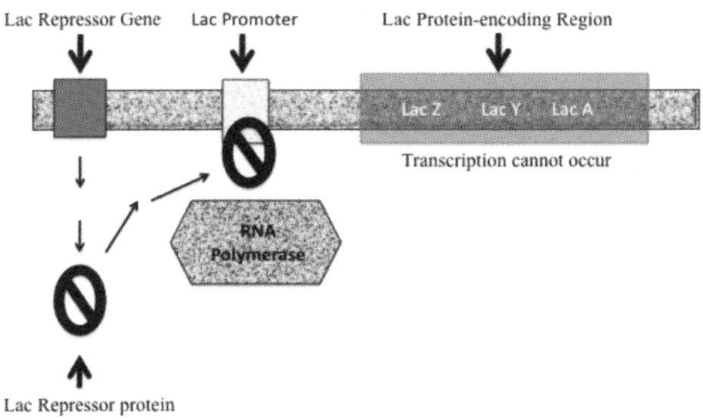

Figure 4A. The lac operon, comprised of the Lac promoter and the Lac-protein-encoding region. In the absence of lactose the lac repressor protein blocks the lac promoter site preventing RNA polymerase from transcribing DNA encoding the lac genes (Z, Y and A) into mRNA. Note there is constitutive expression of the lac repressor protein.

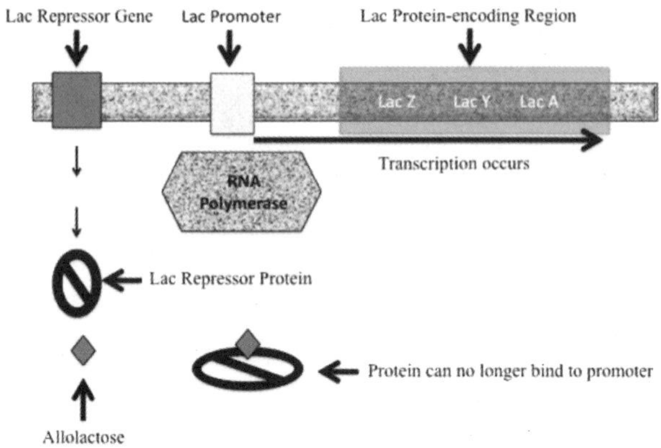

Figure 4B. The lac operon. In the presence of lactose, allolactose is formed. The allolactose binds to the lac repressor protein preventing its binding to the lac promoter allowing RNA polymerase to transcribe the lac genes (Z, Y and A) into the mRNA for β-galactosidase, lactose permease and galactoside transacetylase. The mRNAs will then be translated into the respective proteins.

The above is an incomplete model of the regulation of lactose metabolism since in the presence of glucose and lactose there is only a low level of expression of the lac operon genes while in the presence of lactose and absence of glucose there is a high expression. I will not go into the mechanisms behind this except to note that this is very logical since in the presence of glucose there is no need to expend energy and amino acids for lactose metabolism.

The lac operon is only one example of an operon within *E. coli*. The genetic machinery of *E. coli* enables it to map the world around it in terms of metabolites that are necessary for survival. *E. coli* have receptors of information (e.g., the lac repressor protein), they have a means of responding to this information (e.g., expression of the lac operon genes and the subsequent metabolism of lactose). *E. coli*, thus, have "knowledge" about the environment that is necessary for survival. This is not very different from how we build our knowledge framework: we have information (sensory) receptors, normally we integrate this information (using our nervous system rather than our DNA) and we usually respond to this integrated information by altering our knowledge framework and often by altering our behaviour rather than altering gene expression, although this can occur as well.

2.3 How Animals Build A Knowledge Framework

Multicellular animals are much more complex than bacteria. They can respond to more than just the chemicals in the environment: they can respond to various mechanical aspects of the environment that are interpreted as touch or hearing or equilibrium. Some animals can use vibrations in the air or water to develop very complex three-dimensional maps of their environment (e.g., bats and dolphins). Most animals can respond to parts of the electromagnetic spectrum, i.e., they can "see". Which parts of the spectrum can be "seen" depends very much upon the photon absorption properties of the pigments in the photoreceptor cells. Some animals can build up thermal images of their environment, i.e., they have pin-hole camera type of "eyes" that image small changes in temperature – e.g., pit vipers. Yet others can detect disruptions in the electrical fields around them, either passive detection of electrical fields generated by other animals (e.g., sharks or the monotreme mammal platypus) or by active detection where the animals develop a strong electrical field around them and then detect other creatures or objects by sensing distortions in the electrical fields they have generated – this is very useful if one inhabits murky waters.

We are not born a *tabula rasa*. Our knowledge of the world around is delimited by our sensory receptors and the make-up of our nervous system. What is remarkable are the limited physical features of the environment that can be detected using our sensory receptors. Yet more remarkable is how such simple features can be built into complex mental representations of the world. Of course, now we have built instruments that can increase the number of features of the world around us that can be detected.

The ability to detect multiple aspects of one's environment also necessitates development of a means to integrate this information; hence, nervous systems were developed. Nervous systems essentially develop maps of the sensory information obtained. The nature of the maps depends upon the nature of the sensory information. If the sensory information involves the sense of touch on the body surface, we find maps of the body's surface in different parts of the nervous system. The best known of these maps are present in the cerebral cortex. One example is the map of the body's surface in the primary somatosensory cortex. The area of the cerebral somatosensory cortex devoted to a particular body region is related not to the surface area of the body region, rather it is related to the density of the sensory receptors. Such maps are known as somatotopic maps and the relatively flat somatotopic maps found represented within the cerebral cortex can be rendered into a three-dimensional somatosensory homunculi such as depicted in Figure 5. Upon examining such a homunculus one quickly becomes impressed with the density of receptors on the tongue, lips and hands. It is not a particularly flattering view of ourselves but it does emphasize how we engage with the universe.

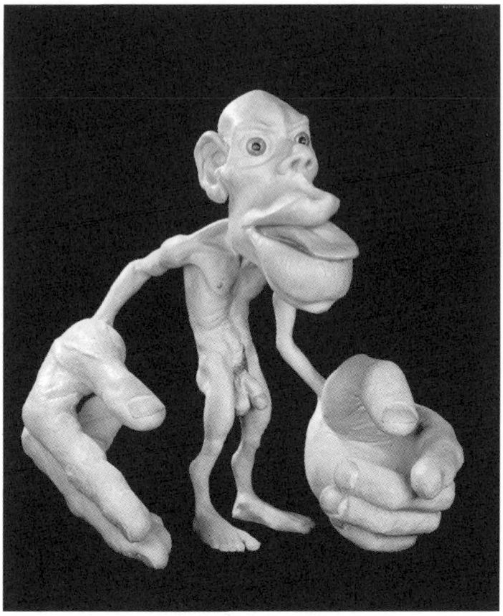

Figure 5. The somatosensory homunculus representing the area of the somatosensory cortex devoted to different regions of the body's surface. ©The Natural History Museum of London, used with permission.

The somatosensory map is developed early and can be considered to be the primary map to which all other maps are referred. We see this in the baby where everything encountered must be touched by hand, by lips and by tongue and then smelled and tasted. The brain knows that chemical information in the form of taste and smell arises from the cranial part of the body since this is where the receptors are located. Although the visual world is mapped initially in bits and pieces in the form of contrast borders, movement and colour, the visual world is then reconstructed in relationship to the body's surface since our body surface is in direct contact with the world around us. Similarly information obtained through sound is mapped relative to our body's surface. We, indeed, occupy the centre of the known universe as seen by our nervous system.

The evolution of the sensory receptors and the brain we have was driven by what specific features of our environment are necessary to detect and analyse to be meaningful for survival. The nervous system is an engine to detect patterns that are meaningful for survival, meaning that is necessary for our survival as an individual as well as a species. All living creatures are meaning-generating machines developing logical connections between the world around us as detected by our receptors and survival.

2.3.1 Knowledge and imagination

As our brain became more complex, our maps of the world around us became more complex where we have superimposition of one map onto another. Such complex multimodal maps become very rich for the mining of meaning. At some point in our evolution the complex maps we developed become less and less tied to the physicality of world around us; that is, there is a dissociation between the world as we directly experience it and the world as we perceive it. We have developed "imagination'. Along with this is that meaning that is being sought is not only for survival in the physical world as it is but survival in the cultural world that we and other animals have developed in order to live in social settings. Imagination or creation of world maps not tightly tied to the physical universe becomes important. We live in an imagined, but not imaginary, universe, an *imago mundi.*

"Imagination is more important than knowledge. For knowledge is limited to all we now know and understand, while imagination embraces the entire world, and all there ever will be to know and understand." – Albert Einstein

The problem with imagination is that it is not tightly tied to the physicality of the universe. Aspects of the world we construct can be imaginary, for example, for many years people have believed in things such as unicorns for which there is no physical evidence. But imagination enables us to be creative: to see the unfamiliar in the familiar.

3. So Again what is Knowledge?

3.1 The Poet and the Scientist

Knowledge is simply a construct of how we 'imagine' the universe is organized. Anyone who has done a bit of reading in cultural anthropology knows that in many senses various peoples live in quite differently constructed universes. Even within a given society aspects of the world the individual lives in can be quite different from the world the individual's neighbour lives in – for example, some are religious, some agnostic while others may be atheists. What everyone has in common is the mechanism whereby their world views are constructed.

So what is the difference between how I, as a scientist, approach understanding parts of the world around me as opposed to I the poet? The approaches are not that different. The essential difference is that the scientist will go out of his/her way to tie the constructed world or imagined world to the physicality of the universe while the poet need not. I will give one example of differences from my history as a researcher.

Soon after I started off as an independent researcher a colleague of mine was diagnosed with amyotrophic lateral sclerosis. Within the next three years his motor functions slowly declined and he finally died from pneumonia. At the time of his death he was carrying out research on an avian form of muscular dystrophy testing the hypothesis that the disease was caused by abnormalities of amino acid transport across the muscle membrane. I agreed to supervise the winding down of his research laboratory and determined the original hypothesis was not upheld by the experimental data obtained. In the meantime I became quite intrigued with this form of muscular dystrophy. It was intriguing since it appeared to be manifesting in only one muscle fibre type, the type IIB fibre and only when the IIB fibre type formed the majority of fibres in a muscle belly. We have essentially 3 main muscle fibre types that are adapted for various muscle functions. For muscles that are almost always active, for example our postural muscles, the muscle fibre type that dominates the muscle belly is the type I fibre. These muscle fibres contract slowly with modest power and are very dependent upon oxidative metabolism of fatty acids for their function. Muscle bellies comprised mainly of type I fibres are richly supplied by blood vessels since they use a lot of oxygen and need to get rid of a lot of waste products derived from their very active metabolism. Type IIA fibres are intermittently very active and can generate much more force than type I fibres: these fibres use both fatty acids and glucose as energy sources and also have a good blood supply for the same reasons as the type I fibres. Type IIB fibres are only sporadically used and can generate the most force of the three fibre types. They are completely dependent upon anaerobic metabolism of glucose (*i.e.*, metabolism that is not dependent upon oxygen), with the glucose obtained from their glycogen stores. They do not require much oxygen for their function and have a relatively poor blood supply. When they are active they have a very high rate of metabolism. Intriguing to me was that type IIB fibres when active generate a lot of heat but because of their poor blood supply they have a poor ability to dissipate heat.

At the time I was interested in a phenomenon known as the heat shock or stress response. This response enables cells to better cope with a variety of stresses including heat stress. It occurred to me that if there were a problem in the heat shock response in these birds one would expect it to become manifest initially in muscle bellies where the predominant muscle fibre was type IIB for the following reasons: 1) when active these muscle fibre types would generate a lot of heat, 2) because of the poor blood supply of these muscles, there would be a poor ability to dissipate the heat generated, 3) this increased heat and absence of a stress response would result in increases in cell membrane permeability and abnormalities of ion homeostasis resulting in prolonged elevation of intracellular Ca^{2+} thereby activating Ca^{2+}-dependent proteases that would cause damage to the muscle fibres exactly as seen in the IIB fibres in the avian form of muscular dystrophy. I thought this map/explanation/hypothesis I had created of avian muscular dystrophy was extraordinarily beautiful since it could explain all the detailed phenomenology of avian dystrophy. I had such marvelous dreams about this beautiful hypothesis and was reminded of a quotation by P.A.M. Dirac

".... it is more important to have beauty in one's equations than to have them fit experiment"

Dirac wrote this to suggest that if Erwin Schrödinger had more confidence in his equations and not paid as much attention to his data he could have published much earlier in a more elegant fashion. Dirac felt that a sense of aesthetics can be more important to detect the patterns of nature than data that often are incomplete.

Unlike the case with Schrödinger, my hypothesis explaining avian muscular dystrophy was not only beautiful but all the data at the time were congruent with the hypothesis. What I had constructed was poetic and a thing of beauty and I suspect that I as a poet would have left the paradigm as it stood. But I, as a scientist, could not. I had to test it to see if it bore any relationship to the universe as it was. The simplest test was to see if there were problems in the heat shock response in the cells from the dystrophic birds. The hypothesis was beautiful, the testing of the hypothesis was simple and I could expect a yes or no response to the question addressed. To my dismay there was no difference in the heat shock response between the cells of the birds with avian muscular dystrophy and the cells of the control birds – the data were unequivocal. I went from such a high to such a low. I was reminded of Thomas Huxley:

"The great tragedy of science - the slaying of a beautiful hypothesis by an ugly fact."

The Poet's beautiful hypothesis or model of the world can never be slain by an ugly fact while the Scientist's beautiful model can be slain.

3.2 What is Creativity?

Creativity is essentially the ability to see connections where none have been seen before, to see the unfamiliar in the familiar. It is the ability to see something in a very different context. It is often a consequence of reflecting on problem Y while having problem X in the back of one's mind – with some aspects of the perspective on problem X being applied to problem Y. The previous section of this essay is a good example of how thinking about one problem while reflecting on another can give rise to a new way of looking at the first problem. This is at the very essence of poetry as well as science. Although my hypothesis that a failure in heat shock response was at the root of avian muscular dystrophy turned out to be false, other experiences by others (as well as myself) have turned out more positively. One well-known example is when Alexander Fleming encountered a *Penicillium sp* mould contamination of his bacterial cultures. Countless microbiologists have certainly encountered *Penicillium sp* contamination of their bacterial cultures and some had even published how this mould killed the bacteria in the cultures. But it was Fleming who saw the potential of this mould in developing drugs to counter bacterial infections. Fleming at the time was interested in antibacterial substances and noting the antibacterial action of *Penicillium* immediately saw some pattern that others had not. Louis Pasteur wrote:

"In the fields of observation chance favours only the prepared mind."

Thus, working on one problem and then encountering a different problem can enable one to see a connection that others may not have seen. Some connections, however, are completely accidental and due to our particular perspective. I will try to illustrate this with the bit of verse on the following page. This poem was created in an attempt to understand someone with a mental disorder characterized by seeing connections in many phenomena where no connections actually existed. In an attempt to give a feeling about the chaotic thought process the writing swirls back and forth, often sweeping across the page. In this verse the bold writing should be read before the writing in italics. But note also that swirling the verse from side to side can create unintended connections where words in bold meet words in italic.

Dys / *connectivity*

Tumbling *Using*
Spinning *Fist*
Uncontrolled *Of*
Isolated *Mind*
Thoughts *Break*
Making *Unreal*
Surreal *Connections*
Connections *Slow*
Sweep *Thoughts*
Through *To*
Mind *Inhibit*
The **Faster**
Chaos **And**
Of **Faster**
The **Developing**
Spiral **Vortex**
Of
Sever **Mad**
The **Dispair**
Brilliance
Of
Mind **Friend**
From **Or**
Reflexive **Family**
Reiterating **Must**
Reflectant **Spin**
To *Refractory*
Catch *Thoughts*
Solitary *That*
Soul *Connect*
Drowning *The*
In *Unconnectedness*
The
Treacle
Of *Liberate*
Past *The*
And *Sanity*
Present *Within*
Con_nec_**tiv**_iti_**es**

In this poem one sees connections between the words in bold and the words in italics resulting in: "**spinning** *fist* **making** *unreal* **mind** *inhibit* **mad** *reflectant* **spin in** *unconnectedness* **present** *within* **con**_nec_**tiv**_iti_**es**" that contradicts what the poem appears to state. The difference in creativity between the minds that are considered sane and those that are considered insane is essentially that sane individuals ultimately identify the pattern of connections that is reflective of our world. Of course, the ability to see connections where none actually exist also appears in the scientific approach. One dramatic example from my own experience is my hypothesis that avian muscular dystrophy was due to an inability of the cells of the dystrophic birds to have a normal heat shock response. All the phenomenologies known at that time were congruent with the hypothesis, but the hypothesis was incorrect.

3.2.1 Serendipity and creativity

Serendipity plays an important role in creativity. If Alexander Fleming had not had a *Penicillium sp* contamination of his bacterial cultures, the discovery of penicillin would have occurred much later. Serendipity has also contributed to my own research in a number of instances, one of which I will relate. I was blessed, although I did not realize this at the time, about two decades ago by the fact that the Facilities Management of the University of Saskatchewan decided to save monies by not air-conditioning the research laboratories on weekends. There may have been sent a notice to this effect but, if so, I was oblivious to this change in policy. At the time I was interested in understanding the cell biological properties of oligodendrocyte precursor cells. These cells give rise to oligodendrocytes that form the insulating myelin sheaths around nerve fibres. I thought that understanding the properties of these precursor cells may give some insight into multiple sclerosis, a demyelinating disease. One approach I took was to isolate the oligodendrocyte precursors from mixed astrocyte and precursor cell cultures (Fig. 6) and then to study them both in isolation as well as in interactions with astrocytes. It turned out to be very difficult to isolate these cells with several months of effort required for my assistant, Connie Wong, and I to perfect the techniques. Once the isolation techniques were perfected it took several weeks from start to finish to obtain cultures highly enriched in the oligodeondrocyte precursors. In contrast to these oligodendrocyte precursor cells, astrocytes and cerebral cortical neurons are very easy to isolate and grow in cell culture.

We did finally manage to get highly enriched cultures of precursor cells with the final step in the two-week isolation procedure occurring on a Friday afternoon during late June. That weekend I was away and came back on Monday morning and was shocked to see that all the oligodendrocyte precursor cells in culture were dead while all the astrocytes and cerebral cortical neurons in my cultures were alive. I was completely baffled but set-backs such as this is not an uncommon experience with experimental scientists. We proceeded to isolate another batch of oligodendrocyte precursor cells and this time I came in on weekends to observe the cultures. To my surprise the lab was an oven with a temperature a little over 40° C. The incubators were even warmer. The tentative conclusion was that the precursor cells were dying in response to temperature elevations. After some discussion Facilities Management agreed to keep the air-conditioning on in certain laboratories during the weekends.

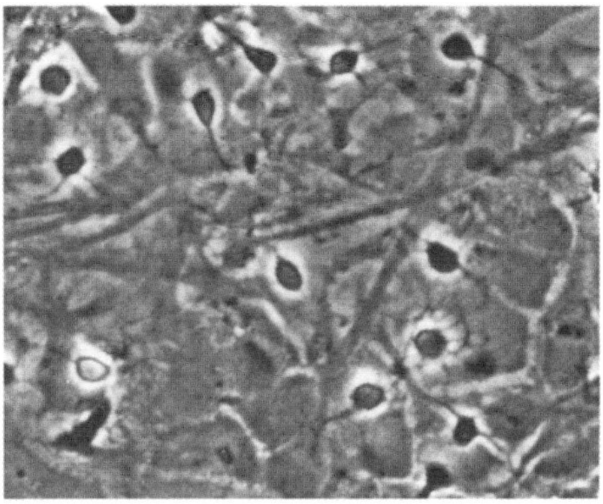

Figure 6. Phase contrast photograph of a mixed glial cell culture obtained from newborn rat cerebral hemispheres. The small process-bearing cells are oligodendrocyte precursor cells. The carpet upon which they rest are the astrocyte cells.

I then started reflecting on the fact that the cortical neurons and astrocytes in culture did not seem affected while the oligodendrocyte precursor cells were negatively affected by this rise in temperature. This led to a series of experiments attempting to delineate what perturbations the precursor cells were susceptible to and what accounted for this susceptibility. This turned out to be a very fruitful area of research since almost every question posed resulted in a response that made sense. We demonstrated that a variety of perturbations resulted in death of the precursor cells including small elevations in temperature and reductions in nutrients and/or oxygen. We learned that these precursor cells were very different from astrocytes in many ways. We demonstrated that these cells had a very high rate of metabolism and because of this they produced an abundance of free radicals and other strong oxidants (Fig. 7). We also learned that the cells had a very high iron content and the consequence of this is that the probability of nasty free radicals being produced from relatively innocuous oxidants is greatly increased. Curiously, the cells had poor means to inactivate the strong oxidants. Astrocytes, in contrast, had a lower rate of metabolism, had a lower rate of strong oxidant production and had a greater ability to inactivate the oxidants produced. We also demonstrated that such differences were present *in vivo* and not just a cell culture artifact.

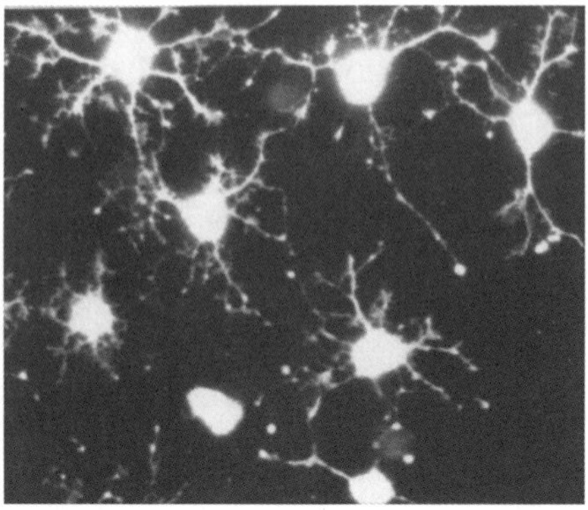

Figure 7. Mixed glial culture subjected to a physiological perturbation. Excessive production of strong oxidants was visualized by means of a compound that became fluorescent when oxidized. It is clear that the oligodendrocyte precursor cells are producing more oxidants than can be inactivated by the endogenous anti-oxidant mechanisms while this is not the case with the underlying astrocytes.

These experiments led me to a deeper understanding on metabolism in that 1) electrons harvested from foodstuffs are used by cells to generate energy stores necessary for cellular function, 2) are used in the synthesis of a number of molecules necessary for cells, or 3) can be used to inactivate strong oxidants. If a cell is synthesizing a lot of membrane as oligodendrocyte precursors do, there is a very high rate of metabolism and strong oxidant production; however, few of the electrons harvested are available for inactivating oxidants since they are being used for synthesis of membrane macromolecules. The consequence is that the cell has only a small margin of safety between producing oxidants and being able to inactivate oxidants. This research also led me to a deeper understanding of the innate anti-oxidant mechanisms of the cell and how to upregulate such anti-oxidant mechanisms.

This very basic research trying to understand why oligodendrocyte precursor cells died in my tissue culture incubators led to a fruitful area of research that can be directly applied to the human condition since an increase in strong oxidant production with no concomitant increase in the ability to inactivate oxidants underlies many disorders. The research publications that arose from this activity cover topics such as approaches to alleviate trauma to the brain (Schültke *et al.* 2005. Neuroprotection following fluid percussion brain trauma: A pilot study using quercetin. *J. Neurotrauma* **22:** 1475-1484) and spinal cord (Kamencic *et al.* 2001. Promoting glutathione synthesis following spinal cord trauma decreases secondary damage and promotes retention of function. *FASEB J.* **15:** 243-250); and age-related disorders such as stroke (Juurlink. 1997. Response of glial cells to ischemia-related insults: Roles of reactive oxygen species and glutathione. *Neurosci. Biobehav. Rev.*, **21:** 151-166; Juurlink, B.H.J. 2001. Anti-oxidant strategies to treat stroke. In G.Z. Feuerstein (ed.), **Progress In Inflammation Research: Inflammation and Stroke**, Springer Verlag, Basel. pp 299-312). Of particular interest my laboratory has shown that a number of compounds found in certain of our foods can upregulate the innate anti-oxidant defence mechanisms resulting in decreased oxidative stress and inflammation, thereby ameliorating aging-related problems (*e.g.*, Wu *et al.* 2004. Dietary approach to attenuate oxidative stress, hypertension and inflammation in the cardiovascular system. *Proc. Natl. Acad. Sci.* USA **101:** 7094-7099; Noyan-Ashraf *et al.* 2008. Phase 2 protein inducers in the diet promote healthier aging. *Journal of Gerontology A: Biological Sciences & Medical Sciences* **63:** 1168-1176). These research activities have also led me to studying epigenetics and fetal determinants of adult health (Noyan-Ashraf *et al.* 2006. Dietary approaches to positively influence fetal determinants of adult health. *FASEB J.* **20:** 371-373; Senanayake *et al.* The dietary phase 2 protein inducer sulforaphane can normalize the kidney epigenome in hypertensive rats and improve blood pressure, Submitted for publication).

Who knows where serendipity may lead? Who knew that so many years ago I would be carrying out research on how diet may ameliorate problems associated with aging and influence fetal determinants of adult health? I was simply interested in trying to understand some of the cell biological properties of one cell type.

3.2.2 Ignorance and creativity

Many of the advances in science are made by the young. The reasons for this is that they have often less fear of displaying ignorance and usually do not have rigidly developed ways of looking at the world. The problem with developing models of the world we live in is that we become accustomed to a particular model and a particular way of looking at things. It becomes difficult to step aside to see the unfamiliar in what is the familiar. At such times a large degree of ignorance about an area can be a blessing. Ignorance allows imagination unfettered by the shackles of what we think is the known. Once hypotheses are generated then it becomes important to fill in the gaps of knowledge. One can do this by reading extensively but often it is easier to consult with colleagues with expertise in the areas in which one is ignorant followed by reading the literature. I have been fortunate at times in not being too frightened to display ignorance on a topic and importantly in having colleagues who would entertain discussions on hypotheses in areas in which I have little knowledge. Time will tell if my latest foray into the unknown will be fruitful, i.e., will my recent hypothesis on the bystander effect of x-rays be upheld.

I have also been fortunate in having wonderful graduate students and post-doctoral fellows sojourning in my laboratory. Many of these individuals have come up with their own hypotheses offering unique insights to problems being pursued in the laboratory and have changed some of the directions of research either directly or indirectly. A very important factor that promotes creativity is the continual interaction with young enquiring minds.

4. Concluding Remarks

Knowledge is a mental reconstruction of the world around us. In Science this reconstruction is based on patterns of information received through our sensory receptors and extensions of these receptors, i.e., the instruments we have developed. What distinguishes science from the arts is that the models of the world around us that we construct must be tested against the information we receive through our senses. The very fact that we, and other creatures, are here instructs that the major axioms we can rely upon are: 1) the information we receive through our senses reflect the world around us, and 2) there is a logic in how the world around us is organized. If this were not the case we could not exist.

The models that are constructed in science may be as creative and fantastic as anything seen in a fantasy novel. For example, the string theory or the multiverse models of the universe. But these are hypotheses only to explain the particular patterns of information we have at the moment about how the universe is organized. It is only when the information we receive through our senses are in overwhelming support of the hypotheses that such hypotheses become the "received" part of our body of knowledge. But even then in science one keeps in the back of one's mind that this "received" body of knowledge may well be incomplete or perhaps not even 'TRUE'. Physicists immediately reflect on the Newtonian Universe in comparison with the Einsteinian Universe or classical mechanics in comparison to quantum mechanics. I, a cell biologist, upon reading "Cells, Gels and the Engines of Life" by Gerald H. Pollack, a speaker at this symposium, am now reflecting whether that the "received" understanding on how the cell functions is actually how the cell functions. For example, the raw data (or information) received through the instruments of cell biology can be interpreted to give rise to two very different models of how the cell membrane potential arises. The classical cell physiological explanation relies upon the presence of differences: i) in ion composition between the inside and the outside of the cell due to actions of ion pumps and ii) cell membrane ion channels of differing permeabilities. The alternative, and very intriguing, explanation relies more upon the properties of water: how water interacts with charged membrane surfaces and how ordered water preferentially excludes one ion over another.

The Universe as viewed through the lens of Science is a tentative construct subject to revision as either more data come in or new perspectives arise. The search for patterns especially patterns of familiarity in the unfamiliar is a major driving force in understanding the world around us. We see this all around us, including in this symposium where a number of speakers explored patterns or equations of quantum mechanics to see if they can be used to better understand the patterns of human economics.

"We shall not cease from exploration, and the end of all our exploring will be to arrive where we started and know the place for the first time."
- T.S. Eliot

Acknowledgements

I am most thankful to my parents whose genetic and cultural gifts resulted in my being continually curious about the world around me, to the students and post-doctoral fellows who have kept me young, to those colleagues who never hesitated to discuss ideas outside the pale and to all the funding agencies who allowed my laboratory to survive. I am also thankful to all of those who provided technical support allowing my laboratory to operate. And finally, to the College of Medicine, Alfaisal University for allowing me to participate in the establishment of a new student-centred, research intensive university in the Kingdom of Saudi Arabia.

ECONO-PHYSICS: MODELS & METAPHORS

PETER M. JACKSON
School of Management, University of Leicester, University Road, Leicester, LE1 7RH, UK
E-mail: pmj1@leicester.ac.uk

"There is really nothing more pathetic than to have an economist or a retired engineer try to force analogies between the concepts of physics and the concepts of engineers. How many dreary papers have I had to referee in which the author is looking for something that corresponds to entropy or to one or another form of energy?" (Samuelson, 1972, p.254).

1. Introduction

At first glance Samuelson's remarks are puzzling. It was Samuelson, who, initially in his, **Foundations of Economic Analysis** (1947), used concepts drawn from classical theoretical physics to inform his analysis of the economic system. This approach is found in his subsequent research and in the research agenda of those whom he influenced (and they are legion, for they are many). Emphasis is placed on systems thinking and the search for a stable equilibrium in both the short-run and the long-run. Neo-classical economics and the neo-classical synthesis, both Samuelson's brain children, owe much to the physics of his day (see Mirowski, 1989 and 2002). And yet in his Nobel Prize acceptance speech Samuelson was making a very important point. There is a tendency for physicists to migrate into economics with a view to using the standard tools of their trade to shed light upon economic problems and indeed to "solve" them. This is a dangerous tendency especially if physicists (or indeed others from the physical sciences more generally defined or the biological sciences) do not take the time and make the effort to understand the nature of economics, as a behavioural science. There are significant fundamental differences between natural/physical systems and economic/social (behavioural) systems that can render theories and models derived from the physical sciences irrelevant and useless if applied to socio-economic phenomena. This cautionary warning does not only apply to the exportation of theories and models from the physical to the social sciences it also applies with equal force to those social scientists who wish to import such ideas.

A recent example is found in Jean-Phillipe Bouchaud's essay published in **Nature** (2008). For a start the title of the essay is provocative, "Economics Needs a Scientific Revolution". Bouchaud (a physicist by training who is head of research at Capital Fund Management) takes economists to task for employing unrealistic axiom based models upon strong assumptions of human actor rationality. This is an age old debate and one which sadly misses the point regarding the nature and purpose of modelling in economics and some of the other social sciences. Bouchaud concludes his misguided essay with the strong demand that those working in economics and financial engineering should change their mindsets and adopt a, "more pragmatic and realistic representation of what is going on in financial markets and to focus on data, which should always supersede perfect equations and aesthetic axioms" (p.1181). Has he never read the vast literature on economics and does he not understand that in order to interpret, understand and explain data basic theory is required. Those, who like Bouchaud, would simply let the data speak for themselves are pitifully unaware that data speak in many different tongues in the Tower of Babel which constitutes empirical science. These tongues need to be interpreted with the assistance of models. Which model is employed will influence what is heard. In one thing, however, it is possible to agree with Bouchand, "to me the crucial difference between modelling in physics and in economics lies rather in how the fields treat the relative role of concepts, equations and empirical data" (p.1181). Such differences will be explored below.

Samuelson's sense of dreariness arises from the uncritical application of physics thinking to non-mechanical systems such as that of an economy, which is a highly complex, unstable and adaptive system made up of markets, institutions and purposeful individual decision makers which are equally complex and adaptive. The complexity of socio-economic systems has been captured and appreciated by many deep thinking physicists such as Bohr, Anderson, Smith and Pines, who do not see scientific endeavour in the crude terms of predicting the behaviour of simple mechanistic system but rather as grappling with the complexity of realities (whatever these might be).

2. The Econo-Physics Turn

Is econo-physics a new direction for economists? For a large number of years economists and other social scientists have allowed concepts drawn from physics to influence their thinking. Daniel Bernoulli (1738) introduced the concept of "utility" to describe individual preferences and Laplace (1812) demonstrated that what appear to be random events can be predictable. The French 19th century sociologist Comte coined the phrase "social physics" as part of his general positive philosophy which was to become "sociology". It has been pointed out that Quetelet, a student of Fourier also referred to social physics. Edgeworth the great 19th century Oxford University economist and statistician laid down the foundation of analytical economics in his **Mathematical Psychics** (1881) and Jevons before him together based their work on "psycho-physics" – mathematical representations of the hedonic calculus and analysis based upon empirical hypotheses about human psychology. Much of modern game theory and the concept of the "core" find their origins in Edgeworth. Irving Fisher, America's first mathematical economist and one of the founding fathers of the Econometric Society, in 1930, put his training in physics to use in developing analytical and statistical economics and in many important respects laid the foundations of modern finance theory. Others who used ideas drawn from the physical and natural sciences to inform their research agenda included Nicholas Georgescu-Roegen who applied the entropy law to his conceptualization of the economic process. Georgescu-Roegen (1966 and 1971) was highly critical of the dominant neo-classical treatment of the economic process as a mechanical analogue and sought to introduce to economics object lessons from physics which included uncertainty; hysteresis; entropy and disorder; determinism and causality. Some of Georgescu-Roegen's ideas were developments of the 19th century economist Herman Heinrich Gossen's (1854) **The Laws of Human Relation and the Rules of Human Action Derived There from**. Gossen was a contemporary of Jevons.

It has already been pointed out that Samuelson's **Foundations** were built upon analogies and ideas drawn from classical physics e.g. the Le Chatelier Principle; comparative statics; stability and dynamics. Later, in the late 1960s and 1970s, Samuelson introduced and developed within the context of economics, the "Brownian dance", of Bachelier, Einstein and Wiener. This line of enquiry spawned a significant and voluminous literature on random walks applied to commodity and financial markets and gave rise to the powerful concept of market efficiency and the central role played by information within economics.

This brief incomplete and superficial history of some economic thought serves but one function and that is to note that there has been a strong symbiotic relationship between physics and economics for well over 100 years, (for a more substantial treatment see Colander, 2000) long before the road was taken to Santa Fe.

A major event which established what can now be thought of as the econo-physics turn took place in 1987. The Santa Fe Institute sponsored a workshop on "Evolutionary Paths of the Global Economy". The workshop brought together a group of natural scientists and economists to introduce each discipline to the work and modes of thinking taking place in the other around the topics of techniques for studying non linear dynamical systems and adaptive paths in evolutionary systems. The organizing group were David Pines and Philip Anderson both of whom are physicists and Ken Arrow a leading economist. Since then a further workshop was organized in 1979 and economists, physicists and computer scientists meet regularly at the Santa Fe Institute as members of an inter-disciplinary research programme. Those associated with the programme represent a galaxy of some of the best minds in economics and physics and include Ken Arrow; W Brian Arthur; Steven Durlauf; Murray Gell-Mann; Stuart Kauffman; William Brock; Alan Kirman; Paul Krugman; Douglas North; Herb Gintis; Sam Bowles; Martin Shubik to name but a few.

Econo-physics can be regarded as a new direction in economics and one which presents many intellectual challenges and much excitement. What then is this new path?

3. Econo-Physics: What Is It?

Econo-physics can, at this stage, only be identified in terms of the problems that its participants study and the techniques that they use to understand these problems. There are too many problems and techniques to review in a short paper such as this. All that can be done is to give a flavour of a few of the issues. A useful survey is found in Farmer, Shubik and Smith (2005). Some of the topics covered include: empirical analysis of regulations in market data; the dynamics of price formation; understanding bubbles and panics; methods for pricing options and derivatives; the distribution of income; implications of symmetry and scaling for market functioning; complexity; self reinforcing mechanisms; increasing returns to scale; non linearity and complex dynamics and so on. Whilst it is not yet a crowded field it is never-the-less a busy one. One major theme running through much of econo-physics is the use and application of ideas drawn from complexity theory (a subject with its own problems of definition). Broadly conceived, complexity theory is used to understand events such as earthquakes and power blackouts i.e. events that are difficult to foresee. It can also be used to understand events that are the product of and outcome of the interactions of interdependent agents behaving in unstable systems that result in very large variations. This kind of behaviour can be described by power laws i.e. the frequency and magnitude of outcomes are explained in a power law of the form $f(x) \rightarrow x^{-a}$, where a is a constant and $f(.)$ is a probability distribution. It is an asymptotic relationship: there are a large number of events which have a small impact (i.e. effect) and a very small number of events that have a massive impact. This is illustrated by earthquakes. A large number of earthquakes have small and insignificant impacts. They are small tremors. Every now and again there is a massive earthquake which does tremendous damage. The power law for earthquakes is described by the Richter scale, which is logarithmic. An important lesson drawn from power laws is that extremely large outcomes are more likely to occur than would be concluded from using a normal bell shaped or Gaussian distribution. The normal distribution implies a relatively even spread around a mean i.e. short thin tails. The use of power laws suggests that events are not normally distributed and that we live in a world that is frequently observed to have fat tails. This idea has been popularised recently in Taleb (2004 and 2007) but has a rich pedigree in statistics and covers a variety of topics including scalable; scale invariant; Pareto-Zipf laws; Yule's law; Paretian stable processes; Levy-stable and fractal laws.

What does all of this tell us about the economy? It suggests that an economy is like any large complex system. It is inherently unstable (a view which Keynes, Schumpeter and many others would readily subscribe to). Economic systems are prone to occasional high failures – as in the global crashes of 1929 and 2009. Systemic failure on this scale is, however, extreme and difficult to predict.

This approach also suggests that the unit of analysis is the system itself and not the individual decision making units that comprise the system (more on this below). The behaviour of a decision making unit is influenced by the system in which it is embedded. It is wrong to assume localized stability in the short-run. Rather a longer-run perspective should be adopted by looking at the systems history and examining its patterns. Focusing on the short-run masks whether or not the short-run events are deviations (outliers) from long-run trends. This also means that point predictions in complex systems are unreliable because of the inherent instabilities.

Power laws had been considered many earlier by George Kingsley Zipf (1932 and 1949). Pareto's distribution is a power law. It was, however, Mandelbrot (a student of John von Neumann) who introduced power laws into the centre of economics. In 1963 Mandelbrot analysed cotton price changes and found that the distribution of cotton price changes follow a power law. In other words, he found fat tails (non Gaussian) in the distribution. Mandelbrot and others examined price changes in many other commodities and also financial markets. In all cases price changes followed a power law implying non equilibrium behaviour (see Mandelbrot, 1963 and 2005).

The observations of power laws, however, create problems rather than solving anything. This is a common feature in many branches of econo-physics. It is an observation extracted from the data that price changes scale. The deeper question is why do they? How is price scaling explained and accounted for? What is the underlying economic behaviour and economic processes that explain this observation? In other words, what theory explains why power laws exist in economic phenomena?

Scaling laws have been used in the study of financial markets. Here the physical analogue is the sand pile and was formally analysed by Bak (1996) and subsequently developed by Brock (1999). The reader is invited to imagine building a sand pile by dropping sand from above. The sand piles up, reaches a critical point (tipping point) where upon an avalanche occurs. The sand then piles up again. This process follows a scaling law. The idea of the sand pile is a means of thinking about situations where some kind of pressure builds up and which is then released through a chain reaction which for a while releases the pressure until it builds up to a critical level once again.

A number of analysts have sought to discover scaling laws in financial markets. If scaling laws underlie patterns in financial time series data then the hope is that it will be possible to predict when the "avalanche" will occur; how big it will be; and how long it will last. Given information about scale laws then those who use it stand to make a lot of money. Predicting turning points is difficult especially when the power law is known to many others. Each grain of sand simply follows the law, however, in an interactive system made up of purposeful actors those who jump ship early because they are risk averse influence the outcome of the game. Where and when the turning point in financial markets will occur is not independent of the behaviours of the constituent elements. For those participating in the game it is very risky. The trick is to get out at the right time before the bubble bursts.

Samuelson, using ideas originating in Bachelier, did much to introduce random walks into the analysis of economic time series. It also forms the basis of the Black-Scholes-Merton model of options pricing. A price change at time t is found to be correlated with a price change at time $(t+k)$. The correlation decays with the power law k^{-a}, where $o<a<1$. These are called long memory processes since it takes a long time to converge. If these long memory processes are applied to financial markets then this violates the fundamental idea of informational efficiency. Much theoretical work needs to be carried out in order to understand what is happening.

The aftermath of a financial crash or some similar major economic catastrophe can be thought of as setting off ripples throughout the economic/financial system like aftershocks following an earthquake. Some researchers have used Omori's law used in seismology to find these ripple effects in the financial system. As in many cases adoption of such "laws" are simply a means of describing data. They do not explain. To obtain understanding requires a deeper analysis of the behaviour underlying the system.

4. Aggregates

It was suggested above that some who specialise in econo-physics prefer to study the behaviour of total systems rather than the behaviour of the constituent parts. Physicists observe patterns of behaviour in aggregate systems e.g. Boyle's law describes the approximate behaviour of aggregates. This raises some very tricky and fundamental issues. What is the relationship between the behaviour of the individual constituent elements of a system and the overall behaviour of the aggregate system? In economics this is the basis of the search for the micro-foundations of macro-economics; in social theory it is the relationship between structure and agency; and in physics it is the link between the observed world of "reality" and the unseen and often imagined world of the quantum.

This problem is illustrated by drawing upon thermodynamics. It has already been pointed out that Boyle's law describes, approximately, the behaviour of an aggregate system that is made up of the interactions of billions of discrete particles. When, however, we move below the aggregate we move into a heavily veiled world of processes and mechanisms that produce these aggregates. It is at this point the challenges begin.

"Every heat engineer knows he can design his heat engine reliably and accurately on the foundations of the second law (of thermodynamics). Run alongside one of the molecules, however, and ask it what it thinks of the second law. It will laugh at us. It never heard of the second law. It does what it wants. All the same a collection of billions and billions of such molecules obeys the second law with all the accuracy one could want". (Wheeler, 1990, p.31).

Part of Bouchaud's criticism of economics is that it is the wrong level of analysis that is being focused on. Like many physicists Bouchaud prefers to analyse the easily captured data of aggregates rather than dive into the murky waters that lie below. Aggregates, however, are not reality. Reality is events. In economics events are the many transactions that produce aggregates such as a "market price". Changes in that aggregate can be described in terms of power laws but not explained by them. Nor does a study of aggregate price behaviour say anything about whether the price is correct price. When considering the launch of a new product or the introduction of a new risky financial instrument there is no aggregate price information for those markets so how is risk to be priced? The Black-Scholes-Merton (BSM) model despite all of its weaknesses represents a major contribution to the pricing of risky derivatives. Those who criticise it do so without coming forward with a reasonable alternative pricing strategy. Physicist such as Derman (2004), who have made a career in the financial services industry, understand and appreciate the significant differences in the methodologies used by economists and physicists (see also, Haven, 2007). It is not a matter of judging BSM in terms of whether or not it is right or wrong. The BSM model is a heuristic device and as such is judged in terms of its usefulness and for many normal conditions it was extremely useful. The weakness of the BSM model is that it is not sufficiently robust in extreme situations. It assumed thin rather than fat tails (i.e. it is Gaussian) – it assumed that the probability of extreme price movements was very small and approached zero. As Mandelbrot (2005) has clearly demonstrated this is not so. These technical issues combined with a high degree of hubris on the part of those who used the model, resulted in the road crash which was observed – in 1987the Dow-Jones index fell by 23% in a single day. Those holding derivatives containing high risk debts lost a massive amount. Similar events occurred in 2008/09. The prices placed on structured financial packages (derivatives etc.) containing toxic, high risk debts (such as sub-prime mortgages) did not take into account the probability of systemic failure i.e. multiple borrowers defaulting simultaneously.

The models used to price risk are clearly flawed and further work needs to be carried out to make them more robust but the point is that this work is done at the micro rather than the macro level.

The quantum theory revolution in physics challenged the classical physics mechanical view of reality held by Descartes, Galileo and Newton. Theirs was a world in which everything in the phenomenal world moves in accord with precise, deterministic, universal laws, which are accessible in principle, by means of suitable measurement. This clockwork view of the universe was displaced and replaced by the quantum view of a probabilistic world. The world of quantum physics is as shadowy as the world of micro economics. It is a world of shadows rather than objects that are directly observable – a world of immeasurables. This is summed up in the work of Bohr and also Kierkegaard, and is encapsulated in Bohr's notion of "complementarity", i.e. the process of coming to know something has the effect of bringing into being some aspect of it whilst at the same time rendering inaccessible some other aspect of it (Bohr, 1963 and Pais, 1991). In other words, measurement provides information about one property at the expense of other properties and as such measurement provides incomplete knowledge about a physical system. Any description of a system is bound up with a particular experimental arrangement (i.e. perspective). Such an experiment focuses on some aspects to the exclusion of others. It is, therefore, incomplete aspects of reality that are brought into consciousness. This is captured by Wolfgang Pauli (1994):

"On the one hand, modern psychology demonstrates a largely objective reality of unconscious psyche; on the other hand bringing into consciousness i.e. observation constitutes an inference with the unconscious contents that is in principle uncontrollable; this limits the objective character of the reality of the unconscious and invests reality with certain subjectivity".

This suggests, as social scientists have argued for many years, that "reality" is constructed. The way we see things is not just constrained by observational measurement techniques but is shaped by a variety of cultural influences. For Bohr complementarity is one of the most fundamental lessons of quantum physics for a physical view of the world. Another basic lesson is that knowledge of reality is indeterminate – all measurements are probabilistic in nature. These assumptions stand in stark contrast to those of the mechanical or clockwork view of classical physics – that a measured outcome is determined by the state of the system (determinacy) and that in principle there is a measurement which is capable of accessing all of the degrees of freedom of a system (transparency).

If indeterminacy and complementarity are accepted then what does this tell about the physical world? Reality is participatory. Individuals have the freedom to choose what questions to ask Nature. They then seek answers to these questions through measurements but Nature responds by giving an answer that is only partly determined by the variables under experimental control. Some aspects of reality become definite but others are not. Reality is temporary and depends on the original choices made.

In the social world these assumptions (complementarity and indeterminacy) play a more intense role. There is great uncertainty in the information obtained from measurement and uncertainty about the processes that generate the information. Processes are simplified in order to make them measurable and measures are full of errors. This is not simply due to the limitations of measuring technologies. It is because of the possibility of multiple possible realities; indefiniteness; complementarity and changing socio-economic realities. It was Popper (1991) who acknowledged that the method of economics was based on situational logic (or situational determinism – see Latsis, 1972). Today both economists and physicists, influenced by John von Neumann's idea of "expected value", which he introduced to both subjects, see the world as probabilistic. Just as physical reality is temporary, social reality is emergent as the situational logic progresses and develops.

5. Complexity & Emergence

Economists have for some time now examined the emergence properties of individual behaviour and the interactions of individuals. This has been done for over 60 years through game theory but more recently models of purposeful agent based behaviour have provided insights. These new tools enable economists to see what they could not see previously. Agent based models are numerical rather than analytical and involve computer simulations. These models are distinct from econo-physics models which simply describe some stochastic process. Agent based models contain purposeful heterogeneous decision makers. One possible future line of research could be to incorporate agent based models and the stochastic process models. In that case purposefully behaving individuals would interact randomly. There are no equations that describe the macro (aggregate) level. It simply emerges from the interaction of the agents. This is a very interesting and productive research agenda which brings together many different strands. Simple rules governing individual decisions and behaviour can produce non-obvious, surprising aggregate outcomes. In some cases these models display a great deal of order (Schelling, 1978); Axtell and Epstein, 1996; Young, 1998).

The significant differences between the subject matters of economics and physics are that the interactions between the atoms/molecules of Brownian motion or the second law of thermodynamics are purely stochastic. Molecules do not learn, they have no memory and they lack passions. On the other hand the individual decision makers that comprise the complex systems that are studied by economists are purposeful, learn, have memories and more especially form expectations. In other words the interactions which are studied by economists are strategic. This is the whole basis of game theory. It is the existence of strategic interactions in the socio-economic sciences which distinguishes them from the physical sciences. Strategic interaction confounds prediction with the result that the behaviours of social systems are emergent.

6. Metaphors & Models

All sciences use a variety of metaphors and models in order to gain an understanding of their realities. Metaphors work through analogies, thus the financial crash of 2009 and the resulting economic depression have been likened to a tsunami or an earthquake. At a superficial level this lends dramatic effect but metaphors tend to be used by the intellectually lazy and the impatient. Those who cannot be bothered to invest the time and effort to construct analytical models think and write in terms of exotic metaphors and those who wish to rush to conclusions without confronting the discipline of careful thinking rush into constructing ill conceived metaphors where more careful analysts would fear to tread. Whilst the result might be colourful and thought provoking writing, never-the-less, the use of metaphors tends to result in research agenda which quickly fizzle out. They lack internal consistency because they have no strong logical foundations. Those who use metaphors quickly become lost in a fog of rhetorical devices.

This does not deny that there is a use for metaphor and rhetoric. Clearly there is a time and place when it is appropriate to think in these ways. When starting to think about a problem metaphors take the properties of something that is familiar and project them onto something that is strange in order to gain some initial understanding and comprehension. Metaphors are like booster rockets that provide the initial propulsion but are not sufficient to complete the journey of discovery. They illuminate by analogy and are, as such, useful as a means of communicating complex ideas to an uninitiated audience (McCloskey, 1985).

Economics is a modelling science. Models are heuristic devices. They (re)present reality. It is through our models that we see the world but models by their very nature leave out many features of reality. "The very act of modelling has the effect of destroying knowledge as well as creating it". (Krugman, 1995, p.72). This in many respects is similar to Bohr's idea of complementarity. Whilst physical scientists seem to have no problems living with the simplifications embodied in models it has, however, resulted in protracted debates in economics and the other social sciences as if the use of simplified representations of a complex reality through the use of models is some kind of sin. One of the problems associated with analytical models in the social sciences is that they are not value neutral. Carefully constructed models and their results often challenge the "beliefs" that are held about the way the socio-economic world works. Rather than re-examining and rethinking their beliefs individuals will attack the model and its simplifying assumptions. Darwin's theory of natural selection and evolution dramatically challenged the beliefs of many Victorian Britons.

Whilst models are incomplete representations they have many advantages. They help to clarify the confusions created by alternative ways of thinking and they reveal the sensitivity of conclusions to the assumptions that are made. Not all models can be judged against empirical observations because they contain so many un-observables. They do, never-the-less, have a powerful role to play. In the social sciences those who occupy the space devoted to giving policy advice frequently make strong policy statements that are founded more on belief than on hard theoretical or empirical analysis. One way of falsifying such statements is to confront them with the highly unrealistic models of the economist such as the Walrasian General Equilibrium model of Arrow and Debreu. These models assume, amongst other things, perfectly competitive markets; complete information sets; future markets; rational expectations etc. etc. Taking one of many policy statements the following question can be asked, "What does the world need to look like for this statement to be true?" If it can be demonstrated that it is only true in the context of the highly artificial world of Arrow and Debreu then that rings bells of caution if such policy prescriptions are to be implemented in the complexity of the real world, (see Hahn, 1984). A similar line of reasoning seems to apply in the world of the quantum. Dirac's equation predicted successfully the existence of unobserved particles. Working backwards from the observations an accurate description of reality in a quantum field can be finessed.

The theories of the second best and Goodhart's "law" in economics have their counter parts in Heisenberg's uncertainty principle. Measurement or intervention changes a system. "The general theorem for the second best optimum states that if it is introduced into a general equilibrium system a constraint which prevents the attainment of one of the Paretian conditions, the other Paretian conditions, although still attainable, are, in general, no longer desirable" (Lipsey and Lancaster, 1956, p.11). The strong implications of this are that policy prescriptions such as, set price equal to marginal cost, which are based on Paretian conditions are not relevant in the real world which is full of policy constraints. Goodhart's "law" simply states that once information about an economic regularity (e.g. the Phillips curve) becomes know that information is fed back into individuals' decision making and thereby destabilizes the original regularity.

Once again the use made of information; especially new information (news), in the context of the strategic interaction of individuals sets economic models apart from models in physics.

7. Whither Econo-Physics

Econo-physics is a new and exciting development within economics but where does that path lead to? As in all situations of uncertainty the destination and the terrain that will be traversed is completely unknown. New knowledge will emerge. Whether or not it is a sustainable development path remains to be seen. If the research agenda is to be viable and not simply be regarded as a fad like catastrophe theory or chaos (see Horgan, 1979) then some caution needs to be exercised as those who take this path grope their way along – a process of tatonnement.

Economics is much more concerned about understanding and explanation than it is with prediction. In economics, along with the social sciences more generally, there are no laws. There are, however, models which, if the researcher is lucky, might be able to distinguish between up or down i.e. positive or negative. There are many competing models each with its own voice: each lets the data speak in a different way. There is no grand meta-theory, though researchers such as Herb Gintis are working on a unification of the behavioural sciences, including biology.

Whilst econo-physics has been very productive in the analysis of data it has been less fruitful in the area of explanation. There is much work to be done in examining non-linear dynamics coupled with models of adaptive learning.

Perhaps an area that will emerge is an exchange of discourses between physicists and economists (or social scientists) on the subject of the deep philosophical issues of ontology. A useful starting point would be Bernard d'Espagnat's (2006) book, **On Physics and Philosophy**. It is not an easy read but is delves deep into the implications of quantum physics for philosophy. His notion of "veiled realism", which is a critique of Kant's demarcation of synthetic a priori knowledge, has much to say to economics. At the same time, however, economists specializing in game theory have much to inform d'Espagnat's notion of "noumenal" reality which under lies empirical regularities without revealing the underlying structure.

Econo-physics does not represent a paradigmatic shift in economics. Rather it is more in the development of a research programme as defined by Lakatos (1976). Whether that research programme develops into something of significance remains to be seen. It could be that with the passing of time the research agenda will simply fade away. As Max Planck described it, "It's not that old theories are disproved, it is just that their supporters die out" (see Mohr, 1977, p.136). A point similarly made by Samuelson, "science advances, funeral by funeral".

References

Axtell, R. and J.M. Epstein, (1996). **Growing Artificial Societies: Social Science from the Bottom Up**. Washington DC, and Cambridge: Brookings Justitution and MIT Press.

Bak, P. (1996). **How Nature Works: The Science of Self Organized Critically**. New York: Copernicus Press for Spinger-Verlag.

Bohr, N. (1963). **Essays 1958-1962: On Atomic Physics and Human Knowledge**. Inter-science Publishers.

Bouchaud, J.P. (2008). "**Economics Needs a Scientific Revolution**". **Nature**, 455; 30th October: p.1181.

Brock, W.A. (1999). "**Scaling in Economics: A Reader's Guide**", **Industrial and Corporate Change** 8: 409-46.

Colander, D. (ed.) (2000). **The Complexity Vision and the Teaching of Economics**. Cheltenham: Edward Elgar.

Derman, E. (2004). **My Life as a Quant**. New Jersey: John Wiley.

Edgeworth, F.Y. (1881). **Mathematical Psychics**. London: C. Kegan Paul.

Espagnat, B.d. (2006). **On Physics and Philosophy**. Princeton: Princeton University Press.

Farmer, J.D., Shubik, M. & Smith, E. (2005). "**Is Economics the Next Physical Science?**" **Physics Today.** pp. 37-42.

Georgescu-Roegen, N. (1966). **Analytical Economics: Issues and Problems**. Cambridge: Harvard University Press.

Georgescu-Roegen, N. (1971). **The Entropy Law and the Economic Process**. Cambridge: Harvard University Press.

Hahn, F. (1984). **Equilibrium and Macroeconomics**. Oxford: Basil Blackwell.

Haven, E. (2007). **A Survey of Possible Uses of Quantum Mechanical Concepts in Financial Economics**. In: Bruza P. D., Lawless W., van Rijsbergen C.J. "Keith", Sofge D. (eds), Proceedings of the Association for the Advancement of Artificial Intelligence (AAAI) Spring Symposium on Quantum Interaction (Held at Stanford University). March 27-29, 2007. AAAI Press, 166-169

Horgan, J. (1997). **The End of Science: Facing the Limits of Knowledge in the Twilight of the Scientific Age**. New York: Broadway Books.

Krugman, P. (1975). **Development, Geography and Economic Theory**. MIT Press.

Latsis, S. (1972). "**Situational Determinism in Economics,**" **The British Journal for the Philosophy of Science**. 23: 207-45.

Lipsey, R.G. & Lancaster, K. (1956). "**The General Theory of Second Best**." **Review of Economic Studies**. 24(1), October, 11-32.

Mandelbrot, B. (1963). "**The Variation of Certain Speculative Prices**," **Journal of Business**, 36(4): 394-419.

Mandelbrot, B. (2005). **The (Mis)Behaviour of Markets**. London: Profile Books.

McCloskey, D.M. (1985). **The Rhetoric of Economics**. Madison: University of Wisconsin Press.

Mirowski, P. (1989). **More Heat than Light: Economics as Social Physics, Physics as Nature's Economics**. Cambridge: Cambridge University Press.

Mirowski, P. (2002). **Machine Dreams. Economics Becomes a Cyborg Science.** Cambridge: Cambridge University Press.

Mohr, H. (1977). **Lectures on the Structure and Significance of Science**. New York: Springer Verlag.

Pais, A. (1991). **Niels Bohr's Times**. Oxford: Oxford University Press.

Pauli, W. (1994). **Writings on Physics and Philosophy**. Springer.

Samuelson, P.A. (1947). **Foundations of Economic Analysis**. Cambridge: Harvard University Press.
Samuelson, P.A. (1972). "**Maximum Principles in Analytical Economics**." **American Economic Review**, 62: 249-62.

Schelling, T.C. (1978). **Micromotives and Macrobehaviour**. New York: Norton.

Taleb, N.N. (2004). **Fooled by Randomness: The Hidden Role of Chance in Life**. New York: Random House.

Taleb, N.N. (2007). **The Black Swan: The Impact of the Highly Improbable**. London: Penguin Books.

Wheeler, J.A. (1990). "**Information, Physics, Quantum: The Search for Links**", in W.H. Zurek (ed.), **Complexity, Entropy and the Physics of Information**. New York: Addison-Wesley.

Young, H.P. (1998). **Individual Strategy and Social Structure: An Evolutionary Theory of Institutions**. Princeton: Princeton University Press.

Zipf. G.K. (1932). **Selective Studies and the Principle of Relative Frequency in Language**. Cambridge: Harvard University Press.

Zipf, G.K. (1949). **Human Behaviour and the Principle of Least Effort**. Cambridge: Addison-Wesley.

METAPHORS AND INTERDISCIPLINARY RESEARCH

E. HAVEN

School of Management, University of Leicester, Leicester, LE1 7RH, UK
E-mail: eh76@le.ac.uk

This paper starts out with a possible definition of what a metaphor can be. It then continues to discuss what a model is, using the approach by Verhulst (1988). We give some examples of methaphorical uses in interdisciplinary settings. We also query if one can always use methaphors in research. The paper continues with a discussion on why we want to pursue interdisciplinary research. We give some examples of such research (focussing on the interface of social science and physics).

1. Why interdisciplinary research?

Some research grant councils are increasingly stressing the importance of research which borrows and actively applies concepts and principles from one discipline into another discipline. The boundaries to such type of research are quite fuzzy: one is neither encouraged nor discouraged to borrow and implement ideas from the sciences into the social sciences or vice versa. In other words, interdisciplinarity may exist i) *within* the sciences; ii) *within* the social sciences, or it may exist iii) *across* the sciences and social sciences. Thus, it is up to the researcher to decide what interdisciplinary road he/she follows. Does such an attitude (which may well vary from country to country) towards interdisciplinary research constitute a solid basis for academic research 'freedom'? We will attempt to answer this question towards the end of this paper.

If such research is being stimulated by major grant agencies, it must be, by default, that there is proven promise for scientific progress following from such type of research. We believe that from the individual researcher's point of view, a multitude of reasons may be lurking behind the decision 'to go interdisciplinary'. The 'progress' ideal may (or should!) be one of them but there may be other explanations such as the thrill of getting read by researchers outside of one's discipline or also the 'novelty effect' associated with the importing of unknown tools of analysis to one's own discipline. Evidently, the novelty effect will be further enhanced when knowledge progress is moved forward. However, this is certainly not always the case. In fact, quite the contrary can also occur: the novelty effect may give the false impression new knowledge is being created when that is patently not the case.

In this paper we will use the premise that a rationale for interdisciplinary research is to augment the impact of model building. This begs the question: what is a model? We will discuss in the sequel of this paper what a metaphor is and we will be using the Verhulst[1] approach to the definition of a model, which incidentally uses the metaphor as a basic ingredient. We will consider some examples of metaphors in interdisciplinary settings and we will then proceed with the query whether we can always use metaphors. We also provide for some examples of interdisciplinary models. We finish the paper with a possible answer to the query we posed at the beginning of this introduction: is the encouragement on performing interdisciplinary research providing for a solid basis towards research 'freedom'?

2. Models: the 'Verhulst way'

Verhulst[1] in a beautiful paper proposes that "A model ...is a metaphor with something added, it has not only qualitative but also quantitative aspects which adds to the precision of description." We would like to devote some paper space to the issue of the input of metaphors in the constitution of a model. The interested reader will want to peruse the paper by Peter Jackson[2] in this proceedings volume, which gives some important cautionary words on the use (and misuse) of metaphors in the social sciences.

What is a metaphor? Ferdinand Verhulst[1] quotes Aristotle who said that a metaphor consists in "giving the thing a name that belongs to something else."

If we follow the Verhulst's definition of a model then can one claim that models are reality? Are metaphors reality? Please see for a nice discussion in the paper by Ferdinand Verhulst.[1]

There exist many interdisciplinary examples of metaphorical usage. The concept of 'memory' is frequently used in time series analysis: 'memory' is measured via the so called auto-correlation function. Clearly, this definition of memory has a very remote connection with the way memory is defined in psychology.

A within-discipline example of metaphorical use may be found in basic quantum physics. The fundamental double-slit experiment in quantum physics commanded the use of a probability devise which could capture interference by superposing probability distributions. Since probability distributions can not be superposed,[3] the idea of 'probability waves' came to the forefront. How metaphorical is a probability wave to a seasoned physicist?

A dramatic example of again, interdisciplinary metaphorical use, is the employment of Brownian motion in finance and economics: the stock price replaces the position of the particle in the glass of water - so to speak.

The above examples show that the metaphorical device can be used beyond the boundaries of rhetoric. In effect, 'time series memory' and 'probability waves' are concepts which have a very precise meaning. Please see Peter Jackson[2] for an excellent discussion on this issue.

3. Can we always use metaphors?

Can a metaphorical object exist as a mathematical object? We believe that a metaphorical object has to be constructed in order to exist. However, is it not reasonable to claim that a mathematical object may exist *without* construction: nature provides for a vault of examples of such mathematical objects? Edward Nelson[4] mentions the intuitionism movement which says that if a "mathematical object exists it means one knows how to construct it." As we indicated above, a metaphorical object for it to exist, needs to be constructed. A mathematical object, if one subscribes to the realism school, which according to Edward Nelson was the movement which gave the "first view of mathematics",[4] may exist independently of whether one knows how to construct it or not. Thus, metaphorical objects seem to share less of a possible common identity with a mathematical object if one adheres to the realism interpretation as opposed to when one opts for the intuitionism interpretation. Thus, in realism when a mathematical object exists, one may know (or not know) how to construct it. Intuitionism says that all mathematical objects, when they exist can be constructed.

There is one interpretation of mathematics where the metaphor is an unusable object: the formalism approach. Consider this quote out of Nelson:[4] "The symbols in (a) formula are marks on paper, and the work of the mathematician is to devise deep and beautiful concatenations of such marks according to strict rules." And Nelson[4] continues: "..what is real in mathematics is the notation, not an imagined denotation." It is precisely this last comment which 'kills off', so to speak, the use of metaphors in a formalism orientation towards mathematics. Indeed metaphors are precisely denotations! If the formalism interpretation is really the pinnacle of truth in interpreting mathematics then how can one define a mathematical object? Is there any point at all in bothering oneself to construct a mathematical object? Constructing such an object implies one gives meaning to it. This is definitely not what formalism is indicating.

In the formalism setting of mathematics, metaphors are thus 'no-go' objects. Therefore, when using the Verhulst[1] approach towards defining the existence of a model, such models will not exist when a formalism setting of mathematics is employed.

Here comes a pragmatic question: is the mathematics we use in economics or physics then still mathematics in the formalism setting? Probably not.

4. An example of a successful interdisciplinary model

Interdisciplinary models can augment model building impact. This is a quite idealistic statement! The impact qualifier in our rationale for using interdisciplinary models, has a lot to do with the knowledge augmentation capacities of such particular research. However, we do also believe that the realities of interdisciplinary research are sometimes far removed from this ideal. We discuss this in the last section of the paper.

In the physics - economics/finance area of research, we can notice a lot of published work which originates from physicists who have applied physics concepts to economics and finance problems. Social scientists have also attempted to import physics concepts to solving problems in economics, finance and psychology. We note however, that the opposite direction of research: i.e. social scientists applying concepts from social science to physics is extremely rare*.

It was in 1973 that Fisher Black (a physicist by training) and Myron Scholes (and economist by training)[5] published their world famous partial differential equation (pde), which allowed for the pricing of option contracts under certain restrictive† conditions. The Black-Scholes pde is part of the so called Kolmogorov pde family. It is a so called backward Kolmogorov pde. Any physicist knows the Fokker-Planck pde (which is a forward Kolmorogov pde).

*The author of this paper must confess he does not know of any such application.

†An important restrictive condition was that the volatility of the underlying asset had to be constant. This condition was relaxed in subsequent research.

The Black-Scholes model has multiple non-perfect traits. We mention some:

- The main ingredient in the set up of the Black-Scholes pde is the geometric Brownian motion on the price of the underlying asset, which in its most common form is a stock price. The geometric Brownian motion for stock prices is an approximation. Stocks are not continuously traded, however the Brownian motion trajectory is continuous and nowhere differentiable.
- Stock prices are assumed to have very little 'memory'. This is also an approximation and the recent credit crunch has probably seriously invalidated that assumption.
- The probability density function that stocks prices follow, when using such motion is of the lognormal type. This is also an approximation. Many studies have shown that probability density functions of asset prices do exhibit fat tails. The paper by Lisa Borland[6] which uses alternative non-Gaussian densities provides for a much richer option pricing model.
- The option pricing portfolio is risk free for a very short period of time and must be re-balanced numerous times. The transaction costs associated with such re-balancing could be a problem.
- A final issue, which may be a problem or not, depending on one's point of view, is that the model may have generated a huge option pricing market. The paper by Millo and MacKenzie[7] provides for a very interesting appreciation of this issue.

The Black-Scholes theory is an incredibly successful theory in social science. It has flaws and it involves many approximations. But this echoes very much what Chary Rangacharyulu[8] says in his paper in this proceedings volume. What type of world do we assume, if we dare to use a geometric Brownian motion for describing the time dependent evolution of stock prices? Are the electrostatic forces in the glass of water which influence the descending particle, equivalent to the effect macroeconomics news has on the price evolution over time? Let us remark that such macroeconomic news may have a serious psychological color! Is the stochastic model the correct one? An electron is not subjected to the same forces like a particle in a glass of water is. Is the stochastic differential equation approach on electrons not also usable in an economics/finance context?

Still, the Black-Scholes model to the standards of this present world, seems to be incredibly successful. The model got away with impunity even though it does model financial objects which are by definition traded by humans. The economist will know that the beautiful secret of the Black-Scholes model is that it can create a portfolio of assets which is risk free for a short period of time. Thus it annuls for a short period of time any issues traders may have in regard to preferences (or phobias) for (against) risk.

This textbook example, we believe, comes to show that one should not always be too over-cautious against using concepts borrowed from physics into the social sciences. There is further evidence of very successful work in the social sciences as proposed by physicists and economists alike. We mention of course the work by one of the main protagonists of the econophysics movement, Boltzmann Medallist, Eugene Stanley.[9] We mention the equally important contributions of Constantino Tsallis.[10] Thomas Lux, an economist[11] has made major contributions, as well as Marcel Ausloos.[12] Please see the contributions of Eugene Stanley; Constantino Tsallis and Marcel Ausloos in this proceedings volume.

5. A wave function as an information function: a usable metaphor?

The applications of quantum physical concepts to macroscopic systems like economics or psychology seem, on prima facie, to be completely contradictory. We would like to insist though that most of the research performed in this emerging area of research might well be less exotic than anticipated. The interested reader may want to take note of the paper by Jerome Busemeyer and Zheng Wang[13] in this proceedings volume which uses to good effect a basic concept of quantum mechanics (interference) within the context of human decision making. Some of the core paradoxes in psychology and economics can possibly be explained by using a quantum probabilistic approach.

Andrei Khrennikov[14] was one of the very first to introduce applications of quantum mechanical principles to macroscopic systems. He also introduced the use of Bohmian mechanics, named after its originator David Bohm,[15] to applications in economics and finance. Louis de Broglie, the French physics Nobel laureate, attributed two roles to the quantum mechanical wave function:[16] i) it aids in determining the likely location of a particle *and* ii) it also influences the location by exerting a force on the orbit. In effect, de Broglie was the precursor to Bohmian mechanics. This idea, which in fact never caught on well in the quantum physics community, can be of use in a macroscopic setting. If the wave function can be likened to an information wave function then that very same function can influence the particle, which is say, an asset price. Consider the very basic idea of quantum physics: i.e. the probability of the likely location of a particle extracted from the wave function, Ψ. Consider a price configuration space of 1 dimensional price vectors. As in Choustova,[17] let Ψ belongs to the space $L^2(\mathbb{R}, dM)$ of square integrable functions with respect to some measure M on $\mathbb{R}$ such that:$\|\Psi\|^2 = \int_{\mathbb{R}} |\Psi(q)|^2 \, dM(q) < \infty$, with q being position (f.i. the price of an asset). Information effects can then be described by Ψ. If $M(q)$ is for instance a Gaussian measure then all positions are distributed by a Gaussian.

In Bohmian mechanics, trajectories exist! This is a radical departure from ordinary quantum physics where trajectories are absent. The Bohm-Newton trajectory is well known to follow: $m\frac{d^2q(t)}{dt^2} = -\frac{\partial V(q)}{\partial q} - \frac{\partial Q(q)}{\partial q}$, subject to the initial conditions that $q(t=0) = q_0$ and $q'(t=0) = q_0^{'}$, where q_0 being the position at $t = 0$ and $q_0^{'}$ is momentum at $t = 0$. The Bohm-Newton trajectory is a smooth trajectory. The function $Q(q)$ in this differential equation denotes the so called quantum potential. Such a potential is unique to Bohmian mechanics. See Holland[16] for an excellent discussion. m denotes mass; $\frac{d^2q(t)}{dt^2}$ is acceleration; $-\frac{\partial V(q)}{\partial q}$ is the partial derivative of the real potential towards position. The variable m and functions V and Q have already been interpreted in an economics setting. See for instance Khrennikov.[18] Note that the quantum potential Q, depends on the wave function (via the amplitude function of the wave function) and therefore the wave function steers the particle.

We can as in Haven[19] apply the information wave function in an non-arbitrage context. We consider also here a so called state price, s, which roughly could be interpreted as an 'insurance' price which one pays to ensure a monetary outcome if a certain state (of nature) occurs. See Neftci[20] for an excellent introduction to this topic. The state price, can be connected to a probability value, as in the famous non-arbitrage theorem. See for instance Etheridge.[21] We can now connect the information wave function to the state price. As we have proposed in Haven,[19] a change in the information wave function $\Psi(q)$, or a change in the information price[‡], q, can induce a change in the state price, s, since the state price draws a probability from $\|\Psi\|^2$. Note that $\Psi(q)$ denotes now information about the information price, q. Thus, in a Bohmian mechanics context, the information steers the

[‡]The information price could refer to for instance the price (or cost) of gathering propietary information. Such information is important in order to execute arbitrage positions for instance.

particle: the information steers the information price, q. The Bohm-Newton trajectory traces out the smooth trajectory of information prices.

For the purposes of this paper, we can go a little further though. We recall that in ordinary quantum mechanics, we have no paths. In the Bohmian mechanics interpretation we do have paths. The existence of those paths can be argued for from the fact there exists a well defined momentum in Bohmian mechanics (which is the gradient of the phase of the wave function). As Basil Hiley[22] remarks, an ensemble of paths exist when the quantum potential is non-negligeable. To come back to the above discussion on information prices, we could claim that from an economics point of view, the multiplicity of paths might call in the notion of *value* of information (as opposed to the precise notion of *price* of information). The value of information has an economic right of existence and, may or may not, coincide with the price of information.

Finally, the title of this section is asking whether the information wave function can be used as a metaphor. We hope that in this brief explanation, we have gone beyond using mere rhetoric in arguing for the use of such metaphor.

6. Encouragement of interdisciplinary research: a solid basis towards research 'freedom'?

The First Interdisciplinary CHESS Interactions Conference generated many beautiful outputs and this Proceedings volume is a testimony to this. One particularly interesting idea was proposed by Bernhard Juurlink[23] who says in his paper that "Ignorance allows imagination unfettered by the shackles of what we think is the known." This is an intuitively appealing idea. Could a duality exist between, on the one hand, important levels of accumulated knowledge which may have a constraining or stifling effect on creativity and, on the other hand, the polar extreme of little knowledge, which is then little constraining and can possibly give way to a lot of creativity (which possibly may have little meaningful content)? An obvious issue may be to remark whether creativity does translate into advancement of knowledge. But is this always a relevant question? What will the syntax based mathematician respond to this? Does his/her formal proof of 'Theorem A' necessarily advance knowledge? Furthermore, is there a richer ground for the gestation of creativity when interdisciplinary research is performed as opposed to when within-discipline research is considered?

Let us tune down the basic criterion by which we judge potential benefits of interdisciplinary research. Let us instead be slightly more pragmatic (and maybe less idealistic). Consider the following case scenario. The head of department in the School of Social Science at University X has to decide whether to support a tenure application of an assistant professor who has a record of research output published in journals which would rather belong to the research output of a different School. From a pure research perspective, what criterion (or criteria) can he/she use to support such candidate? The relevance and prestige of the journals the candidate published in will be relevant and may be indicative of the potential for knowledge advancement from such research. Levels of prestige from journals which sit outside one's discipline can be difficult to judge though. Hence, the candidate may be set back. In the case where a tenure candidate would have published solely within his discipline, his/her contribution towards knowledge advancement could be more clearly judged. If the young, as Bernhard Juurlink[23] says "have often less fear of displaying ignorance and usually do not have rigidly developed ways of looking at the world" and if those same young have no fear to display ignorance through the conduct of inter-disciplinary research, then the cost of a set back on a promotion may be even more resting on the shoulders of those very young. In that sense, it may well be very good that interdisciplinary thinking is encouraged by some research councils. The benefit of the large grant can indeed stem the 'losses' associated with a research output which is 'hard' to judge.

To close down then on the question which rests in the title of this section, we could make the following argument. What is the value associated to the freedom of doing a particular type of research? We have highlighted the core (and most difficult) criterion which is knowledge advancement. If increased freedom of doing inter-disciplinary research can lead to increased success of knowledge advancement then such freedom is highly valuable. But the world is less ideal than that. Some inter-disciplinary research seems to be solely satisfied with importing a technique from discipline A and adapt it (albeit in sometimes very shallow ways) to a problem context in discipline B, for which such application was never developed in the first place. We note the existence of a multitude of journals which cater to this type of endeavour. Research freedom in that very case may well be less valuable.

If ignorance can be conducive for augmenting imagination (and therefore creativity and potentially knowledge creation), it is probably not so unrealistic to claim that the very potential for imagination creation could be quite discipline specific. We believe that ignorance is probably more 'taxing' within a syntax based mathematics (as opposed to other disciplines (or other approaches towards mathematics)) because of the very absence of metaphors in such an approach towards mathematics. As we have indicated before in this paper, such absence of metaphors also does command an absence of models if we follow the Verhulst approach.

A researcher who has a doctoral degree in social sciences and who conducts research in social science with the help of tools from the exact sciences, will have, by default, a level of ignorance of knowledge in the exact sciences. This ignorance can be fueling imagination and the metaphor is a very handy 'conduct' by which such imagination can be taking shape. In the absence of metaphors, the level of imagination is that much harder to generate. Hence, metaphors and ignorance of knowledge seem to be related. After all, it was Aristotle (cited in[1]), who said that a metaphor was equivalent to "giving a thing name that belongs to something else".

In conclusion, in the absence of metaphors there may well be little scope to argue for the existence of interdisciplinarity. There is furthermore not much of a point to discuss the premise we followed in this paper about the rationale behind the conduct of interdisciplinary research. Augmenting model building impact is probably irrelevant if metaphors do not exist. Hence, any discussion about research freedom associated with such form of research would be probably futile.

References

1. F. Verhulst (1988); The Validation of Metaphors; Mathematisch Instituut; Utrecht (www.math.uu.nl/publications/preprints/1078.ps.gz)
2. P. Jackson (2010); Models and Metaphors; Proceedings of the First Interdisciplinary CHESS Interactions Conference; World Scientific.
3. M. Morrison (1990). Understanding Quantum Physics. Prentice-Hall.
4. E. Nelson (2010). Confessions of an Apostate Mathematician: http://www.math.princeton.edu/~nelson/papers.html.
5. F. Black; M. Scholes (1973), The Pricing of Options and Corporate Liabilities, Journal of Political Economy,81 637-654.
6. L. Borland (2002); A Theory of Non-Gaussian Option Pricing; Quantitative Finance; 2; 415-431.
7. D. MacKenzie; Y. Millo; Negotiating a Market, Performing a Theory: the Historical Sociology of a Financial Derivatives Exchange; American Journal of Sociology; 109; 107-145
8. C. Rangacharyulu (2010); Structures and Symmetries in Physics; Proceedings of the First Interdisciplinary CHESS Interactions Conference; World Scientific.
9. V. Plerou; E. Stanley (2007) ; Test of scaling and universality of trade size and share volume: evidence from three distinct markets; Physical Review E; 76 (4); 046109 (10 pages)

10. M. Gell-Mann; C. Tsallis (Eds) (2004); Nonextensive Entropy: Interdisciplinary Applications; Oxford University Press
11. Th. Lux Th.; M. Marchesi (1999); Scaling and criticality in a stochastic multi-agent model of a financial market; Nature; 397; 498-500
12. R. Lambiotte; M. Ausloos, Uncovering collective listening habits and music genres in bipartite networks, Physical Review E 72 (2005) 066107 (11 pages)
13. J. Busemeyer; Z. Wang (2010), Quantum Probability Applied to Social and Behavioral Sciences; Proceedings of the First Interdisciplinary CHESS Interactions Conference; World Scientific.
14. A. Yu. Khrennikov (1999); Classical and quantum mechanics on information spaces with applications to cognitive, psychological, social and anomalous phenomena; Foundations of Physics 29, 1065-1098.
15. D. Bohm (1952); A suggested interpretation of the quantum theory in terms of "hidden" variables, Part I and II; Physical Review 85, 166-193.
16. P. Holland (1993); The quantum theory of motion: an account of the de Broglie-Bohm causal interpretation of quantum mechanics. Cambridge University Press, Cambridge.
17. O. Choustova (2006). Quantum Bohmian model for financial markets. Physica A 374, 304-314.
18. A. Yu. Khrennikov (2004). Information dynamics in cognitive, psychological and anomalous phenomena, Series in the Fundamental Theories of Physics, Kluwer, Dordrecht.
19. E. Haven (2008), The Variation of Financial Arbitrage via the Use of an Information Wave Function. International Journal of Theoretical Physics; 47; 193-199
20. S. Neftci (2000). An Introduction to the Mathematics of Financial Derivatives, Academic Press, New York.
21. A. Etheridge (2002). A Course in Financial Calculus, Cambridge University Press, Cambridge.
22. B. Hiley (2008). Quantum reality unveiled through process and the implicate order; Proceedings of the Second Quantum Interaction Symposium; Oxford University.
23. B. Juurlink (2010); Fumbling Towards Ecstasy*: A Journey To Understand A Small Corner Of The Universe; Proceedings of the First Interdisciplinary CHESS Interactions Conference; World Scientific. (author's note: title used from a song of Sarah McLachlin)

THE IMPOSSIBILITY OF A THEORY OF EVERYTHING

TOM G. PORTER
Bioresource Policy, Business and Economics, University of Saskatchewan, Saskatoon, Saskatchewan, S7N 5A8, Canada
E-mail: tom.porter@usask.ca

Interdisciplinary work is a philosophical necessity to advance human understanding. This will be demonstrated through a review of ancient and recent literature about why a theory of everything is impossible. Interdisciplinary work that includes theoretically separate perspectives about how the world works is needed to advance understanding. Any one discipline will improve through interactions with other disciplines. This is different from a convergence of theories, because interdisciplinary work fails when independence among the theories is absent. A non-trivial example of the consequence is the $5 billion financial crisis caused in 1998 by Long Term Credit Management (LTCM) hedge fund. The LTCM example shows how the use of analogy and metaphor that facilitates interdisciplinary work is also the means where theoretical independence is undermined. Advancement of human understanding requires collaboration among theoretically separate perspectives.

1. Philosophical Basis for Interdisciplinary Work

1.1. *The Pursuit of Knowledge*

The idea of a Theory of Everything is a powerful motivation for the pursuit of knowledge. This motivation has been responsible for the advancement of human understanding in many disciplines. There is, however, a limit to the benefits of this pursuit. Human understanding also advances, perhaps in more profound ways, through differing perspectives and the resulting recognition of limits to any one theory or model.

The advantage of exposing a theory and its models to different perspectives is in the discovery of limitations of the theory. Theories often advance in explanatory power by extension – that is, through assessment and improvement of the internal logic of the theory. Extension of theory happens through experimentation, modeling and deduction among other means. The processes of extension will identify logical inconsistencies even though it cannot solve the problem of incompleteness. A differing perspective and the associated theoretical propositions are needed to reach beyond the limits of a given theory.

Thus there are two key propositions that lead to the conclusion that interdisciplinary studies are fundamental to the advancement of human understanding. First, there are limits on all theories and models. A given theory cannot provide answers, or even a common methodology, to answer all questions. Second, independence among theories is required to reach beyond these limits. The value of interdisciplinary work is realized only if there is independence in the philosophical bases of the different theories and their structures.

The first proposition regarding limitations appears in two arenas of discussion: first, in low explanatory power; and second, in low recognition of the limitations. A limitation is easily recognized when a theory is unable to adequately explain a phenomenon. The greater problem is when a theory appears to explain a phenomenon, but it is incomplete in its capacity to recognize a limitation. This is to say that a theory does not always identify its own limitations. This first proposition is the problem that interdisciplinary work is able to address.

The second proposition is important because it offers a condition for interdisciplinary work to be successful in addressing limitations – *i.e.*, in advancing human understanding. The limitations of a theoretical construct can be addressed through interdisciplinary studies if there is philosophical independence in the additional perspective. When presented with a problem that one theory cannot successfully address, a differing perspective may offer understanding. This differing perspective is more important if it helps expose limitations heretofore where unrecognized by the original theoretical structure. The role of interdisciplinary work creates a type of safety net to identify and perhaps resolve explanatory limitations. The key assumption is that interdisciplinary work must occur among theoretically separate disciplines.

1.2. *Entscheidungsproblem*

All disciplines are faced with difficult and often intractable problems. The term *Entscheidungsproblem* refers to the question of creating a formal algorithm to determine the validity of any given formal logic statement. In other words, is it possible to develop a logical treatment for all logical problems? As it turns out, the answer is no. Thus, even logic has a limit. The necessary reality of indeterminate questions that do not have logical solutions is why a theory of everything is not possible.

The existence of intractable problems cannot be eliminated, but many seemingly intractable problems can be solved with algorithms and mathematics. This is the function that logic serves both in reducing internal inconsistencies and in completing theoretical structures. Nonetheless, intractable problems and incompleteness are necessarily part of all theories. These realities form the core of the argument about why a Theory of Everything is impossible. This is not meant to deride attempts to pursue a Theory of Everything; rather, it is meant to demonstrate the fundamental motivation for interdisciplinary studies. Interdisciplinary work can serve to identify and perhaps solve difficult problems and incompleteness.

The proof that a Theory of Everything is impossible originates in Georg Cantor's work on infinities in the late 1800s and is most recently and completely demonstrated by David Wolpert in 2008 [1] as described in P. Binder [2]. The fundamental notion from Cantor's results is that there are separate sets of infinity between zero and one. Even if one sets out to capture all numbers in one infinite set, it will still be incomplete. This is an interesting point because it reminds us that the number line is undefined. Even if a theory can be expressed in mathematical terms it is limited by the inherent limitations of the number line. This is a seemingly obscure result, but it is fundamental to the theorems of incompleteness.

Kurt Gödel proved that for any numerically generated theory, it cannot be both consistent and complete [3]. Furthermore, a system cannot determine the consistency of its own axioms (or, initial premises). The interpretation is that any theory will invariably reach indeterminate questions including the existence of unprovable statements. The most common example of an unprovable statement is: "this statement is false". This general statement seems trivial, but any specific numerical theory will find a question that it intends to solve without success. Non-trivial examples depend on the specific context and are shown in following sections.

Alan Turing produced the same logical result by demonstrating that a set of algorithms could be developed to effectively duplicate the results of any other algorithm. This is now known as the Turing Machine that has universal processes that are able to duplicate all other machines [4]. This result means an algorithm can be developed to determine any decidable function and is encouraging for the idea of a Theory of Everything.

Turing's result holds for all decidable problems. The numerical problems may be extended to any logical problem so long as any concept or decision can be expressed in unambiguous terms. An unambiguous concept can be expressed in binary terms – *e.g.* an "either/or" statement – and all numbers can be expressed using 0 and 1. This result has been known from the early mathematicians Leibniz, Bertrand Russell, and Whitehead [5]. The result also implies a limit to numerical theories because, as Cantor demonstrated, there is a limit to the number line.

The other relevant result of Turing's Machine is the limit of the Machine's predictive power and the "halting problem" of the Turing Machine. The halting problem is that where the Machine encounters an indeterminate question or intractable problem. The Machine cannot answer the question, so it "halts". Whether through Cantor or Gödel, any numerically generated theory faces undecidable problems. Further, an algorithm cannot predict whether a Turing Machine will halt. In essence, a theory can determine its internal consistencies, but not its completeness or external inconsistencies

Wolpert extends the argument to discuss the logic of science. There is a necessary incompleteness in mathematics because it is a logic statement; as well as in computation because it is built on algorithms. The method of diagonalization that Cantor used to show uncountable infinities is also used by Wolpert to show that any one theory will necessary miss some portion of "knowledge space" [6]. The significance of the result is that there is philosophical certainty that any given theory offers an incomplete understanding – *i.e.*, the impossibility of a Theory of Everything.

To summarize: there are necessarily intractable problems that cannot be eliminated through analysis or theoretical expression. Intractable problems may be reduced through logic and rigorous analysis, but not eliminated. Wolpert used two Turing Machines to demonstrate his results. The argument will be made in the next section that more than one discipline is needed to advance human understanding – the motivation for interdisciplinary studies.

1.3. *Limits in the History of Science*

There are many examples of limitations in models and theories in the recent history of science. The limitations appear in several forms and are not always recognizable. The first of three forms covered here is a barrier, such as an intractable problem, that is recognized in theory. The second form involves a theoretical construct with no explanatory power. The propositions have been broadened to the point that these are tautologies. In the third form, a limit is revealed through accident or unanticipated observation. This is the most interesting and important because it is in the failure to recognize limits where the great failures and as well as the great discoveries are found.

The first example of the limitations in theories and models is with undecided or intractable problem. In these cases a theory or model cannot answer a question. When faced with such a limit various solutions are sought using logical deduction, an expansion of definitions, or a different construct. One of the great examples in the modern history of science is found in René Descartes. He was trying to determine the philosophy of nature – what we might now call the laws of nature. In order to achieve the outcome he had to find an objective point of reference – an Archimedean Point. *"I think therefore I am"* is the famous statement that gives each of us a reference point for reality; however Descartes could not link mind and body. The link to our bodies would then enable the theory to make a link from one person to another and thus provide an Archimedean Point as a perspective from which to understand reality [7]. Several philosophical explanations have emerged in an attempt to resolve this question (*e.g.*, Correspondence Theory), but these fail to create objective points of reference and mark the limit of Descartes' theory as a whole [8].

Tautologies illustrate a second limitation in models and theories. These occur where the definitions, axioms and propositions are defined so broadly that they no longer have explanatory value. Propositions frequently become broadened in scope in order to explain new observations or satisfy qualifications and maintain the consistency of a theory: "the death of a thousand qualifications" [9]. The best, although seemingly trivial, example is found in defining a straight line as an arc. A straight line is part of a circle with an infinitely large radius. While true, it is a useless concept when trying to go from point A to B! The example is worth mentioning because it reminds us that the number line is undefined at infinity.

A second and more artistic example of tautology is found in the work of M.C. Escher. In his famous work *The Gallery* the viewer sees four perspectives that are subsets of each other.

The impossible result was expressed in mathematical terms by Hendrik Lenstra. Lenstra successfully used the logic of mathematics to extend *The Gallery* into its centre – Escher left the centre blank. Lenstra demonstrates that Escher's perspective can be extended inward and that the centre could be a smaller reflection of the original print. Indeed the pattern could be repeated indefinitely to such a small scale that it becomes a singularity. A singularity is an undefined mathematical concept [10]. Although Escher did recognize that the pattern could be repeated in increasingly smaller scales, he did not construct a centre precisely because the centre had no interest or meaning [11]. The centre of *The Gallery* represents a limit.

The third and most important form of the limits of a theory is serendipity or catastrophe – depending on the outcome. Many limits are not known or acknowledged until after a seminal event. Several examples exist in engineering where there were tragic failures in bridges (*e.g.*, Tacoma Narrows) and supertanker construction. The models for the Tacoma Narrows suspension bridge did not account for negative air pressures over the road deck during high winds. The negative pressure caused the road deck to lift much like the wing of an airplane. This created such violent movements in the road deck that it destroyed itself. Suspension bridges now have holes in the deck that prevent negative pressure.

Supertankers are built to withstand 15 meter waves. Every prior model of ocean waves and statistical models demonstrated that this height ensures safety in all but the most rare weather condition. These models relegated reports of "rogue waves" to myth. A rogue wave is 25 meters in height and, statistically, is a once in 10,000 year occurrence. Once the phenomenon was studied using different methods, 10 waves were identified inside of 2 weeks. Over 200 super-carriers have disappeared in the 1980s and 1990s that were attributed to unspecified weather events. The idea that rogue waves could be responsible was not possible using the existing models. Understanding required non-linear physics methods [12]. These examples demonstrate the limits of modeling in engineering because models do not predict their own limits. The limits were found outside of the models.

These are examples of the limits of models, but examples are common for the limits of theories as well. Two examples can be found in physics regarding materials transition phases and the conservation of energy. The first example concerns properties of Charge Density Waves in metals: "When metals are cooled, they often undergo a phase transition to a state exhibiting a new type of order" [13]. There are a number of unusual properties that are described using analogies as opposed to mechanisms: "...metals...'remember' electrical pulse lengths" [14]. The properties related to Charge Density Waves remain "mysterious and controversial" [15]. The most famous of these properties that was discovered as a result of changing states is that of superconductivity, and it was discovered serendipitously [16]. The implication is that the models of material phases retain intractable problems.

The second example concerns the concept of the conservation of energy, which has evolved many times over the past 150 years. Revisions to the concept had to be developed to accommodate unexpected results in experimentation [17]. The revisions that started 150 years ago continue within the fields of subatomic physics. As P. Mirowski has written: "In the late 1920s and early 1930s, Paul Dirac was attempting to apply the theory of relativity to quantum mechanics in order to improve the description of the energy of a free electron. He was stunned to find that the equation he had derived admitted of two solutions: one, the conventional positive solution; the other, a *negative* solution for kinetic energy." The solution to the unexplained phenomenon was the discovery of positrons. "The floodgates of particle production were open, and to this day, the whole motley menagerie of subatomic particles has yet to be definitively taxonomized" [18]. The independence of the theory of quantum mechanics and the theory of relativity was needed to identify limits within each and to expand our understanding of the subatomic universe.

Limits of theories and their models are found throughout the history of science. The limits are always overcome and human understanding has grown as a result of new discoveries and new evidence gained from outside the theory or model. Any one example could be dismissed as an oversight or as incomplete observation rather than to the philosophical limits inherent in and scientific theory. The belief that there are no limits to scientific theory is itself a theoretical proposition. The evidence supporting – but not proving – this proposition is found in the power of deduction. The objectives of this paper are not to denigrate the power of deduction, but to add to it. Wolpert logically proves that there is a limit on all theories; including those developed through deduction. The means of reaching beyond these limits is through engagement with philosophically separate theories – *i.e.*, interdisciplinary work.

1.4. *Reaching Beyond the Limits*

Given the logical limits of all theories, our ability to extend understanding for engineering, economics, medicine or other fields requires engaging different perspectives. Thus the value of interdisciplinary work, but the value can only be realized if the multiple disciplines represent fundamentally different theories. If multiple disciplines only represent variations on a theme or philosophy, then the work is necessarily limited because reaching beyond these limits requires theoretically independent propositions.

There are a several examples of the successful application of knowledge based on multiple perspectives. The following three examples demonstrate that successful application of knowledge to specific non-trivial problems is only possible with two philosophical approaches:

1. Geophysics is developing rapidly as a science with computer modeling and three dimensional simulations. As a consequence, geophysical science often occurs in an office rather than in a mine site. However, the conclusions of modeling can only be verified *in situ*.
2. Evidence-based medicine is the practice of using the best available scientific evidence for health decisions. An algorithm could be a physician if this were sufficient. The evidence is not always sufficient in itself and physicians are legally obligated to exercise their best clinical judgment when treating real people. Judgment is required because one set of symptoms implies multiple conditions and deciding on a diagnosis may depend on judging the degree of each symptom as it might apply to several conditions.
3. Enterprise software incorporates the best available means of managing vast amounts of information. Yet, enterprise software creates a space for “tacit knowledge”. Algorithms operate with explicit knowledge, but geographic applications often present intractable problems because of the variations at each locality. The tacit understanding of these local conditions is integral to successful management of information.

Each of these examples represents successful applications of knowledge using two fundamentally independent methodologies. The limits of one are overcome by the capabilities of the other. The following section will examine these same benefits and limits in economics – particularly in financial economics.

2. Extending Natural Science Results to the Social Sciences

Modern economics is a social science that originated with Adam Smith's philosophical question about the "just price". The question is about a price high enough for producers to sell their goods and low enough for people to buy the goods. For Smith, The dynamics of the market place operated as if there were an "invisible hand" guiding the market to an appropriate price. The rigor of Neoclassical Economics a century later, in effect, convert Smith's analogy into a "mechanism", which was assumed to behave like "mechanisms" in natural science. The scientific model brought the rigor of natural science methods helped make great advances in economic science. The limit, however, is that economics is a social science rather than a natural science. And, I will argue, the nature of intractable problems in one discipline does not translate into those of the other.

There are non-trivial and historic failings as a result of inappropriate extension of natural sciences in the social sciences. The limits of one theory were ignored and bad solutions (limited or inappropriate solutions) to human problems were developed for intractable social questions. The advantages of interdisciplinary studies were not realized because independence was not maintained when transposing the lessons from one discipline onto the questions of another.

2.1. *Reaching Beyond the Limits of Understanding*

According to P. Mirowski:

> *Neoclassical economists made savvy use of the resonances between body, motion and value by engaging in brazen daylight robbery: The [neoclassical economists] appropriated the mathematical formulations of mid-nineteenth century energy physics,...made them their own by changing the labels on the variables, and then trumpeted the triumph of a truly "scientific economics" [19]*

Economics advanced the discipline through interdisciplinary studies, but it inherited certain limitations in using the natural science model to analyze human societies as well. One consequence was that the original theoretical base of the classical economists started to fade from the discipline. The distinction between Classical and '"scientific" economics can be illustrated through the relative approaches of Francis Edgeworth and Vilfredo Pareto.

Edgeworth applied mathematical concepts to sociology in this seminal work, *Mathematical Psychics* [20]. In this work, Edgeworth delineates the conditions for perfect competition, and in the process explains the contract and indifference curve concepts in economics. He also stipulates the conditions for

"Pareto Optimality" and the idea that imperfect competition yields an indeterminate range of solutions for maximum social welfare [21]. Humans had not evolved, according to Edgeworth, to the point that interpersonal comparisons of utility could be measured. This odd view of human limits shows that Edgeworth recognized the limits of mathematics and stated that in order to maximize social welfare, an ethic must be chosen on faith [22].

Pareto, like Edgeworth, accepted the utilitarian ethic, but rather than admit that he accepted his ethic on the basis of faith he avoided the contentious issues in theology and philosophy by dismissing the issues as not scientific. "Strange disputes about predestination, about the efficacy of grace, etc.…show that men have not freed themselves from these daydreams which people have gotten rid of in the physical sciences" [23]. Pareto's argument is that the equations work in economics simply because they work in mechanics [24]. As we have seen, however, philosophical and theological issues about ethics and morality represent intractable problems for mathematics and were viewed as irrelevant. By ignoring the problem, Pareto helped economics adopt energy physics and replaced the theoretical basis of classical economics. The objective was to make economics a natural science rather than a social science. The interdisciplinary success evident in Edgeworth's work, by adding rigor to the greatest extent possible, was taken too far by Pareto. The key difference was Edgeworth maintained the theoretical basis of the social science while Pareto tried to wholly incorporate economics into the theoretical basis of natural science. Thus his approach ceased being truly interdisciplinary.

2.2. *Refitting Fact to Theory*

The development of economics as a discipline, building upon Pareto and others, has enhanced the mathematical aspects of social science. Economics depends on observation of social activity rather than on experimentation. This is a significant challenge because validation through experimentation is critical in advancing human understanding through science. The means to demonstrate natural laws in economics that are equivalent to those in physics is a difficult task.

Thorstein Veblen addressed the problem of observation given the close analogy between economics and physics:

> *…economic theory is formulated as an analysis of the 'natural' course of life of the community…which might, not unfairly, be stated as some sort of law of the conservation of economic energy…*

> *But, since a strict uniformity is nowhere to be observed at first hand in the phenomenon...it has to be found by a laborious interpretation...and a diligent abstraction and allowance for disturbing circumstances...where causal continuity is denied [25].*

Veblen further states that the "investigator proceeds on a conviction of the orderliness of natural sequence". A proposition is assumed based on conviction or faith because controlled observation is impossible. Veblen's approach would be disturbing for natural sciences because some observations are accepted and other observations are ignored. These methods are the normal course of activity in social sciences and humanities. These are also the methods that offended Pareto. Developing arguments for or against the efficacy of theory and philosophy is time consuming and difficult in the absence of controlled experimentation. This makes interdisciplinary studies difficult.

Mistakes have been made by trying to adopt the methods of other disciplines – *i.e.*, using energy physics or mechanics in place of economic propositions. One of the best examples of the consequence of inappropriate adoption of a scientific mechanistic model is found in the descriptions of "economic man" as a narrowly self-interested agent for utility maximization. Then Game theory experimentation by economists of individuals' actions found purported anomalies in human behavior that did not fit the mechanistic model. The first and most important "anomaly" of human behavior is cooperation [26]. The experimentation further revealed that participants in game theory scenarios are more likely to cooperate if they are allowed to talk to one another [27]! These trivial results were reported as surprising only because the original theoretical propositions of human behavior had been replaced to satisfy the adoption of mechanics methods as economic methods.

The above game theory experiment shows that economics, and other social sciences, are developing the means to conduct experimentation in order to validate various theories and hypotheses. Using a proper scientific method is certainly an advancement for the social sciences and is a great example of learning from interdisciplinary perspectives.

The game theory experiments yielded other outcomes. Most significant for our purposes is the finding that people will continue to buy and sell assets until the prices are bid past the point of maximum value. This means that people normally create speculative bubbles [28]. The more liquid a market, the higher speculation can drive up the price. These are some of the results of Vernon Smith's Nobel Prize work on economic behavior. The analogy between human behavior in markets and mechanics has a limit. One of the consequences of not recognizing the limits of the model is financial collapse.

2.3. *Long Term Capital Management*

Long Term Capital Management (LTCM) was a financial company in the 1990s that enjoyed a spectacular rise over five years but suffered a historic five months collapse in its last year. This brief history is based on Roger Lowenstein, *When Genius Failed*, New York (2001). LTCM was a hedge fund that utilized models to predict financial market prices and conducted its trades in pairs in order to eliminate market uncertainty. LTCM is of historical importance for several reasons:

- LTCM successfully converted $1 billion in 1994 into $4 billion in 1998, controlled leveraged assets of over $100 billion, and through additional instruments had a market exposure of over $1 trillion.
- LTCM was one of the first arbitrage companies. Arbitrage has since become a common trading strategy on Wall Street and around the world.
- LTCM used models based on the best financial economics of Nobel Prize winning theories.
- LTCM was liquidated for $600 million five months after reaching a peak value of over $4 billion.

LTCM was led by John Meriwether, an extremely successful bond market trader, in partnership with two Nobel Prize winning economists, Myron Scholes and Robert C. Merton. The firm was essentially a small group of highly successful traders conducting trades using the results of models built on the Black-Scholes financial formula. Scholes and Merton won the Nobel Prize for the theoretical models (Fischer Black died prior to the award of the Nobel Prize). LTCM, in short, was a firm built with the best knowledge and staffed by the best traders [29].

The nature of arbitrage is to buy one financial instrument low and sell it high at the same time as selling another instrument high and then buying it low. See Lowenstein's account for a full description. In very general terms, prices move in response to two separate circumstances: either due to overall market conditions or due to conditions specific to the instrument. By trading in pairs – usually bonds – LTCM could eliminate risk associated with general market movements.

The significant contribution of the Black-Scholes model is in identifying past volatility as an unbiased predictor of future volatility [30]. Bond prices fluctuate daily, but the mean price of a bond can be deduced with an estimate of its variation. This is similar to arguing that the price will revert to its mean. This is important because the Black-Scholes model introduced an innovation in economics that allowed modelers to predict the price of a bond. Despite the daily volatility, models could determine when bond prices deviated from their long term price. The deviation from the long term price created the trading opportunity.

LTCM did not make money on every trade and did not make money in every month, but its successes were spectacular. The fund earned a 28% return in the first year; a year when the average investor lost money. The more interesting part of the firm was its belief in knowing the odds of market transactions – the firm calculated that it would lose 5% of its money 12% of the time, or 12 times in 100 years. Like flipping a coin, they were certain that a series of trades would eventually create a return if they watched the odds. They were repeated rewarded for this belief. The firm earned 59% in 1995, 57% in 1996, and 25% in 1997.

All of these returns were lost in 1998. The failure is attributable to a number of internal and external factors. The internal factors all related to the dependence on a single theoretical foundation. The four key internal factors were the hubris of the traders, the assumption of independence of world financial markets, the assumption of a normal distribution of errors, and the assumption that the past is a good predictor of the future. The first internal factor was the excess confidence in the genius of the models. In the early years of operations, the firm's traders did not depend exclusively on the models. They debated trades and probable market conditions. In short, traders used additional perspectives to make decisions. In the last years, the complete confidence in the models dominated all debates about potential trades and the firm's exposure to risk. Additional perspectives were no longer part of the analysis.

Second, the firm had developed such a large portfolio that they needed to find new trades and they entered highly leveraged positions in most financial markets around the world. The highly leveraged position that allows for larger returns also creates an opportunity for larger losses. Being in multiple markets did not diversify risk as they expected because financial markets had become closely linked over time. One of the key propositions of arbitrage is that by trading in pairs they eliminate general market risk.

Third, LTCM price predictions errors were modeled using a normal distribution. This is often the single most important critique of the LTCM models. The implication of normal distributions is that the probability of extreme events becomes increasingly remote the further that a price deviates from its mean. The result for LTCM was that as its models started to make incorrect predictions the firm put more money into the markets rather than stabilizing its positions as other arbitrage firms started to do.

Fourth, the predictions of volatility were based on past price fluctuations. This is the key feature of the Black-Scoles model. The past fluctuations were used as the best estimate of future fluctuations. The past is the best predictor of the future, but it is still not a good predictor of the future. Further, the past does not fully account for extreme events. The past provided a sufficient estimate for normal volatility. The only guidance that the past provides for future crises is that crises occasionally occur. This is a limit for the model.

Within the context of financial markets, the models accurately predicted prices and produced fantastic returns to the firm and its investors. In the context of politics, the firm was wholly ill-equipped. In August 1998, Russia defaulted on its bonds. The default was not an isolated event; it was part cause and fully a consequence of a series of political and financial crises that were occurring in most geographic regions. From a modeling point of view, the markets were acting irrationally. LTCM relied on its models as other investors were reducing their exposure to uncertainty. LTCM put more money into failing markets because the potential for profit was growing according to the models.

The significance of the Russian bond default is that the bond markets ceased to operate. Volatility, the predictor that LTCM depended upon, had no meaning. Indeed the models of LTCM were producing recommendations that were the opposite of the reality in front of the traders. The possibility of default was too remote an event for the model to consider relative to other possibilities.

The bulk of the LTCM story has not been duplicated here. For example, significant events regarding derivatives, leveraging, and internationalization have not been discussed. The significance of the portion reported here is that the LTCM models were incomplete. John Maynard Keynes made the same point, although not in a theoretic manner: "Markets can remain irrational longer than you can remain solvent" [31]. This is an expression about intractable events that occur more frequently than can be modeled.

3. Conclusion

The temptation in analyzing LTCM is to assume that the models of LTCM had errors or that the modelers did not develop a sufficiently sophisticated algorithm. Given Wolpert's recent work on limits on the explanatory power of theories and Gödel's and Turing's work as well, a complete analysis of the failure of LTCM must consider the position that financial models, regardless of sophistication, will fail to predict events at some point. In LTCM's case, the possibility of default by the Russian government could not be treated like an assignment of a probability; the need to recognize the possibility was an intractable problem requiring additional perspectives from outside the model.

The random noise that occurs in the market place is indeed random and normally distributed. The possibilities of extreme events are non-market events, and, therefore, cannot be modeled using market information. The extreme events are intractable problems that the models cannot accommodate. Nonetheless, the models continue to be used because these still have tremendous explanatory power. The Nobel Prizes awarded to Merton and Scoles were earned and deserved.

The nature of the market remains beyond the scope of any one model. This is not an indictment of the models. This may be an indictment of relying solely on modeling to address all market events. Financial crises are common in the history of markets. Each crisis was precipitated by a different event. The financial crashes of Black Monday in October 1987, the Mexico Peso devaluation in 1994 and the recent collapse of AIG in 2008 are each examples of extreme events that financial models could not predict.

Turing discussed the halting problem of a universal machine. The machine invariably faces a question for which there is no answer. This is the limit of the machine – *i.e.*, the theory. The financial crises represent extreme events that are beyond the scope of the models used to manage normal financial markets. The means to deal with these limits is by finding theoretically independent perspectives to identify and reach beyond the limits. This is the value of interdisciplinary studies.

References

1. D.H. Wolpert, *Physica D* **237** (2008)
2. P.-M. Binder, *Nature*, **455** (2008), p. 884-885.
3. S. Kleene, *Mathematical Logic*, 1967
4. D. Deutsch, *The Fabric of Reality*, Penguin Books (New York, 1997)
5. P. Drucker, *The Essential Drucker*, Harper-Collins (2001), p. 345.
6. Binder, p.884.
7. E.K. Hunt, "Freedom, Necessity and Human Values in the Social Sciences", *Recent Approaches to the Social Sciences* (1979), p. 37.
8. E.N. Zalta (ed.), *Stanford Encyclopedia of Philosophy*, http://plato.stanford.edu/ (2002).
9. D. Gordon, *Rev of Aus. Econ.*, **5**, 1 (1991).
10. W.L. Hosch, *Encyclopedia Britannica*.
11. S. Robinson, *SIAM News*, **35**, 8 (2002).
12. B. Stutz, *Discover*, July (2004).
13. R.E. Thorne, *Physics Today*, May (1996).
14. *Ibid*, p.42.
15. D.E. Feng, *Phys. Rev. Lett.* **99**, 216404 (2007).
16. R.B. Ouboter, *Scientific American*, March (1997).

17. P. Mirowski, *More Heat than Light*, Cambridge (1989), p. 53.
18. *Ibid*, pp.86-87.
19. *Ibid*, p. 9.
20. F. Y. Edgeworth, *Mathematical Psychics*, London (1881), p. 1.
21. M. Blaug, *Economic Theory in Retrospect*, Cambridge (1992).
22. Edgeworth, p.82.
23. V. Pareto, quoted in Mirowski, p.221.
24. Mirowski, p.272.
25. T. Veblen, quoted in Mirowski, p. 139.
26. R.M. Dawes and R.H. Thaler, *J Econ. Pers,* **2** 3 (1988), p. 187.
27. *Ibid*, p. 193.
28. V.L. Smith, *Am. Econ. Rev.* **66** 2 (1976).
29. R. Lowenstein, *When Genius Failed*, New York (2001).
30. *Ibid*, p.29.
31. J.M. Keynes, quoted in Lowenstein, p. 123.
32. 4.V. Baran, M. Colonna, M. Di Toro and V. Greco, *Phys. Rev. Lett.* **86**, 4492 (2001).

QUANTUM PROBABILITY APPLIED TO SOCIAL AND BEHAVIORAL SCIENCES*

J. R. BUSEMEYER

Psychological and Brain Sciences, Indiana University, Bloomington, In, 47405, USA
E-mail: jbusemey@indiana.edu

ZHENG WANG

Department of Communications, Ohio State University, Columbus, USA
E-mail: wang1243@osu.edu

Social and behavioral sciences rely on probabilistic models to describe the behavior of people. In the past, these models have been based on the classic (Kolmogorov) theory. We argue that it may be useful to develop models using quantum (Von Neumann) probabilities. Quantum theory provides a coherent understanding of interference produced by one measure disturbing another measure, which is a common problem that occurs with human judgments.

1. Why Use Quantum Probability Theory in Social and Behavioral Sciences?

Quantum theory was originally invented by physicists to explain findings that seemed paradoxical from a standard physical view point. Later Von Neumann[1] provided an axiomatic foundation for quantum theory, and by doing so, he discovered that it implied a new type of logic and probability theory. Consequently, there are now two general theories for assigning probabilities to events: classic (Kolmogorov[2]) theory and quantum (Von Neumann) theory. For many years social and behavioral scientists have relied solely on classic probability theory to understand the results of their experiments. However, social and behavioral scientists also face dramatic findings that remain paradoxical from a classic probability viewpoint. These paradoxes suggest that measurements involving human judgments interfere in a manner that violates the classic law of total probability.[3] Quantum

*This work is supported by NSF 0817965

probability theory is viewed as an alternative mathematical approach for generating probability models that can explain these interference effects found in the social and behavioral sciences.

No assumptions about the biological substrates are made here, and instead this is an exploration into new mathematical tools for constructing social and behavioral science theories. Quantum principles have been used recently in a number of psychological applications including perception,[4] conceptual structure,[5] information retrieval,[6] and human judgments.[7]

2. How to use Quantum Probability in Social and Behavioral Sciences?

Consider the following simple example. Suppose a survey is conducted and individuals are asked two different types of survey questions before an election. Question A concerns the persons voting preference, and for this question, the person must choose one of three answers: Democrat, Republican, or Independent. Question B concerns the persons political orientation and in this case the person must choose one of three answers: Moderate, Liberal, Conservative. Question A produces three mutually exclusive and exhaustive events (labeled D,R,I); question B also produces produces three mutually exclusive and exhaustive events (labeled M,L,C). These questions must be asked in some order and so half of the individuals are asked question A first (the AB condition), and the other half are asked question B first (the BA condition). In general, the order of presentation has a large influence on the probabilities of responses.[8] The answer to the first question disturbs the response to the second question. Now let us compare classical and quantum formulations of probability models for these probabilistic events. In the end, we shall see an advantage for using quantum probabilities to understand order effects caused by interference of one measurement on another when human judgments are involved.

2.1. *Classical Probability Representation*

Classic probability theory represents events as subsets of a sample space. A critical question is how to form the sample space for the two different order conditions. This depends on whether or not there are order effects. For simplicity, let us first consider the case in which there are no order effects.

2.1.1. *No order effects*

In this case, we ignore the difference between two conditions so that we can use a single sample space defined as a set containing nine distinguishable (ignoring order) elements $S = \{D \cap M,\ R \cap M,\ I \cap M,\ D \cap L,\ R \cap L,\ I \cap L, D \cap C,\ R \cap C,\ I \cap C\}$ which are mutually exclusive and exhaustive. For example, the element $D \cap M = M \cap D$ represents observing the answer 'democrat' on a voting question and the answer 'moderate' on a political orientation question, where we do not distinguish the order of the questions. Other events are formed by union, so that for example $D = \{\ D \cap M,\ D \cap L,\ D \cap C\}$ and $M = \{D \cap M,\ R \cap M,\ I \cap M\}$.

Classic probability postulates a *state* vector that is used to assign probabilities to all of the events. The state P assigns a probability (a real number between zero and one) to each of the nine elements of the sample space S, and all these probabilities sum to one. For example, $P(D \cap M)$ assigns a probability to the event $D \cap M \in S$. For later comparisons with quantum theory, it will be useful to define the classic state P as a 9×1 column vector containing the 9 probabilities for the nine elements of the sample space S, and the sum of these probabilities equals one. An event then corresponds to a projector, which is a 9×9 indicator matrix that projects the state onto a vector that only contains the probabilities of elements contained in the event. For example, the indicator matrix $M_{D \cap M}$ is a projector used to select the probability for the event $D \cap M \in S$; this 9×9 matrix contains all zeros except for a one in the diagonal entry corresponding to the row in P that contains the probability for the event $D \cap M$. The matrix product, $M_{D \cap M} \cdot P$, is a projection that produces a 9×1 vector containing the probability $P(D \cap M)$, and zeros otherwise. The event D corresponds to an indicator matrix M_D with zeros everywhere except ones on the diagonals corresponding to events $D \cap M, D \cap L,\ D \cap C$; the matrix product, $M_D \cdot P$, is a 9×1 vector containing the three probabilities $P(D \cap M)$, $P(D \cap L)$, $P(D \cap C)$, and zeros otherwise. Similarly, the event M has a corresponding indicator matrix M_M and the projection $M_M \cdot P$ contains the three probabilities $P(M \cap D)$, $P(M \cap R)$, $P(M \cap I)$, and zeros otherwise. The nine projectors corresponding to the nine elements in S form a resolution of the identity: they are pair-wise orthogonal and they sum to unity. Also we note that the projectors for events defined on S always commute, so that for example $M_D \cdot M_M = M_M \cdot M_D$.

The probability of an event equals the sum of the elements of the corresponding projection. We introduce the notation $|X|^k = \sum |x_i|^k$ for a vector X with coordinates x_i, $i = 1, n$. Then the probability of the event $D \cap M$ equals $P(D \cap M) = |M_{D\cap M} \cdot P|^1$, and $P(D) = |M_D \cdot P|^1 = P(D \cap M) + P(D \cap L) + P(D \cap C)$, and $P(M) = |M_M \cdot P|^1 = P(M \cap D) + P(M \cap R) + P(M \cap I)$. Note that the identity matrix I corresponds to the universal event so that $|I \cdot P|^1 = 1$. The complement of the event D is represented by $(I - M_D)$, and the probability that event D does not occur equals $P(\tilde{}D) = |(I - M_D) \cdot P|^1$. Similarly, the probability that event M does not occur equals $P(\tilde{}M) = |(I - M_M) \cdot P|^1$.

The conditional probability of event M given that we have already observed D is defined as $P(M|D) = P(M \cap D)/P(D)$. Similarly, the conditional probability of D given M equals $P(D|M) = P(D \cap M)/P(M)$. These conditional probabilities can be described using the state vector notation as follows: $P(M|D) = |M_M \cdot M_D \cdot P|^1/P(D)$ and $P(D|M) = |M_D \cdot M_M \cdot P|^1 /P(M)$. The joint probability of events D and M is therefore equal to

$$\begin{aligned} P(D \cap M) = P(D) \cdot P(M|D) &= |M_M \cdot M_D \cdot P|^1 \\ &= |M_D \cdot M_M \cdot P|^1 = P(M \cap D). \end{aligned}$$

2.1.2. *Order effects*

Now we suppose that there are large order effects so that the first question disturbs the answer to the second question. In this case, a separate sample space is used for each condition. The sample space for condition AB is $AB = \{DM, RM, IM, DL, RL, IL, DC, RC, IC\}$ and the sample space for condition BA is $BA = \{MD, LD, CD, MR, LR, CR, MI, LI, CI\}$ and the order now matters. Using rules of set theory, we can form new events by union of elements within each sample space. For example, for condition AB, we can define event $D1 = \{DM, DL, DC\}$ and event $M2 = \{DM, RM, IM\}$. Similarly, for condition BA we can define events $M1 = \{MD, MR, MI\}$ and $D2 = \{MD, LD, CD\}$. The state P_{AB} for condition AB assigns a probability (a real number between zero and one) to each of the elements of the sample space AB and all these probabilities sum to one. For example, $P_{AB}(DM)$ assigns a probability to the event $DM \in AB$. Similarly, the state P_{BA} assigns a probability to each of the elements of the sample space BA. For example, the probability of the event $MD \in BA$ equals $P_{BA}(MD)$. The two probability vectors, P_{AB} and P_{BA} are allowed to be different to account for order effects. However, classic probability does not provide an principles for relating the two probability distributions obtained from the two conditions. We shall see that this problem is solved by using quantum probabilities.

2.2. *Quantum Probability Representation*

Quantum probability theory represents events as subspaces of a vector space (defined on a complex field). However, there are two ways to form this vector space depending on whether or not there are order effects. If there are no order effects, then a compatible representation can be formed; but if order effects are present, then an incompatible representation is needed. Let us first assume no order effects and consider the compatible representation, and later assume order effects and consider the incompatible representation. We shall see that when the events are all compatible, there is no difference between classic and quantum probabilities. The most interesting implications arise when the events are incompatible.

2.2.1. *Compatible Events*

For the compatible representation, order of question does not matter, and so there are nine distinguishable events $S = \{D \cap M,\ R \cap M,\ I \cap M,\ D \cap L,\ R \cap L,\ I \cap L, D \cap C,\ R \cap C,\ I \cap C\}$, which are mutually exclusive and exhaustive. Events are represented as subspaces of a vector space, and so this implies that we require nine mutually orthogonal subspaces that span the entire vector space to represent these events. The minimum representation is obtained by assuming each subspace representing each distinguishable event is a one dimensional ray. This implies that the vector space is a nine dimensional Hilbert space denoted H. Corresponding to each subspace is a projector that projects vectors in the vector space onto the subspace. For example, the unique event $D \cap M$ is represented by a ray defined by the unit length vector $|D \cap M\rangle \in H$, and the projector for this subspace is the outer product $M_{D \cap M} = |D \cap M\rangle \langle D \cap M|$. Other events can be defined by the *span* of the union of elementary events. For example the event D is represented by $\text{span}\{|D \cap M\rangle, |D \cap L\rangle, |D \cap C\rangle\}$ which corresponds to the projector $M_D = M_{D \cap M} + M_{D \cap L} + M_{D \cap C}$. As another example, the event M is represented by $\text{span}\{|D \cap M\rangle, |R \cap M\rangle, |I \cap M\rangle\}$ which corresponds to the projector $M_M = M_{D \cap M} + M_{R \cap M} + M_{I \cap M}$. The nine projectors corresponding to the nine elements in S form a resolution of the identity: they are pair-wise orthogonal and they sum to the identity. Also we note that all the projectors for all the events defined on S always commute, so that for example $M_D \cdot M_M = M_M \cdot M_D$.

Quantum theory postulates a state vector that is used to assign probabilities to all of the events. The unit length state vector $|\psi\rangle \in H$ assigns probabilities to events as follows: First we project the state onto the subspace that represents the event in question, and then we compute the squared length of the projection. For example, the quantum probability for the event $D \cap M$ is equal to $Q(D \cap M) = |M_{D\cap M} \cdot |\psi\rangle|^2$, and the quantum probability for the event D equals $Q(D) = |M_D \cdot |\psi\rangle|^2 = |(M_{D\cap M} + M_{D\cap L} + M_{D\cap C}) \cdot |\psi\rangle|^2$. Furthermore, the orthogonal property of the projectors implies that $|(M_{D\cap M} + M_{D\cap L} + M_{D\cap C}) \cdot |\psi\rangle|^2 = |M_{D\cap M} \cdot |\psi\rangle|^2 + |M_{D\cap L} \cdot |\psi\rangle|^2 + |M_{D\cap C} \cdot |\psi\rangle|^2 = Q(D \cap M) + Q(D \cap L) + Q(D \cap C)$. Note that the identity I corresponds to the universal event so that $|I \cdot |\psi\rangle|^2 = 1$. The complement of the event D is represented by $(I - M_D)$, and the probability that event D does not occur equals $Q(\tilde{}D) = |(I - M_D) \cdot |\psi\rangle\,|^2$. Similarly, the probability that event M does not occur equals $P(\tilde{}M) = |(I - M_M) \cdot |\psi\rangle|^2$.

The conditional probability of event M given that we have already observed D is defined as $Q(M|D) = |M_M \cdot M_D \cdot |\psi\rangle\,|^2/Q(D)$ and similarly $Q(D|M) = |M_D \cdot M_M \cdot |\psi\rangle|^2 / Q(M)$.The joint probability of events D and M is therefore equal to $Q(D\cap M) = Q(D)\cdot Q(M|D) = |M_M \cdot M_D \cdot |\psi\rangle|^2 = |M_D \cdot M_M \cdot |\psi\rangle|^2 = Q(M \cap D)$.

Finally, we can define

$$|\psi\rangle = \tag{1}$$
$$\sqrt{P(D\cap M)} \cdot |D\cap M\rangle + \sqrt{P(D\cap L)} \cdot |D\cap L\rangle + \sqrt{P(D\cap C)} \cdot |D\cap C\rangle$$
$$+\sqrt{P(R\cap M)} \cdot |R\cap M\rangle + \sqrt{P(R\cap L)} \cdot |R\cap L\rangle + \sqrt{P(R\cap C)} \cdot |R\cap C\rangle$$
$$+\sqrt{P(I\cap M)} \cdot |I\cap M\rangle + \sqrt{P(I\cap L)} \cdot |I\cap L\rangle + \sqrt{P(I\cap C)} \cdot |I\cap C\rangle\,.$$

In this case, we see that the quantum model with compatible events can assign exactly the same probabilities to all events as the classic model that assumes no order effects.

2.2.2. *Incompatible Events*

Now we assume that the order of measurement matters. Once again, we start by proposing a single vector space denoted H, and the subspaces of this vector space are used to represent the events for both questions and for both conditions. For each condition, the three events D, R, and I are represented by three subspaces with corresponding projectors M_D, M_R, and M_I respectively; similarly the three events M, L, C are represented by three subspaces with corresponding projectors M_M, M_L, M_C. The three events D, R, I are mutually exclusive and exhaustive, which implies that the three projectors are mutually orthogonal ($M_D \cdot M_R = M_D \cdot M_I = M_R \cdot M_I = 0$) and sum to the identity $M_D + M_R + M_I = I$. In other words, the set of three projectors $\{M_D, M_R, M_I\}$ describe a set of three orthogonal subspaces that together span the vector space H. The same properties hold true for the three events M, L, C and the corresponding three projectors M_M, M_L, M_C. Thus the set of three projectors $\{M_M, M_L, M_C\}$ also describe a set of three orthogonal subspaces that together span the same vector space H. Therefore we have two sets of projectors that describe two sets of orthogonal subspaces, and each set spans the same vector space H.

Furthermore, the three projectors M_D, M_R, M_I all commute ($M_D \cdot M_R = M_R \cdot M_D$, $M_D \cdot M_I = M_I \cdot M_D$, $M_R \cdot M_I = M_I \cdot M_R$); and the other three projectors M_M, M_L, M_C also all commute ($M_M \cdot M_L = M_L \cdot M_M$, $M_M \cdot M_C = M_C \cdot M_M$, $M_L \cdot M_C = M_C \cdot M_L$). However, an order effect implies that the question about voting preference interferes with the question about political orientation and visa versa. Therefore we must assume that the events D, R, I are *incompatible* with the events M, L, C; which implies that the three projectors M_D, M_R, M_I do not commute with the three projectors M_M, M_L, M_C (e.g., $M_D \cdot M_M \neq M_M \cdot M_D$).

Now we can consider the probabilities of the events obtained from the two conditions. First consider the condition AB. The probability of answering 'democrat' to the first question is simply $Q(D) = |M_D \cdot |\psi\rangle|^2$. If the person does in fact decide to vote 'democrat', then the probability of selecting the 'moderate' political orientation equals $Q(M|D) = |M_M \cdot M_D \cdot |\psi\rangle|^2 / Q(D)$. The probability of first voting 'democrat' and then selecting 'moderate' equals $Q(D) \cdot Q(M|D) = |M_M \cdot M_D \cdot |\psi\rangle|^2$. The total probability of selecting 'moderate' for condition AB, when asked about voter preference first, equals $Q(D)Q(M|D) + Q(R)Q(M|R) + Q(I)Q(M|I)$.

Next consider condition BA. The probability of answering 'moderate' to the first question is simply $Q(M) = |M_M \cdot |\psi\rangle|^2$. The probability of first selecting 'moderate' and then voting 'democrat' equals $Q(M)Q(D|M) = |M_D \cdot M_M \cdot |\psi\rangle|^2$, which is not equal to the other order, $Q(D) \cdot Q(M|D) = |M_M \cdot M_D \cdot |\psi\rangle|^2$, because the projectors do not commute. Furthermore, an interference effect occurs because

$$
\begin{aligned}
Q(M) &= |M_M \cdot |\psi\rangle|^2 = |M_M \cdot I \cdot |\psi\rangle|^2 \qquad (2)\\
&= |M_M \cdot (M_D + M_R + M_I) \cdot |\psi\rangle|^2 \\
&= |M_M \cdot M_D \cdot |\psi\rangle + M_M \cdot M_R \cdot |\psi\rangle + M_M \cdot M_I \cdot |\psi\rangle|^2 \\
&\neq |M_M \cdot M_D \cdot |\psi\rangle|^2 + |M_M \cdot M_R \cdot |\psi\rangle|^2 + |M_M \cdot M_I \cdot |\psi\rangle|^2 \\
&= Q(D)Q(M|D) + Q(R)Q(M|R) + Q(I)Q(M|I).
\end{aligned}
$$

The inequality follows from the fact that the three vectors being summed before taking the squared length are not orthogonal and their inner products produce interference effects. Thus the law of total probability is violated because the total probability of selecting 'moderate' after first deciding about voter preference is not equal to the probability of selecting 'moderate' when political orientation was asked first.

In the simplest case, the three events D, R, I are represented by unidimensional rays, corresponding to three orthonormal vectors $|D\rangle, |R\rangle, |I\rangle$ all belonging to H, which form the three orthogonal projectors $M_D = |D\rangle\langle D|$, $M_R = |R\rangle\langle R|$, $M_I = |I\rangle\langle I|$ which form a resolution of the identity. Under this simplifying assumption, H is a three dimensional space spanned by the three orthonormal eigenvectors $\{|D\rangle, |R\rangle, |I\rangle\}$:

$$\begin{aligned} |\psi\rangle &= I \cdot |\psi\rangle \\ &= (|D\rangle\langle D| + |R\rangle\langle R| + |I\rangle\langle I|) \cdot |\psi\rangle \\ &= \langle D|\psi\rangle \cdot |D\rangle + \langle R|\psi\rangle \cdot |R\rangle + \langle I|\psi\rangle \cdot |I\rangle . \end{aligned} \tag{3}$$

In other words, any state $|\psi\rangle$ can be expressed in terms of coordinates with respect to the voter preference question basis as follows:

$$\alpha = \begin{bmatrix} \langle D|\psi\rangle \\ \langle R|\psi\rangle \\ \langle I|\psi\rangle \end{bmatrix} .$$

Again in the simplest case, the three events M, L, C are represented by unidimensional rays, corresponding to three orthonormal vectors $|M\rangle, |L\rangle, |C\rangle$ all belonging to H, which form the three orthonormal projectors $M_M = |M\rangle\langle M|$, $M_L = |L\rangle\langle L|$, $M_C = |C\rangle\langle C|$ which form another resolution of the identity. In this case, H is a three dimensional space spanned by another set of three orthonormal eigenvectors $\{|M\rangle, |L\rangle, |C\rangle\}$:

$$\begin{aligned} |\psi\rangle &= I \cdot |\psi\rangle \\ &= (|M\rangle\langle M| + |L\rangle\langle L| + |C\rangle\langle C|) \cdot |\psi\rangle \\ &= \langle M|\psi\rangle \cdot |M\rangle + \langle L|\psi\rangle \cdot |L\rangle + \langle C|\psi\rangle \cdot |C\rangle . \end{aligned} \tag{4}$$

In other words, any state $|\psi\rangle$ also can be expressed in terms of coordinates with respect to the political orientation question basis as follows:

$$\beta = \begin{bmatrix} \langle M|\psi\rangle \\ \langle L|\psi\rangle \\ \langle C|\psi\rangle \end{bmatrix} .$$

The two sets of eigenvectors $\{|D\rangle, |R\rangle, |I\rangle\}$ and $\{|M\rangle, |L\rangle, |C\rangle\}$ span the same vector space H and they are related by a unitary transformation:

$$\begin{aligned} |D\rangle &= \langle M|D\rangle \cdot |M\rangle + \langle L|D\rangle \cdot |L\rangle + \langle C|D\rangle \cdot |C\rangle \\ |R\rangle &= \langle M|R\rangle \cdot |M\rangle + \langle L|R\rangle \cdot |L\rangle + \langle C|R\rangle \cdot |C\rangle \\ |I\rangle &= \langle M|I\rangle \cdot |M\rangle + \langle L|I\rangle \cdot |R\rangle + \langle C|I\rangle \cdot |C\rangle \,. \end{aligned} \tag{5}$$

Note that an eigenstate of the voter preference question, for example $|D\rangle$, is in a *superposition* state with respect to the political orientation question. The inverse unitary transformation is given by

$$\begin{aligned} |M\rangle &= \langle D|M\rangle \cdot |D\rangle + \langle R|M\rangle \cdot |R\rangle + \langle I|M\rangle \cdot |I\rangle \\ |L\rangle &= \langle D|L\rangle \cdot |D\rangle + \langle R|L\rangle \cdot |R\rangle + \langle I|L\rangle \cdot |I\rangle \\ |C\rangle &= \langle D|C\rangle \cdot |D\rangle + \langle R|C\rangle \cdot |R\rangle + \langle I|C\rangle \cdot |I\rangle \,, \end{aligned} \tag{6}$$

Note once again that an eigenstate of the political orientation question, for example $|M\rangle$, is in a *superposition* state with respect to the voter preference question.

The unitary matrix U transforms the coordinates (α) of the state expressed in the basis of the voting preference question into coordinates (β) of the state expressed in the basis of the political orientation question:

$$\beta = \begin{bmatrix} \langle M|\psi\rangle \\ \langle L|\psi\rangle \\ \langle C|\psi\rangle \end{bmatrix} = \begin{bmatrix} \langle M|D\rangle & \langle M|R\rangle & \langle M|I\rangle \\ \langle L|D\rangle & \langle L|R\rangle & \langle L|I\rangle \\ \langle C|D\rangle & \langle C|R\rangle & \langle C|I\rangle \end{bmatrix} \cdot \begin{bmatrix} \langle D|\psi\rangle \\ \langle R|\psi\rangle \\ \langle I|\psi\rangle \end{bmatrix} = U \cdot \alpha. \tag{7}$$

The columns of the U matrix provide the coordinates of the $|D\rangle, |R\rangle, |I\rangle$ vectors when described in terms of the political orientation basis. The cell entries of U indicate the transition amplitudes from the voter preference states to the political orientation states; for example, $\langle M|D\rangle$ is the transition amplitude from state $|D\rangle$ to state $|M\rangle$. The inverse matrix $U^\dagger$ transforms the coordinates (β) of the state expressed in the basis of the political orientation question into coordinates (α) of the state expressed in the basis of the voter preference question:

$$\alpha = \begin{bmatrix} \langle D|\psi\rangle \\ \langle R|\psi\rangle \\ \langle I|\psi\rangle \end{bmatrix} = \begin{bmatrix} \langle D|M\rangle & \langle D|L\rangle & \langle D|C\rangle \\ \langle R|M\rangle & \langle R|L\rangle & \langle R|C\rangle \\ \langle I|M\rangle & \langle I|L\rangle & \langle I|C\rangle \end{bmatrix} \cdot \begin{bmatrix} \langle M|\psi\rangle \\ \langle L|\psi\rangle \\ \langle C|\psi\rangle \end{bmatrix} = U^\dagger \cdot \beta. \tag{8}$$

The columns of the $U^\dagger$ matrix give the coordinates of the $|M\rangle, |L\rangle, |C\rangle$ vectors when described in terms of the voter preference basis. The cell entries of $U^\dagger$ indicate the transition amplitudes from the political affiliation states to the voter preference states. For example, $\langle D|M\rangle$ is the transition amplitude from state $|M\rangle$ to state $|D\rangle$.The transformation between coordinate systems must be unitary in order to preserve lengths and inner products (orthogonality).

Order effects are easier to examine in this simple case. The probability of answering 'democrat' followed by 'moderate' equals

$$\begin{aligned} Q(D)Q(M|D) &= |M_M \cdot M_D \cdot |\psi\rangle|^2 \\ &= |(|M\rangle\langle M|)\cdot(|D\rangle\langle D|)\,|\psi\rangle|^2 \\ &= ||M\rangle\langle M|D\rangle\langle D|\psi\rangle|^2 \\ &= |\langle M|D\rangle|^2 \cdot |\langle D|\psi\rangle|^2 \end{aligned} \tag{9}$$

and the probability of answering in the opposite direction equals

$$\begin{aligned} Q(M)Q(D|M) &= |M_D \cdot M_M \cdot |\psi\rangle|^2 \\ &= |(|D\rangle\langle D|)\cdot(|M\rangle\langle M|)\,|\psi\rangle|^2 \\ &= ||D\rangle\langle M|D\rangle\langle D|\psi\rangle|^2 \\ &= |\langle D|M\rangle|^2 \cdot |\langle M|\psi\rangle|^2 \end{aligned} \tag{10}$$

The first term in the product, $|\langle M|D\rangle|^2 = |\langle D|M\rangle|^2$ is the same for the two probabilities. The difference between the two orders is produced by the second term in the product

$$|\langle D|\psi\rangle|^2 = |\langle D|M\rangle\langle M|\psi\rangle + \langle D|L\rangle\langle L|\psi\rangle + \langle D|C\rangle\langle C|\psi\rangle|^2 \neq |\langle M|\psi\rangle|^2 . \tag{11}$$

The law of total probability is violated again, which is seen by expanding the probability of selecting the 'moderate' answer under the BA condition:

$$\begin{aligned} Q(M) &= |M_M \cdot (M_D + M_R + M_I)\cdot|\psi\rangle|^2 \\ &= |M_M \cdot M_D \cdot |\psi\rangle + M_M \cdot M_R \cdot |\psi\rangle + M_M \cdot M_I \cdot |\psi\rangle|^2 \\ &= ||M\rangle(\langle M|D\rangle\langle D|\psi\rangle + \langle M|R\rangle\langle R|\psi\rangle + \langle M|I\rangle\langle I|\psi\rangle)|^2 \\ &= ||M\rangle|^2 \cdot |\langle M|D\rangle\langle D|\psi\rangle + \langle M|R\rangle\langle R|\psi\rangle + \langle M|I\rangle\langle I|\psi\rangle|^2 \\ &= |\langle M|D\rangle\langle D|\psi\rangle + \langle M|R\rangle\langle R|\psi\rangle + \langle M|I\rangle\langle I|\psi\rangle|^2 \\ &\neq |\langle M|D\rangle\langle D|\psi\rangle|^2 + |\langle M|R\rangle\langle R|\psi\rangle|^2 + |\langle M|I\rangle\langle I|\psi\rangle|^2 . \end{aligned}$$

3. Discussion

Quantum probability is considered by many to be a very specialized probability theory that is only useful in Physics. We question this idea and believe there are useful applications of this theory outside of Physics. One of the founding fathers of quantum theory, Niels Bohr,[9] speculated on this possibility and so did David Bohm.[10] In fact, quantum and classic probability theories are highly similar (once they are both expressed using a common projection notation as we have tried to do here). Both classic and quantum theory can be defined by a state vector, and probabilities of events are determined by projections of the state vector. There are two key differences. One is that classic probability restricts the elements of the projections to be probabilities and measures the probability of a projection by the sum of elements in the projection; whereas quantum theory allows the elements of the projection to be complex and measures the probability of a projection by the sum of the squared magnitudes of the elements in the projection. The second is that quantum theory makes use of a unitary transformation that allows one to linearly map one set of events into another set of events producing incompatible events.

The problem with the classic probability model is that it does not provide any principle for relating the state P_{AB} from order condition AB to the state P_{BA} from order condition BA when there are order effects on human judgments. In the past, researchers have developed cognitive models,[11] that derive the probabilities for the observed events in sets AB and BA from a common set of cognitive/affective states that are not unobservable. A quantum model provides a simpler representation that could facilitate the development of new cognitive models.

The quantum probability model, based on incompatible events, provides a coherent theory for relating the probabilities across the two order conditions. The unitary matrix U is the key for understanding this relationship between the probabilities of events obtained from the different order conditions. The psychological intuition is that a person requires one viewpoint, described by the set of eigenvectors $\{|D\rangle, |R\rangle, |I\rangle\}$, for evaluating questions about voter preference; but they need to change to a new viewpoint, described by a different set of eigenvectors $\{|M\rangle, |L\rangle, |C\rangle\}$, for evaluating questions about political orientation. These two viewpoints reside in the same vector space and they are related by a unitary transformation. The person cannot evaluate both of these questions at the same time, and they have to evaluate these questions sequentially, one at a time. In other words, they can only examine on perspective or viewpoint at a time. The state of the individual before being asked about voter preference will not be the same as the state of the individual after answering the question about voter preference. Before answering any question at all, the individual is in an initial state represented by $|\psi\rangle$ and views the world from this point. If the person clearly settles on an answer for one question, say decides in favor of 'democrat,' the the state transfers from state $|\psi\rangle$ to state $|D\rangle$ and the person now views the world from a different point. Furthermore, the person cannot be certain about both questions at the same time. Once the person becomes certain that he/she will vote, say 'democrat,' then this person must be uncertain about the political orientation question. This occurs because the eigenvector $|D\rangle$ corresponding to 'democrat' is a superposition state with respect to the political orientation eigenvectors $\{|M\rangle, |L\rangle, |C\rangle\}$. Thus making a voter preference decision will disturb and produce uncertainty about the political orientation question and visa versa.

References

1. J. Von Neumann, *Mathematical Foundations of Quantum Theory* (Princeton University Press, 1932/1955).
2. A. N. Kolmogorov, *Foundations of the theory of probability* (N.Y.: Chelsea Publishing Co., 1933/1950).
3. J. R. Busemeyer, Z. Wang and A. Lambert-Mogiliansky, *Journal of Mathematical Psychology* **xx**, xx (2009).
4. H. Atmanspacher, T. Filk and H. Romer, *Biological Cybernetics* **90**, 33 (2004).
5. D. Aerts and L. Gabora, *Kybernetes* **34**, 192 (2005).
6. K. Van Rijsbergen, *The geometry of infomation retrieval* (Cambridge University Press, 2004).
7. A. Y. Khrennikov, *Information dynamics in cognitive, psychological, social and anomalous phenomena* (Kluwer Academic, 2004).
8. D. W. Moore, *Public Opinion Quarterly* **66**, 80 (2002).
9. N. Bohr, *Atomic theory and human knowledge* (Wiley, 1958).
10. D. Bohm, *Quantum Theory* (Dover, 1979).
11. J. R. Busemeyer and A. Diederich, *Cognitive Modeling* (Sage, 2009).

MARKET EFFICIENCY AND THE RISKS AND RETURNS OF DYNAMIC TRADING STRATEGIES WITH COMMODITY FUTURES

LORNE N. SWITZER

Associate Dean, Research and Van Berkom Endowed Chair of Small Cap Equities, Finance Department, John Molson School of Business, Concordia University, 1455 De Maisonneuve Blvd. W., Montreal, Quebec, CANADA H3G 1M8
E-mail: switz@jmsb.concordia.ca

HUI JIANG

John Molson School of Business, Concordia University, 1455 De Maisonneuve Blvd. W., Montreal, Quebec, CANADA H3G 1M8
E-mail: jasmine1224@gmail.com

This paper investigates relationships between profits from dynamic trading strategies, risk premium, convenience yields, and net hedging pressures for commodity futures. As a market efficiency study, it crosses a number of disciplines, including traditional finance, behavioral finance, and behavioral psychology. The term structure of oil, gold, copper and soybeans futures markets contains predictive power for the corresponding term premium. However, only oil futures and soybean futures lead their spot premium. Significant momentum profits are identified in both outright futures and spread trading strategies when the spot premium and the term premium are used to form winner and loser portfolios. Profits from active strategies based on winner and loser portfolios are conditioned on market structure and net hedging pressure effects. Dynamic trading strategies based on contracts with extreme backwardation, extreme contango, and extreme hedging pressures are also tested. On average, spread trading outperforms outright futures trading in capturing the term structure risk and hedging pressure risk. For such strategies, long-short the long-term spread offers the greatest and most significant return and it offers the only exploitable trading profits built on the past hedging pressure. The existence of profits from active trading strategies based on winners is consistent with behavioral finance and behavioral psychology models in which market participants irrationally overreact to information and trends.

Keywords: Commodity futures; market efficiency; dynamic trading strategies.

1. Introduction

In recent years, commodity market booms and busts have inspired significant interest in this asset class amongst academics and practitioners. As an individual asset class, commodities show several features that distinguish themselves from financial assets, including:

1. Commodities are real assets that can be used for consumption. They can be packaged as derivative securities for generating returns for investors and for transferring real consumption risk. Therefore, their performance is subject to more factors than financial assets, including but not limited to business cycles, local and global supply-and-demand, technology development, substitute and complementary products.
2. Based on economic theory, commodity returns in the absence of shocks should mean-revert to the equilibrium marginal rate of production.
3. Unlike financial assets that are cash-settled at expiration, commodity futures need to be settled by physical delivery; this entails unique storage and shipping costs. For those who want to avoid physical exposure at expiry, contract roll-over is essential. According to Feldman and Till (2006), this roll yield drives the overall yield over the long-term horizon.
4. Each commodity should be treated as an individual asset class instead of one asset class for all. Erb and Harvey (2006) associate commodity index performance with the performance of its components.
5. The equilibrium CAPM model does not work well for the commodity markets because some of its critical assumptions are violated: such as an insufficiently diversifiable base of participants in the commodity markets pointed out by Hirshleifer (1988) and the exclusion from capital assets by Erb and Harvey (2006).

Given this background, much of the extant theoretical and empirical research has looked at commodities apart from financial assets. One strand of literature relates commodity returns to inventory level storage costs as reflected in convenience yields and the cost of carry (e.g. Kaldor (1939), Working (1948), Brennan (1958), Fama and French (1988), Dincerler, Khoker and Simin (2005)). These yields determine the state-time opportunity set for consumption. A second strand focuses on hedging pressures, which constrain the risk transference function (e.g. Working (1953), Cootner (1960), Chang (1985), Carter, Rausser and Schmitz (1983)). Bessembinder (1992), and De Roon Goorbergh and Nijman (2000, 2004)). A third strand looks at market efficiency from a returns predictability perspective (Fama (1991)). Evidence of abnormal returns from momentum or contrarian strategies in strategies based on ***past performance*** or historical data would be inconsistent with the efficient markets paradigm. For example, Miffre and Rallis (2007) test several momentum strategies based on ***past returns*** that on average generate significant excess returns of 9.38% per annum. Their contrarian strategies, proposed to capture the long-term reversal are not successful. However, based on the theories of storage and hedging/risk transference, one could expect that the returns and risk premia for commodity futures to be based on a richer set of variables than ***past performance***. This paper proposes to explore this issue, extending Miffre and Rallis (2007) by identifying variables based on the theory, such as the convenience yield and net hedging pressure that might be considered for generating active trading strategies as tests of market efficiency.

Our focus is on the four commodity futures contracts; NYMEX crude oil, COMEX gold, COMEX copper and CBOT soybean contracts. These contracts are chosen for the following reasons: (i) their liquidity; (ii) their diverse historical term structure and hedging experiences; and (iii) their different sensitivities to the business cycle.

We find that both the term structure and hedging pressure variables based on theory are significant determinants of commodity returns, based on VAR Granger Causality Models. Furthermore, these variables are found to contain information that can be used to construct profitable trading strategies.

The remainder of the paper is organized as follows. Section 2 describes the data. In section 3 we present the methodology for the tests. The empirical results follow in Section 4. Section 5 performs provides the results for the active trading strategies. The paper concludes with a summary in section 6. For readers unfamiliar with commodities, some terminology is provided in Appendix 1.

2. Data

Daily closing prices of NYMEX crude oil, COMEX gold, COMEX copper and CBOT soybean from January 1, 1990 to December 31, 2006 are obtained from Bloomberg. Continuous time-series of futures prices are constructed in the order of the first -nearby futures, second-nearby futures, etc. up to one-year maturity or the last contract month before the end of a calendar year, such as November for the soybean contract. The one-year maturity is chosen because it covers a sufficiently long forward period and also has substantial liquidity to deal with. Table 1 provides some summary statistics of the four commodity futures contracts.

Table 1: Futures contract information

Contract	Exchange	Delivery months	Last trading day
WTI crude oil	New York Mercantile Exchange	All months	The third business day prior to the 25th calendar day of the month preceding the delivery month.
Gold	Commodity Exchange Inc.	2 4 6 8 10 12	The third to last business day of the maturing delivery month.
Copper	Commodity Exchange Inc.	All months	The third to last business day of the maturing delivery month.
Soybean	Chicago Board of Trade	1 3 5 7 9 11[1]	The business day prior to the 15th calendar day of the contract month.

The first - nearby futures contract is constructed from the series of prices of the next expiring contract until one week prior to the last trading day of it, at which point the contract is rolled over to the next expiring contract. We adopt the common practice of treating the first-nearby futures price, as the spot price, since a united spot market for each commodity does not exist. By extension, the first-nearby futures is literally the second-nearby futures and by analogy, all the next nearby futures contracts are mapped to their subsequent one interval lagged contracts.

For calendar spreads, only the first-nearby futures and futures maturing in six months and in one year are considered because these contracts usually have the longest trading life, normally available for trading 18 months before maturity. These are the most widely used contracts for spreads by practitioners.

Figure 1 shows the term structures and average term structures (based on the average futures prices across maturities) of the four commodities.

[1] Soybean futures are also traded for delivery in August. We neglect the August contract when constructing the time-series data in order to make data sets evenly spaced.

Figure 1: Term Structure of Commodity Futures Prices, June 1990- January 2006

Panel A: Contract Prices, daily data

Oil

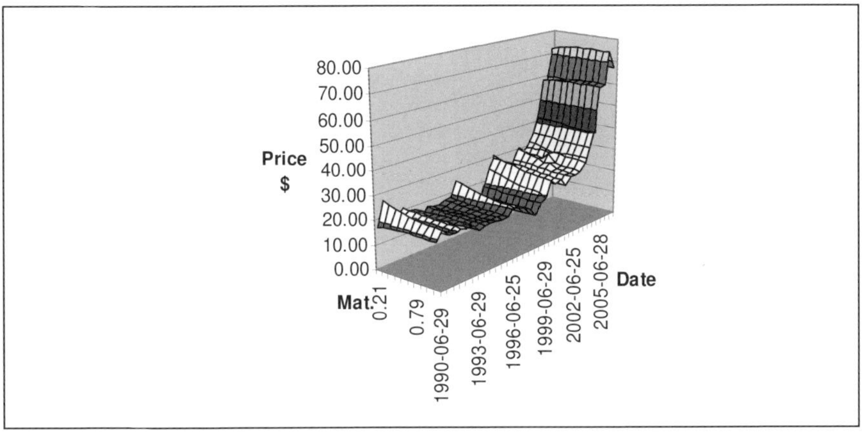

Gold

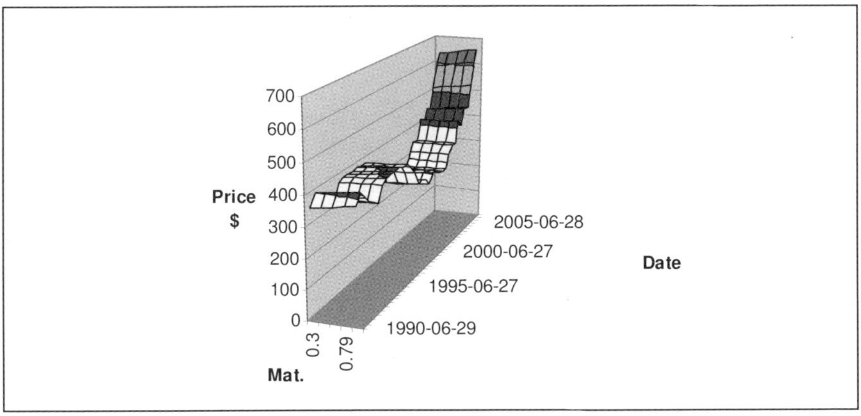

Oil and copper markets are in backwardation on average over the period of the sample. In contrast, the gold market and the soybean market is also generally in contango with its sharpest slope appearing in the pre-harvest third quarter—from July to September.

Table 2 provides descriptive statistics of the daily returns based on the spot price and the first-nearby futures price. All returns are calculated against the one-day price lag of the same time series and not across contracts with different maturities. Returns of the CRB index (the oldest tradable as well as the most comprehensive commodity index), the Russell 3000 index (a proxy for the U.S.

(Figure 1 cont.)

Copper

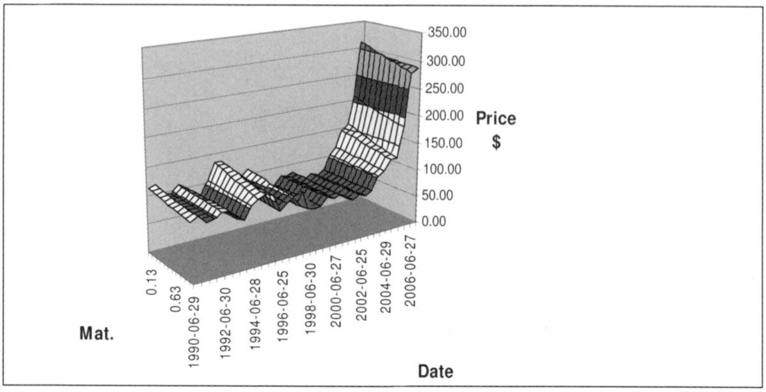

Soybeans

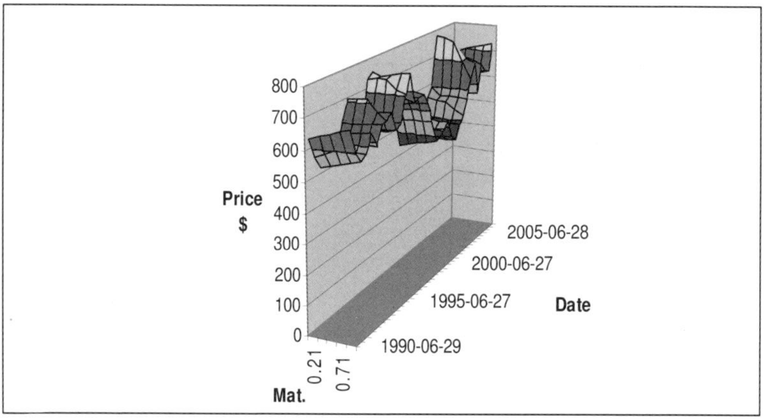

listed equity market portfolio) and the 10-year U.S. Treasury bond (the industry norm of measuring long-term interest rate) are also shown as benchmarks of commodity indices vs. financial asset returns.[2]

All commodity returns show significant departures from the normal distribution, based on the Jarque-Bera estimates. Oil and copper futures exhibit the highest returns, Among commodities, oil futures exhibit the highest volatility while gold futures exhibit the least volatility which is only marginally higher than that of the CRB commodity index.

---

[2] The annualized standard deviations of commodity returns are possibly elevated since commodity returns are normally autocorrelated and asymmetrically distributed.

(Figure 1 cont.)

Panel B: Term Structure of Average Commodity Futures Prices, daily data

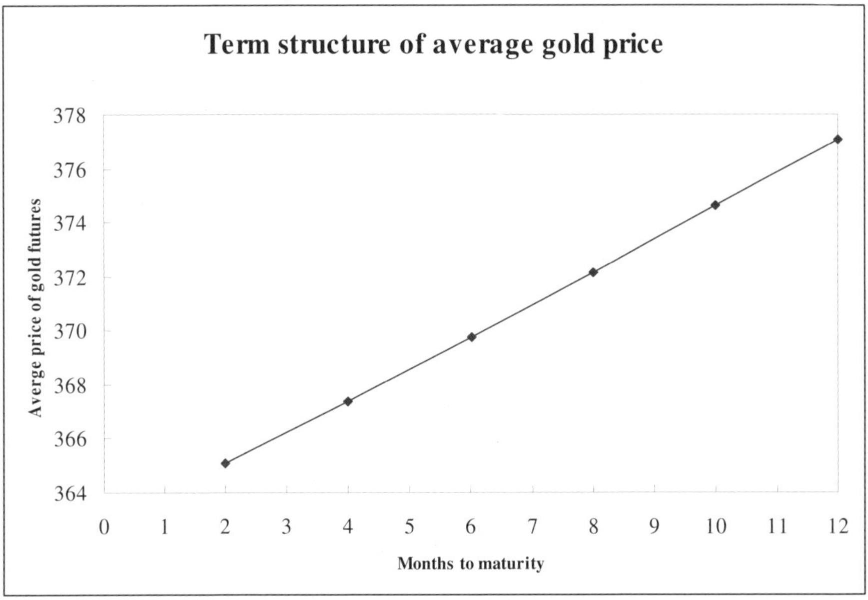

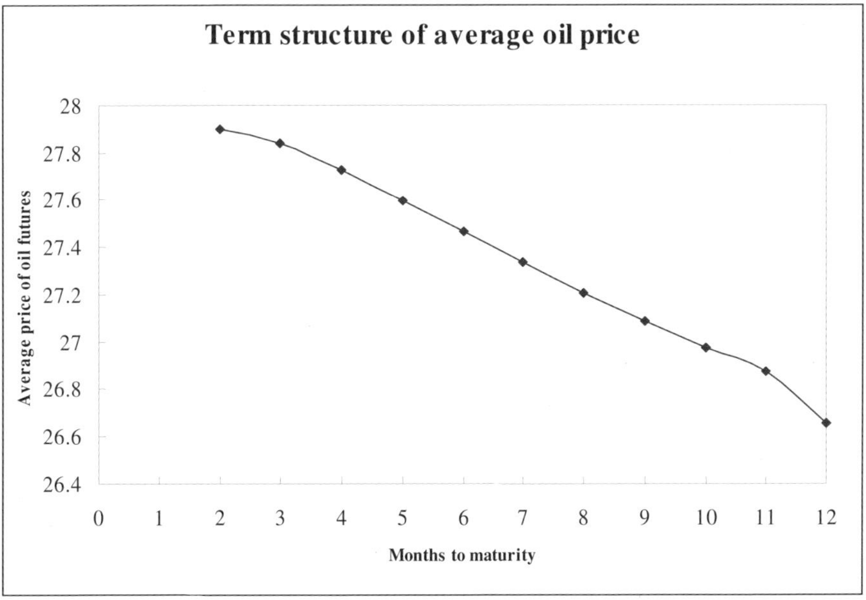

(Figure 1 cont.)

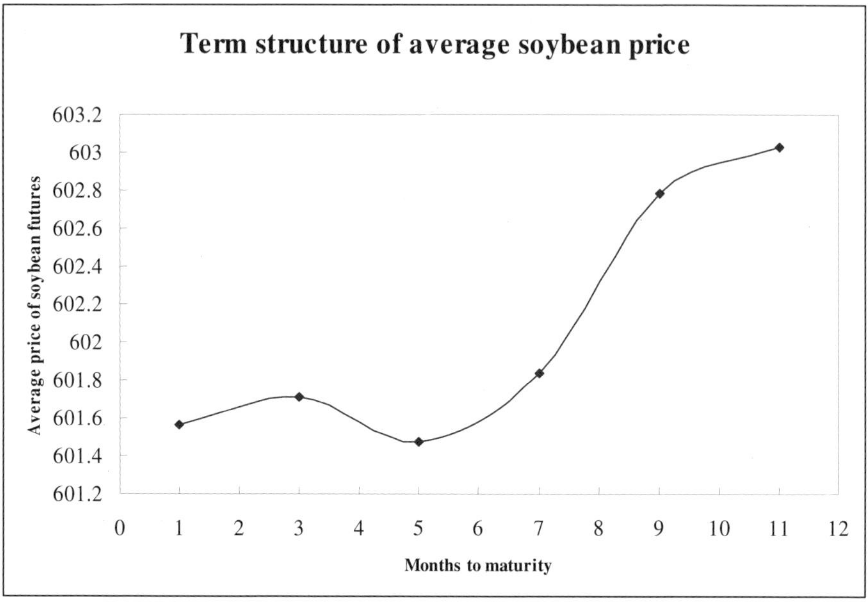

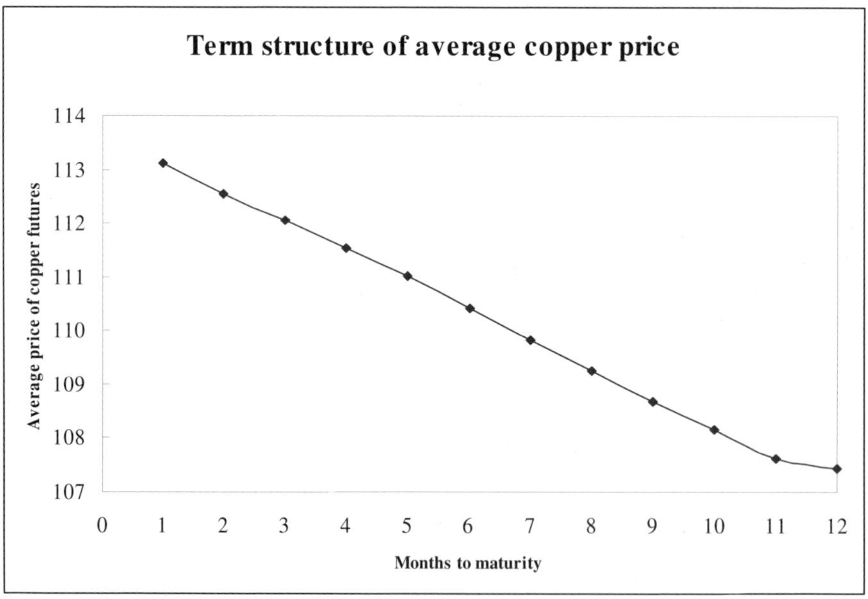

Oil and copper markets are in backwardation on average over the period of the sample. In contrast, the gold market is characterized and the soybean market is also generally in contango with its sharpest slope appearing in the pre-harvest third quarter—from July to September.

Table 2 provides descriptive statistics of the daily returns based on the spot price and the first-nearby futures price. All returns are calculated against the one-day price lag of the same time series and not across contracts with different maturities. Returns of the CRB index (the oldest tradable as well as the most comprehensive commodity index), the Russell 3000 index (a proxy for the U.S. listed equity market portfolio) and the 10-year U.S. Treasury bond (the industry norm of measuring long-term interest rate) are also shown as benchmarks of commodity indices vs. financial asset returns.[3]

All commodity returns show significant departures from the normal distribution, based on the Jarque-Bera estimates. Oil and copper futures exhibit the highest returns. Among commodities, oil futures exhibit the highest volatility while gold futures exhibit the least volatility which is only marginally higher than that of the CRB commodity index.

3. Methodology

3.1 Decomposition of futures returns

Following Roon, Nijman and Veld's (1998), we decompose futures returns into spot premia and term premia by simulating the returns of a strategy of (1) buying a *k*-day contract and (2) buying a *k*-day contract, selling a *n*-day contract and holding the spread for *k* days ($k<n$) where *k* and *n* represent number of days until maturity of the first-nearby futures and of distant futures respectively.

From the cost-of-carry model, in a structurally contangoed market, a certain yield $y_t^{(n)}$ based on futures prices, *f* and spot prices, *s* [4] can be locked in by longing an asset in the spot market and simultaneously shorting it in the futures market to be delivered at time $t+n$.

$$y_t^{(n)} \equiv \frac{f_t^{(n)} - s_t}{n} \quad (1)$$

[3] The annualized standard deviations of commodity returns are possibly elevated since commodity returns are normally autocorrelated and asymmetrically distributed.

[4] The prices are in log form. The prices are in log form.

Table 2: Summary Statistics on Returns Based on the Spot price and the First-nearby futures price

Panel A. Daily Returns Results

Return of	N	Mean(%)	Median(%)	Std Dev	Skewness	Kurtosis	Max	Min	Jarque-Bera
Oil SP	4262	0.02	0.06	0.02	-1.34	25.15	0.14	-0.38	88390.31
Oil F_1	4260	0.02	0.06	0.02	-1.24	21.98	0.12	-0.33	65016.13
Gold SP	4258	0.01	0.00	0.01	-0.24	13.12	0.09	-0.08	18207.74
Gold F_1	4254	0.01	0.00	0.01	-0.27	12.58	0.09	-0.08	16333.49
Copper SP	4260	0.02	0.00	0.02	-0.23	7.19	0.11	-0.12	3149.4
Copper F_1	4256	0.02	0.00	0.02	-0.27	7.30	0.12	-0.12	3322.24
Soybean SP	4255	0.00	0.00	0.01	-3.32	74.80	0.07	-0.34	921711
Soybean F_1	4255	0.00	0.00	0.01	-1.24	21.34	0.07	-0.21	60724.67
CRB	4250	0.01	0.02	0.01	-0.05	4.61	0.04	-0.03	463.62
R3000	4250	0.03	0.06	0.01	-0.12	6.81	0.05	-0.07	2576.452
10y T-bond	4172	0.00	0.00	0.00	0.35	5.14	0.00	0.00	883.251

Return of SP refers to return of spot markets; return of F_1 refers to return of the most nearby futures contracts. Return of CRB commodity index, return of Russell 3000 index and return of 10-year U.S. Treasury bond are listed here as benchmarks.
Returns are calculated with daily data from January 1st 1990 to December 31st 2006. All returns are calculated against the one-day price lag of the same time series and not across contracts with different maturities.

Panel B. Annualized Mean and Standard Deviation of Spot and First Nearby Future Returns

Return	Oil SP	Oil F_1	Gold SP	Gold F_1	Copper SP	Copper F_1	Soybean SP	Soybean F_1	CRB	R3000	10-y T-bond
Annual mean (%)	5.75	5.93	2.58	2.49	5.83	6.03	0.90	0.90	1.50	8.63	-0.16
AnnualStd Dev (%)	36.22	32.09	14.38	14.29	24.40	24.05	23.27	21.97	10.01	15.51	0.92

Similarly, the forward yield $h_t^{(k,n)}$ can be earned from time $t+k$ to $t+n$ by taking a long position in the contract to mature at $t+k$, shorting its equivalent to mature at $t+n$ ($k<n$), and holding the spread for k days.

$$h_t^{(k,n)} \equiv \frac{f_t^{(n)} - f_t^{(k)}}{n-k} = \frac{ny_t^{(n)} - ky_t^{(k)}}{n-k} \quad (2)$$

Re-arranging (2) we obtain:

$$\begin{aligned}(n-k)h_t^{(k,n)} &= ny_t^{(n)} - ky_t^{(k)} \\ &= E_t\left[(n-k)y_{t+k}^{(n-k)}\right] + \Theta_t^{(k,n)}\end{aligned} \quad (3)$$

$$\Theta_t^{(k,n)} \equiv \sum_{i=0}^{k-1} \theta_t^{(n-i)}$$

where $\Theta_t^{(k,n)}$ is the estimate of the *term premium* and can be rewritten in terms of log spot price and log futures price as:

$$\Theta_t^{(k,n)} = \left(s_{t+k} - f_t^{(k)}\right) - \left(f_{t+k}^{(n-k)} - f_t^{(n)}\right) \quad (4)$$

It is reasonable to assume that for a naked long position in the first nearby futures contract that matures in k days, the term premium $\pi_{y,t}^{(n)}$ is negligible. In this case, the maturity risk premium is explained by spot premium $\pi_{s,t}$, which is, by definition, the expected spot return in excess of the one-period yield.

$$E_t\left[r_{s,t+k}\right] = E_t\left[s_{t+k} - s_t\right] = ky_t^{(k)} + \pi_{s,t} \quad (5)$$

Rearrange (5) to get the expression of spot premium:

$$\pi_{s,t} = \left(s_{t+k} - f_t^{(k)}\right) \quad (6)$$

Table 3 provides descriptive statistics for the spot and term premia. In most cases, they are found to be significantly different from zero at conventional levels. Average term premia decline as maturities increase in oil and copper markets and rise in longer-dated gold and soybean markets, implying backwardation (contango) of oil and copper (gold and soybean). A structurally backwardated market carrying an inverse-charge from nearby to distant delivery renders a long spread unprofitable—thus, a positive term premium is needed for remuneration, vice versa for a contangoed market.

Table 3. Summary statistics of spot premium and term premium

Panel A. Daily Returns Results (as percentages)

Daily Mean	***Spot premium*** $\pi_{s,t}=\left(s_{t+k}-f_t^{(k)}\right)$	***Term premium*** $\Theta_t^{(k,n)}=\left(s_{t+k}-f_t^{(k)}\right)-\left(f_{t+k}^{(n-k)}-f_t^{(n)}\right)$									
	K	***p=3***	***p=4***	***p=5***	***p=6***	***p=7***	***p=8***	***p=9***	***p=10***	***p=11***	***p=12***
Oil	1.99		0.34	0.32	0.30	0.28	0.27	0.25	0.23	0.23	0.23
Gold	0.24				-0.59		-0.59		-0.58		-0.57
Copper	1.91	0.26	0.24	0.22	0.21	0.20	0.19	0.18	0.18	0.18	0.12
Soybean	-0.07			-0.19		-0.16		-0.17		-0.19	

Daily Std Dev	***Spot premium*** $\pi_{s,t}=\left(s_{t+k}-f_t^{(k)}\right)$	***Term premium*** $\Theta_t^{(k,n)}=\left(s_{t+k}-f_t^{(k)}\right)-\left(f_{t+k}^{(n-k)}-f_t^{(n)}\right)$									
	K	***p=3***	***p=4***	***p=5***	***p=6***	***p=7***	***p=8***	***p=9***	***p=10***	***p=11***	***p=12***
Oil	15.84		2.85	3.77	4.60	5.35	5.99	6.56	7.11	7.58	8.01
Gold	6.83				0.41		0.50		0.59		0.70
Copper	12.08	1.53	1.91	2.19	2.51	2.80	3.13	3.46	3.78	4.10	9.89
Soybean	12.59			4.99		5.97		5.97		6.37	

Returns are calculated with daily data from January 1st 1990 to December 31st 2006.
Spot premium is obtained by longing the 1st-nearby futures mature in k days and term premium obtained by longing the same futures maturing at $t+k$ and shorting distant futures maturing at $t+n$ ($k<n$) and holding the spread for k days.
p refers to number of months until maturity for contract being shorted. n and k refer to number of days until maturity for contract being shorted and for the 1st-nearby futures being longed in both strategies respectively. In this paper, the 1st-nearby futures are mature in different months across four markets, i.e. 3 months, 4 months, 2 months and 3 months for oil, gold, copper and soybean respectively.

Table 3 (Cont'd)

Panel B. Annualized Mean and Standard Deviation of Spot and Term Premium (as percentages)

AnnualMean	***Spot premium*** $\pi_{s,t} = \left(s_{t+k} - f_t^{(k)}\right)$	***Term premium*** $\Theta_t^{(k,n)} = \left(s_{t+k} - f_t^{(k)}\right) - \left(f_{t+k}^{(n-k)} - f_t^{(n)}\right)$									
	K	***p=3***	***p=4***	***p=5***	***p=6***	***p=7***	***p=8***	***p=9***	***p=10***	***p=11***	***p=12***
Oil	7.95		4.08	1.91	1.20	0.84	0.64	0.49	0.40	0.34	0.30
Gold	0.72				-3.54		-1.76		-1.16		-0.86
Copper	11.45	3.07	1.42	0.88	0.62	0.47	0.38	0.32	0.27	0.24	0.14
Soybean	-0.28			-1.15		-0.49		-0.34		-0.28	

Annual Std Dev	***Spot premium*** $\pi_{s,t} = \left(s_{t+k} - f_t^{(k)}\right)$	***Term premium*** $\Theta_t^{(k,n)} = \left(s_{t+k} - f_t^{(k)}\right) - \left(f_{t+k}^{(n-k)} - f_t^{(n)}\right)$									
	K	***p=3***	***p=4***	***p=5***	***p=6***	***p=7***	***p=8***	***p=9***	***p=10***	***p=11***	***p=12***
Oil	31.69		9.88	9.22	9.20	9.26	9.28	9.28	9.31	9.28	9.25
Gold	11.84				1.00		0.86		0.83		0.85
Copper	29.58	5.30	4.67	4.39	4.34	4.34	4.42	4.52	4.63	4.73	10.83
Soybean	25.18			12.22		10.34		8.44		7.80	

T-test results for Significance of Premia:

(1) Oil: spot premium and term premiums up to *p=8* are significantly different from zero at 1% level; term premia *p=9* and *p=10* are significant at the 5% level; the rest are all significant at 10%.

(2) Gold: spot premium is significantly different from zero at the 5% level; the remaining premia are all significant at 1%.

(3) Copper: term premium *p=12* is not significantly different from zero; all other premia are all significant at the 1% level.

(4) Soybean: spot premium is not significantly different from zero; term premium *p=5* is significant at the 5% level; all other premia are significant at 10%

Spot price trading risk is high based on volatility of returns. For oil and soybean markets, term volatilities are also high (annualized volatility of 7% and 12% respectively).

3.2 Estimation of the convenience yield and net hedging pressure variables

We estimate the convenience yield proxy using the Fama and French (1988) approach, as the negative of the interest-adjusted basis.[5] Three most widely used contracts—the 3-month,[6] the 6-month and the 12-month futures are used to calculate the short-, intermediate- and long-term convenience yields for each commodity.

The metric used to capture net hedging pressure, H_t, is the difference between the short and long hedge positions of commercial traders divided by their total hedge positions[7], with all positions' information from the semi-monthly[8] report of *Commitments of Traders in Commodity Futures* ("CFTC report" hereafter). A positive H_t means a net short hedging market whereas a negative ratio a net long hedging market.

The convenience yield is calculated as the negative of the interest-adjusted basis.

Oil and copper exhibit relatively high average convenience yields, on average.[9] Oil exhibits the highest unconditional convenience yield across commodities. Gold has the lowest and least volatile convenience yield; with the exception of copper, volatility increases at a decreasing rate with maturity,

---

[5] $$[F(t,T)-S(t)]/S(t)-R(t,T)=[W(t,T)-C(t,T)]/S(t)$$

where F and S are prices of futures and spot markets, R is interest foregone, W is storage cost and C is convenience yield. Interest-adjusted basis in the left-hand-side can be expressed as difference between relative storage cost and relative convenience yield. For a constant storage cost, the variation of relative convenience yield naturally dominates that of the interest-adjusted basis, so that convenience yield can be approximately expressed as the negative of the interest-adjusted basis.

[6] Since gold futures only mature in even months and soybean futures only in odd months, different contracts other than the three maturities have to be chosen for both. The 4-month futures is used to calculate the short-term convenience yield for gold, and the 7-month and 11-month contracts are used to calculate the intermediate-term and long-term convenience yields for soybean.

[7] $$H_t=\frac{\text{short hedge positions - long hedge positions}}{\text{total hedge positions}}$$

[8] This report was released bi-weekly until September 1992. From October 1992 on, was released on a weekly basis.

[9] Detailed results are available on request.

consistent with the Samuelson (1965) hypothesis for all commodities except copper.

For the entire observation period, all the four commodity futures are net short hedging markets, supporting Keynes' insurance perspective hypothesis. Net hedging pressure does vary across time and markets. Copper and soybean futures are net short hedging the majority of time, and oil and gold markets vacillate more often. Net short hedging pressure characterizes the gold market from 2001 to 2006.

4. Structural Granger-Causality VAR Estimation of Commodity Returns

This study employs a direct approach[10] to capture the effects of the autoregressive components of commodity returns, which would underpin momentum investment strategies, as well as other variables based on theory which would structurally impact on commodity returns, and estimate a nested Vector Autoregressive (VAR) system. The system includes as *joint variables* the commodity risk premia, convenience yields, and net hedging pressure. Since the spot premium, term premium, convenience yield and net hedging pressure variables are stationary in nature, unrestricted VAR Granger-causality models are appropriate. [11] The testing period is from the issue of the first weekly CFTC report[12] on October 6th 1992 to its last issue in 2006 on December 26th, 2006. The data encompass from 670 to 730 weekly observations across the four markets.

Table 4 provides estimates of VAR Granger-causality models for risk premia, convenience yields (with different maturities), and net hedging pressure for spot premia across markets.[13]

The most striking result of the estimation is that spot premia serve as leading (predictor) variables for all of the variables, both convenience yields and net hedging pressure across all commodities. As such, momentum trading strategies

---

[10] Bessembinder (1992) uses the Fama and MacBeth (1973) two step approach to capture the effect of net hedging pressure on residual risks of commodity futures

[11] Based on Augmented Dickey-Fuller (ADF) and Phillips-Perron (PP) unit root tests, the null hypothesis of a unit root is rejected by all the tested time series under all settings at a significance level of 5%. These results are available on request.

[12] The CFTC report was once published every two weeks until the end of September 1992. Ever since October 1992, this report has become a weekly issue and has been disclosed on each Wednesday. To keep as much information intact as possible, weekly data of risk premiums, net hedging pressure and convenience yield are re-calculated and used in this test.

[13] We use the FPE (final prediction error) criterion, the Akaike information criterion, the Schwarz criterion (SC); and the Hannan & Quinn (HQ) criterion and in all cases find an optimal lag lengths of one.

Table 4. VAR estimates—Spot Premia(RF_1), convenience yield (CY_p) and net hedging pressure (QT), weekly data, from October 1992 to December 2006

OIL						GOLD					
	RF_1	CY_1	CY_2	CY_3	QT		RF_1	CY_1	CY_2	CY_3	QT
$RF_1(-1)$	0.89	0.01	0.03	0.04	0.03	$RF_1(-1)$	0.90	0.00	0.00	0.00	0.20
t-value	*54.32****	*3.27****	*5.60****	*6.56****	*3.90****	*t-value*	*55.01****	*-2.17***	*-1.79**	*-1.14*	*4.24****
$CY_1(-1)$	-1.50	0.75	0.16	0.33	-0.04	$CY_1(-1)$	-0.89	0.11	-0.70	-0.74	-1.86
t-value	*-2.80****	*11.75****	*1.12*	*1.47*	*-0.19*	*t-value*	*-0.59*	*0.78*	*-5.10****	*-5.05****	*-0.43*
$CY_2(-1)$	1.25	0.10	0.75	-0.35	0.09	$CY_2(-1)$	1.04	0.68	1.46	0.57	1.86
t-value	*3.00****	*1.96***	*6.50****	*-1.99***	*0.53*	*t-value*	*0.57*	*4.16****	*8.88****	*3.22****	*0.36*
$CY_3(-1)$	-0.43	-0.02	0.09	1.10	-0.05	$CY_3(-1)$	-0.47	-0.10	-0.06	0.89	-0.82
t-value	*-2.86****	*-1.10*	*2.14***	*17.42****	*-0.76*	*t-value*	*-0.95*	*-2.20***	*-1.38*	*18.34****	*-0.57*
QT(-1)	0.03	0.00	0.00	0.00	0.91	QT(-1)	0.00	0.00	0.00	0.00	0.95
t-value	*0.76*	*-0.32*	*0.19*	*0.28*	*53.21****	*t-value*	*0.19*	*-4.61****	*-5.15****	*-5.83****	*73.54****
C	0.01	0.00	0.00	0.00	0.00	C	0.00	0.00	0.00	0.00	0.02
t-value	*2.73****	*2.02***	*-0.77*	*-0.67*	*0.88*	*t-value*	*1.48*	*8.15****	*8.25****	*8.24****	*1.76**
COPPER						**SOYBEAN**					
	RF_1	CY_1	CY_2	CY_3	QT		RF_1	CY_1	CY_2	CY_3	QT
$RF_1(-1)$	0.93	0.00	0.01	0.02	0.08	$RF_1(-1)$	0.90	-0.02	0.00	0.01	0.07
t-value	*64.69****	*0.87*	*2.20***	*3.74****	*4.02****	*t-value*	*57.66****	*-2.88****	*0.36*	*1.27*	*5.01****
$CY_1(-1)$	0.05	0.41	-0.27	-0.25	-1.26	$CY_1(-1)$	0.22	0.73	-0.26	-0.26	-0.20
t-value	*0.10*	*5.78****	*-2.77****	*-1.74**	*-1.94**	*t-value*	*3.18****	*26.71****	*-8.13****	*-6.97****	*-3.04****
$CY_2(-1)$	0.09	0.32	1.01	0.02	1.75	$CY_2(-1)$	-0.22	0.10	1.00	0.10	0.02
t-value	*0.20*	*4.56****	*10.41****	*0.16*	*2.75****	*t-value*	*-3.05****	*3.47****	*29.54****	*2.61****	*0.32*
$CY_3(-1)$	-0.05	-0.04	0.04	1.02	-0.62	$CY_3(-1)$	0.08	-0.02	0.02	0.92	0.00
t-value	*-0.35*	*-1.71**	*1.44*	*22.80****	*-3.11****	*t-value*	*1.43*	*-1.00*	*0.88*	*31.01****	*0.03*
QT(-1)	0.01	0.00	0.00	0.00	0.93	QT(-1)	0.00	0.00	0.00	0.01	0.98
t-value	*0.79*	*-2.41***	*-1.52*	*-1.08*	*68.57****	*t-value*	*-0.51*	*0.28*	*1.18*	*1.97***	*126.48****
C	0.00	0.00	0.00	0.00	0.01	C	0.00	0.00	0.00	0.00	0.00
t-value	*0.00*	*2.42***	*1.26*	*1.47*	*3.76****	*t-value*	*0.64*	*1.41*	*1.34*	*1.82**	*1.55*

based on recent performance may be well grounded. Convenience yields show market-dependent results. In particular, they lead spot premiums in the oil and soybean markets. On the other hand, net hedging pressure does not lead the spot premium in any market. Bi-directional causality is detected between term slopes of different horizons in all markets, showing that some intertwining of information is contained in each part of the term structure. Causality between the term structure and net hedging pressure is observed only for the metal markets studied. For gold, net hedging pressure uni-directionally leads the term structure while for copper, the overall term structure leads net hedging pressure and net hedging pressure leads the short-term slope only.

We also estimate VAR Granger-causality models for term premia (spread trade returns), convenience yields and net hedging pressure, for the various commodities.[14]

4.1 Metal futures

Aside from the autoregressive components, significant causality effects of term premia and convenience yields for spread returns on term structure and net hedging pressure are observed in the gold spread using the 6-month contract and in all the three copper spreads. Negative (positive) coefficients of gold (copper) spreads can be naturally associated with its contangoed (backwardated) convenience yield shape. Causality between the term structure and net hedging pressure is observed in both spread rates similar to that which obtains in spot returns: Net hedging pressure uni-directionally drives gold's term structure while the term structure leads copper's net hedging pressure.

Perhaps the most important finding in the spread market in these metal markets comes from the significant leading effect of the entire term structure on the term premium. In contrast, no significant relationship between the term structure and the corresponding spot premium exists. This result is consistent with Roon, Goorbergh and Nijman (2004) who conclude that the term premium should reflect term structure risks. Similar to the spot premium, variation of the term premium explained by term structure decreases with the contract horizon. Hence, the latest short-term convenience yield contains the most relevant information for predicting future term premiums.

---

[14] The detailed results are not included, in order to conserve space. These results are available on request.

4.2 Oil futures

For all three oil spreads, a leading effect from the term premium to the overall term structure and net hedging pressure appears in the 12th and 13th lags, showing that the spread premiums contain information that can be used to forecast the term structure and net hedging pressure in approximately three months away.

The overall term structure improves forecast of term premium across time, but at slightly different lags. In general, the predictive power of term structure shows up in various term premia in most cases up to the third month in the future.

4.3 Soybean futures

For the second soybean spread,[15] all three term premia clearly lead the term structure by one week, one month and three months ahead, respectively.

In sum, both premia Granger-cause the overall term structure and net hedging pressure. While the term structure contains information useful in predicting spot premia in oil and soybean markets, it does demonstrate predictive power with respect to term premiums in all four of the commodity markets studied. This result demonstrates that the term structure risk should be better captured by the term premium than by the spot premium. No significant causality from net hedging pressure to either premium is observed.[16]

5. Returns from Active Trading Strategies

In this section, we use the findings of the previous section on the predictive power of term structure and hedging pressure variables to construct active trading strategies, as tests of market efficiency from a returns predictability perspective (Fama (1991)). Three active trading strategies are introduced: the momentum/contrarian strategy, the convenience yield strategy, and the net hedging pressure strategy.

---

[15] The reason to focus on the results of the second soybean spread is that the term structure effect should be most pronounced in the term slope derived from the July contract, since it is the only contract that expires just prior to the upcoming harvest season starting in September. Test results show that it is the case.

[16] As a robustness test, we have also performed the VAR Granger Causality tests using bi-weekly and monthly data. These results are qualitatively similar to those reported here. They are available on request.

5.1 Momentum/contrarian strategies

5.1.1 Portfolio construction

Momentum/contrarian portfolios are established using bi-weekly data from January 1990 to December 2006. To construct the portfolios, past spot premia and term premia for all four commodity futures are ranked at the end of each period. Based on these rankings, we go long (and short) futures or futures spreads in those commodities categorized as "winners" ("losers") with the highest (lowest) premium. After a designated period of time, the positions are rebalanced or unwound. We continue the process of enter-hold-exit during the testing period. To facilitate presentation, each strategy is named after its ranking (R) and holding (H) periods as the *R–H* strategy, with ranking and holding periods set as one month, three months, six months and twelve months. [17]

Miffre and Rallis (2007) build momentum portfolios using an overlapping trading strategy, following Moskowitz and Grinblatt (1999) and Jegadeesh and Titman (2001). Our approach extends this strategy, with some notable differences. First, rather than performing the rankings based on returns (i.e. changes in the log settlement prices), we perform our rankings using the spot and term premia, as defined in section 3.1. Hence, we extend both the Roon et al calendar spread strategies as well as the Miffe and Rallis (2007) long-only strategy. In addition, Miffre and Rallis (2007) form equally-weighted portfolios. Our study uses a value weighted approach of investing one dollar in both winner and loser portfolios in each period. This method avoids capital-allocation risk introduced by large differences in unit contract prices across commodity markets.

---

[17] It should be noted that trading results are comparable only between outright futures strategies or spread strategies but not across both groups, since negligible costs are incurred when initiating spreads. Spread returns are calculated as:

$$R_{cs} = \frac{P_{endf} - P_{openf}}{(|P_{openl}| + |P_{opens}|)/4}$$

where P_{endf} and P_{openf} are ending and opening balances of spread portfolio, $|P_{openl}|$ and $|P_{opens}|$ are opening balances of long and short sides when spread is initialized. Their absolute sum is then divided by 4 to take account of the offsetting effect of a long-short strategy. The revised calculation solves the "infinitesimal denominator" problem and makes different calendar spreads comparable across time and markets.

Finally, in this study, returns of winner or loser portfolios are based on the holding-period returns of taking long positions in the corresponding portfolios. The momentum (contrarian) trading return is defined simply as the difference between the winner (loser) portfolio return and loser (winner) portfolio return.

5.1.2 Empirical results

Table 5 provides summary statistics for momentum trading returns for outright positions that use the first nearby contract.[18] The momentum (contrarian) strategy is deemed a success when the "Momentum" column of the respective table is positive (negative) The purpose of using multiple contracts or spreads to build momentum trading portfolios is two-fold: to check robustness and to empirically test Miffre and Rallis's (2007) assertion that the term structure drives the momentum effect in commodity markets.[19]

On the whole, the momentum strategy performs well in most cases, particularly when the holding period is under six months. For the one month holding period and one-month ranking period, the annual return of the strategy is 9.24%, which is slightly lower than that found by Miffre and Rallis (2007) (10.87%). However, in contrast to Erb and Harvey (2006) and Miffre and Rallis (2007) the "one-month holding period and 12 month ranking" strategy does not persist as a dominant strategy for our sample, which extends to 2006, with the return and Sharpe ratio of this strategy of 3.12% and 30.69% respectively.

We also examine momentum returns based on a calendar spread that longs the first nearby futures contract and shorts the 12-month futures contract.[20]

On the whole, the spread trading strategies compare favorably with those of the outright strategies. The average annual return for the spread strategy with a one month holding period one month ranking portfolio is 7.99% with a Sharpe Ratio of 70.1. For a month holding period the annual return is 4.87% with a Sharpe Ratio of 39.97%. This is noteworthy since spread positions involve less risk.

---

[18] We also perform the analysis for contracts that expire in six months and twelve months. They are not included in order to conserve space and are available on request. The results for the longer maturity contracts are largely similar to those reported here, except as noted in the text.

[19] Miffre and Rallis (2007) suggest that momentum profits are related to the market structure, trading in long-term futures could generate more profits.

[20] Detailed results are available on request. We also perform the analysis for two other calendar spread positions: 1) long the first nearby futures contract and short the second nearby futures contract; 2) long the first nearby futures contract and short the six month futures contract. The results for these alternative positions are largely similar to those reported here, except as noted in the text, and are available on request.

Table 5. Summary statistics of momentum winner, loser, and momentum (MMT) trading returns, long-only the 1st-nearby futures, bi-weekly data.

	Holding period of 1 month			Holding period of 3 month			Holding period of 6 month			Holding period of 12 month		
	Winner	Loser	MMT	Winner	Loser	MMT	Winner	Loser	MMT	Winner	Loser	MMT
Panel A: ranking period of 1 month												
Mean	0.0474	-0.0499	0.0924	0.0337	-0.0250	0.0599	0.0162	-0.0083	0.0239	0.0096	-0.0031	0.0124
t-stat	11.8075	-12.7826	18.3919	8.5752	-7.4514	13.1905	4.0704	-2.3720	4.8397	2.4963	-0.9544	2.7643
Std Dev	0.0804	0.0782	0.1006	0.0787	0.0671	0.0909	0.0798	0.0702	0.0990	0.0768	0.0655	0.0897
Sharpe R	0.5896	-0.6383	0.9184	0.4282	-0.3721	0.6587	0.2033	-0.1185	0.2417	0.1247	-0.0477	0.1380
Panel B: ranking period of 3 months												
Mean	0.0412	-0.0374	0.0761	0.0196	-0.0125	0.0329	0.0102	-0.0021	0.0119	0.0060	0.0009	0.0051
t-stat	10.1120	-9.4505	14.2127	4.6339	-3.4596	6.3375	2.4857	-0.6007	2.3035	1.5871	0.2935	1.1103
Std Dev	0.0816	0.0793	0.1073	0.0845	0.0722	0.1038	0.0821	0.0693	0.1031	0.0761	0.0639	0.0919
Sharpe R	0.5050	-0.4719	0.7098	0.2314	-0.1728	0.3165	0.1241	-0.0300	0.1150	0.0793	0.0147	0.0554
Panel C: ranking period of 6 months												
Mean	0.0250	-0.0125	0.0399	0.0112	-0.0042	0.0146	0.0071	0.0037	0.0035	0.0028	0.0063	-0.0031
t-stat	5.4554	-3.5515	7.3043	2.5454	-1.2662	2.8253	1.6174	1.2090	0.7256	0.7326	2.0733	-0.7283
Std Dev	0.0916	0.0704	0.1095	0.0884	0.0658	0.1032	0.0876	0.0620	0.0976	0.0768	0.0613	0.0858
Sharpe R	0.2724	-0.1774	0.3648	0.1271	-0.0632	0.1411	0.0808	0.0604	0.0362	0.0366	0.1035	-0.0364
Panel D: ranking period of 12 months												
Mean	0.0212	-0.0125	0.0312	0.0104	-0.0042	0.0133	0.0050	0.0037	0.0012	0.0000	0.0076	-0.0073
t-stat	5.0708	-3.5515	6.1464	2.5626	-1.2419	2.7554	1.1977	1.1825	0.2622	0.0000	2.5210	-1.6559
Std Dev	0.0838	0.0704	0.1016	0.0813	0.0671	0.0967	0.0835	0.0634	0.0953	0.0761	0.0603	0.0880
Sharpe R	0.2532	-0.1774	0.3069	0.1280	-0.0620	0.1376	0.0598	0.0591	0.0131	0.0000	0.1259	-0.0827

Mean and standard deviation are annualized by multiplying by 26 and $\sqrt{26}$ respectively.

Extending the ranking and holding period reduces momentum profits. For the outright futures strategy, the significant momentum effect disappears when both periods exceed six months; significant contrarian profits are observed for holding periods of twelve months. Momentum trading with the shortest three-month futures dominates all strategies. For the calendar spread strategy, significant contrarian profits dominate all scenarios when ranking and holding periods are longer than three months. Profits decline as we move to shorter (i.e. six month followed by three month) spread positions. Miffre and Rallis's (2007) conjecture of a linkage between the momentum effect and the term structure are thus empirically supported through these trading results.

On the whole a short-term price continuation, analogous to that identified by Jegadeesh and Titman (1993, 2001) for equity markets is also observed for the spread momentum positions, as in the outright positions. For these markets, momentum profits are maximized by longing the most outperforming futures or calendar spread and shorting its most underperforming counterpart in the previous month, holding the pair for one month and continuously rolling it over to the next pair selected with the same criterion at each rebalancing point. This strategy works equally well for outright contracts and calendar spreads with different maturities.

5.2 Momentum strategy return determinants, backwardation and contango

Miffre and Rallis (2007) suggest that high momentum profits are not just based on compensation for risk but are related to backwardation and contango. The results of the previous section support a relationship between the momentum effect and the term structure. A direct test for a momentum—backwardation link, which incorporates other structural determinants of returns based on the theories of storage and hedging is provided by the following regression for both winner and loser portfolios:

$$R_{f,t} = \alpha_0 + \beta_1 cy_t + \beta_2 H_t + \beta_{3i} \sum_{i=1}^{11} D_i + \varepsilon_t \quad (7)$$

where $R_{f,t}$ is the return of longing (shorting) the winner (loser) portfolio, cy_t the portfolio convenience yield, H_t the portfolio net hedging pressure; D_i are monthly dummy variables, while ε_t is the error term. The regressors cy_t and H_t are constructed from the convenience yield and net hedging pressure of the most outperforming and underperforming commodities included in winner and loser portfolios at each period. Returns are regressed on convenience yield derived from corresponding maturity, i.e., 3-month convenience yield is regressed against returns of long-only the 1st-nearby futures and the shortest calendar spread and so on and so forth.[21]

From (7) a linkage between momentum and backwardation (Miffre and Rallis (2007)) would be supported if:

1. The regression estimate of β_1 is significantly positive (negative) for the long winner (short loser) portfolio for a holding period of 12 months. A positive correlation suggests that winner (loser) wins (loses) more when market becomes more backwardated (contangoed). This is the simplest case since no rebalancing is performed during the one year horizon, and as a consequence the roll yield can be ignored.
2. Positive or negative estimates of β_1 for portfolio returns under other ranking and holding periods because of the dual forces offsetting each other. However, if the winner (loser) portfolio does have more exposure to backwardated (contangoed) contracts or both, β_1 should, on average, be much larger in absolute value for the winner portfolio than that for the loser portfolio because inventory influences price more in a backwardated market than in a contangoed market.
3. Since β_1 is determined by the joint correlation between the convenience yield, the spot return and the roll return, the sign of β_1 demonstrates the dominant return explained. For a longed portfolio, in backwardation or contango, spot return always declines with a rising convenience yield while roll return always rises with it, and vice versa for a shorted portfolio. Therefore, a positive β_1 points to an increased term structure effect on roll return (spot return) in a longed (shorted) portfolio and so on and so forth.

Empirical estimates of (7) are shown in Table 6 for the momentum trading returns using the first nearby futures contract. Table 7 shows the results for calendar spread strategy that longs the first nearby futures contract and shorts the 12 month futures contract.[22]

---

[21] The effect of backwardation on futures return is complex since the futures excess return is roughly made up of the spot return and the roll return, which both correlate with the term structure, but in opposite directions (see Till (2007)). Increasing backwardation (contango) should diminish (enhance) the spot return and increase (decrease) the roll return. Thus, the overall effect of term structure on the portfolio return should be a balance of both forces, either in backwardation or in contango. As per Till (2007) the spot return is defined as distant futures price divided by nearby futures price, and roll return is the difference between futures return and spot return.

[22] The regressions are also performed for the other outright and spread positions outlined in this paper. The results are very similar to those reported here and are available on request.

Table 6. Regression results of momentum trading returns with the 1st-nearby futures on convenience yield, net hedging pressure and monthly dummies, bi-weekly data; p-values in parentheses

Holding period of	1 month		3 months		6 months		12 months	
	Long Winner	Short Loser	Long Winner	Short Loser	Long Winner	Short Loser	Long Winner	Short Loser
Panel A: ranking period of 1 month								
Variable								
Intercept	0.02	0.05	0.03	0.08	0.05	0.06	0.08	0.05
	-0.13	(<.0001)***	-0.14	(<.0001)***	-0.11	(0.0249)**	(0.0748)*	-0.13
CY3MONTH	0.41	0.47	0.52	0.39	0.95	-0.19	1.32	-0.46
	(0.0225)**	(<.0001)***	(0.0906)*	(0.0248)**	(0.0348)**	-0.47	(0.032)**	-0.19
HP	0.04	-0.10	0.14	-0.14	0.15	-0.13	0.16	-0.14
	(0.014)**	(<.0001)***	(<.0001)***	(<.0001)***	(<.0001)***	(0.0004)***	(0.011)**	(0.0049)***
Panel B: ranking period of 3 months								
Variable								
Intercept	0.00	0.05	-0.01	0.04	0.00	0.02	0.01	0.01
	-0.90	(<.0001)***	-0.65	(0.0308)**	-0.89	-0.40	-0.86	-0.81
CY3MONTH	0.47	0.42	1.47	0.34	2.20	0.33	1.49	-0.58
	(0.0153)**	0.00	(<.0001)***	(0.0553)*	(<.0001)***	-0.18	(0.0202)**	(0.073)*
HP	0.03	-0.07	0.09	-0.15	0.11	-0.17	-0.05	-0.10
	(0.0625)*	(<.0001)***	(0.0066)***	(<.0001)***	(0.015)**	(<.0001)***	-0.45	(0.047)**
Panel C: ranking period of 6 months								
Variable								
Intercept	-0.02	0.02	-0.04	0.03	0.00	0.02	-0.07	-0.08
	-0.12	(0.0378)**	-0.11	(0.0593)*	-0.91	-0.46	-0.11	(0.0175)**
CY3MONTH	0.42	0.04	0.95	0.16	2.08	0.12	2.77	-0.62
	(0.048)**	-0.75	(0.0071)***	-0.44	(<.0001)***	-0.68	(<.0001)***	-0.13
HP	0.04	-0.09	0.09	-0.17	0.11	-0.18	0.01	-0.18
	(0.0286)**	(<.0001)***	(0.0063)***	(<.0001)***	(0.0193)**	(<.0001)***	-0.91	(0.0003)***
Panel D: ranking period of 12 months								
Variable								
Intercept	0.00	0.02	0.01	0.04	0.00	-0.01	-0.07	-0.10
	-0.79	(0.0973)*	-0.80	(0.0368)**	-0.94	-0.66	(0.0905)*	(0.0007)***
CY3MONTH	0.00	0.01	0.59	-0.86	1.85	-0.32	2.65	-0.83
	-0.98	-0.98	(0.0654)*	(0.0185)**	(<.0001)***	-0.51	(<.0001)***	-0.18
HP	0.06	-0.07	0.06	-0.10	0.01	-0.20	-0.13	-0.40
	(0.0015)***	(<.0001)***	(0.0505)*	(0.0004)***	-0.83	(<.0001)***	(0.018)**	(<.0001)***

Numbers in parentheses are *p-value*. Coefficients with *p-value* highlighted by ***, ** and * are significant at 1%, 5% and 10% levels, respectively.

Table 7. Regression results of momentum trading returns of calendar spread that longs the 1st-nearby futures and shorts the 12-month futures on convenience yield, net hedging pressure and monthly dummies, bi-weekly data; p-values in parentheses

Holding period of	1 month		3 months		6 months		12 months	
	Long Winner	Short Loser	Long Winner	Short Loser	Long Winner	Short Loser	Long Winner	Short Loser
Panel A: ranking period of 1 month								
Variable								
Intercept	-0.01	0.05	0.03	0.05	0.05	0.04	0.10	-0.03
	-0.68	(0.0002)***	-0.31	(0.0085)***	(0.0619)*	-0.15	(0.0081)***	(0.0003)***
CY12MONTH	0.30	0.00	0.33	0.01	0.54	-0.16	0.88	-0.09
	(<.0001)***	-0.92	(<.0001)***	-0.83	(<.0001)***	(0.0611)*	(<.0001)***	(<.0001)***
HP	0.00	-0.03	0.07	-0.05	0.17	-0.04	0.07	-0.04
	-0.82	-0.12	(0.0862)*	(0.0619)*	(0.0001)***	-0.36	-0.17	(<.0001)***
Panel B: ranking period of 3 months								
Variable								
Intercept	-0.03	0.02	-0.03	0.03	0.00	-0.01	-0.04	0.03
	(0.0534)*	(0.0494)**	-0.25	(0.0072)***	-0.88	-0.58	-0.35	-0.21
CY12MONTH	0.27	0.03	0.72	-0.10	0.66	-0.24	0.89	-0.27
	(<.0001)***	-0.47	(<.0001)***	(0.0044)***	(<.0001)***	(0.0078)***	(<.0001)***	(0.005)***
HP	0.00	-0.04	0.02	-0.04	0.08	-0.09	-0.03	-0.15
	-0.92	(0.0182)**	-0.62	(0.0061)***	(0.0645)*	(0.0111)**	-0.53	(<.0001)***
Panel C: ranking period of 6 months								
Variable								
Intercept	-0.03	0.02	-0.04	0.02	-0.03	0.00	-0.14	-0.02
	(0.0791)*	(0.0126)**	(0.0928)*	(0.0631)*	-0.30	-0.93	(0.0004)***	-0.47
CY12MONTH	0.18	0.00	0.45	-0.04	0.67	-0.11	0.80	-0.27
	(0.0002)***	-0.92	(<.0001)***	-0.43	(<.0001)***	-0.17	(<.0001)***	(0.0285)**
HP	0.05	-0.03	0.10	-0.08	0.11	-0.09	-0.04	-0.11
	(0.0481)**	(0.0105)**	(0.0061)***	(<.0001)***	(0.0141)**	(0.0006)***	-0.46	(0.0037)***
Panel D: ranking period of 12 months								
Variable								
Intercept	-0.01	0.02	0.02	0.04	-0.04	-0.02	-0.14	-0.10
	-0.37	(0.0106)**	-0.37	(0.0009)***	-0.19	-0.30	(<.0001)***	(0.0015)***
CY12MONTH	0.15	-0.17	0.36	-0.38	0.59	-0.44	0.83	-0.36
	(0.0022)***	(<.0001)***	(<.0001)***	(<.0001)***	(<.0001)***	(<.0001)***	(<.0001)***	(0.0458)**
HP	0.04	-0.02	0.00	-0.05	-0.05	0.00	-0.23	-0.24
	(0.0883)*	(0.0434)**	-0.91	(0.014)**	-0.33	-0.96	(<.0001)***	(<.0001)***

Coefficients with *p-value* highlighted by ***, ** and * are significant at 1%, 5% and 10% levels, respectively.

It is clear that the convenience yield has significant explanatory power for momentum trading returns under most of the ranking and holding periods. Consistent with hypothesis one, β_1 is significantly positive and significantly negative for longed winner and shorted loser portfolios that are being held for and beyond six months. Hypothesis two is also supported in most cases, since β_1 is larger in size for winner portfolios than for loser portfolios. Consistent with hypothesis three, both the 6-month and 12-month convenience yields drive the futures roll return. However, the 3-month convenience yield has a greater influence on the roll return of winner portfolios and on the spot return of loser portfolios. Net hedging pressure is another significant factor of momentum returns in most cases. β_2 is positive for winner portfolios and negative for shorted loser portfolios, which means on average, a rising short hedging pressure boosts both winner and loser returns and a declining short hedging pressure reduces both. The only exception occurs in a scenario with a one-year ranking and one-year holding periods, which could be due to a large number of short hedgers choosing one year as the hedging span and their collective unwinding at the end of one year boosting portfolio returns.

In sum, both convenience yield and net hedging pressure are significant determinants of the returns from momentum strategies. However, the intercept terms of these regressions remain significant, particularly for short term losers for holding periods under six months.[23] Hence, the persistent high returns from momentum strategies cannot be wholly attributable to structural factors. Since convenience yield and net hedging pressure contain information useful for momentum strategies, can they be useful in generating profitable trading strategies on their own? To address this question, we explore two such strategies:

a) taking a long position in a portfolio with the highest convenience yield and short portfolio with the lowest convenience yield.[24] For spread trading, this strategy involves going long the short spread with the highest convenience yield and short the short spread with the lowest convenience yield.

b) long portfolio (short spread) with the highest hedging pressure and short portfolio (short spread) with the lowest hedging pressure.

[23] Note however that the intercept term is only mildly significant for the one month holding period/12 month ranking period strategy.

[24] This coheres with Miffre and Rallis's (2007) suggestion of consistently trading the most backwardated and contangoed contracts.

These two strategies are implemented to examine whether information contained in convenience yield and net hedging pressure are exploited into abnormal profits. Specifically, relative convenience yield and net hedging pressure are used to assign commodities into portfolios. Relative convenience yield (net hedging pressure) is defined as last period's convenience yield (net hedging pressure) divided by average convenience yield (net hedging pressure) in the ranking period. In each period, these two ratios are calculated and ranked to form two portfolios that have the highest and lowest ratios. A relative ratio is used instead of the average convenience yield (net hedging pressure) because the four commodities have very different historical levels of both factors, which invalidates a direct comparison between each other. Four strategies are simulated: two outright futures and two spreads with the shortest and longest maturities. On the whole these trading strategies are less profitable than those based on outright and spread positions presented above.[25]

The net hedging pressure strategy is less profitable than the convenience yield strategy. Long-short the long-term spread outperforms all other strategies based on net hedging pressure, and is the only one with exploitable profits. This result is consistent with Roon et al (2004), whereby hedging pressure effect can be inferred through spreads. The low hedge ratio portfolio drives the majority of trading profits. For both strategies, trading profits are inversely related to the holding period.

7. Conclusions

This paper explores some structural determinants of profits from momentum based strategies using commodity futures, as tests for market efficiency from a returns predictability perspective. High returns are observed using momentum strategies not only using outright futures positions but also spread trading strategies, even while the latter might be deemed less risky. On average, spread trading outperforms outright futures trading in capturing the term structure risk and hedging pressure risk.

Convenience yields and hedging pressures, variables consistent with the theories of storage and risk transference for commodity futures are examined as possible determinants of returns. High returns from a popular momentum trading strategy based on a ranking period of 12 months and a holding period of one month dissipate after accounting for hedging pressure effects, consistent with the rational markets model.

---

[25] Detailed results for the convenience yield and net hedging pressure trading strategies are available on request.

Appendix: Terminology

Term structure of commodity yield:
The term structure of commodity shows the relationship between the maturity of a futures contract and the futures price at a specific point in time.

Backwardation and contango:
These terms are used to describe the shape of term structure. Backwardation (contango) refers to a downward (upward) sloping term structure in which futures prices decline (rise) with maturity. According to Keynes (1930) and Hicks (1939), futures markets are used predominantly by short hedgers (producers in underlying commodities) so that futures price should be biased down to remunerate speculators for being long. Current futures price will rise to converge to the spot price as maturity approaches.

Carrying charge and inverse carrying charge:
Also called the "market-determined price of storage", these terms refer to the difference between two contract prices with different maturities. It is evident that a backwardated market anticipates an inverse carrying charge and a contangoed market a carrying charge from nearby to distant futures.

Long spread and short spread:
In commodity trading, a long (short) spread is to long (short) a nearby contract and short (long) a distant contract of the same commodity.

Outright futures strategy:
Outright futures position is a position in a futures contract that is not offset, in contrast to spread trading strategy.

References

Bessembinder, H., (1992), "Systematic Risk, Hedging Pressure, and Risk Premiums in Futures Markets", *The Review of Financial Studies* 5, No. 4, 637-667.

Brennan, M.J., (1958), "The supply of storage," *American Economic Review* 48, 50-72.

Carter C., Rausser G., and Schmitz A., (1983), "Efficient Asset Portfolios and the Theory of Normal Backwardation," *Journal of Political Economy* 91, 319-331.

Chang E., (1985), "Returns to Speculators and the Theory of Normal Backwardation", *Journal of Finance* 40, pp 193-208.

Cootner, P., (1960), "Returns to Speculators: Telser vs Keynes," *Journal of Political Economy* 68, August, 396-404.

De Roon, F., Nijman Theo E., and Veld Chris, (1998), "Pricing Term Structure Risk in Futures Markets,"*Journal of Financial and Quantitative Analysis* 33, No.1, 139-157.

De Roon, F., R. van den Goorbergh, and T. Nijman, (2000), "Hedging pressure Effects in Futures Markets," *Journal of Finance* 55, 1437-56.

De Roon, F., R. van den Goorbergh, and T. Nijman, (2004), "An Anatomy of Futures Returns: Risk Premiums and Trading Strategies", WO Research Memoranda 757, Netherlands Central Bank, Research Department.

Dincerler,C., Khokher, Z. and T. Simin, (2005), "An Empirical Analysis of Commodity Convenience Yields," Working Paper, Pennsylvania State University.

Erb, C. and C. Harvey, (2006), "The Strategic and Tactical Value of Commodity Futures," *Financial Analysts Journal* 62(2), 69-97.

Fama, E. F. and K. R, French, (1988), "Business Cycles and the Behavior of Metals Prices," *Journal of Finance* 43, 1075-1093.

Fama, E. F. and J. MacBeth (1973), "Risk, Return and Equilibrium: Empirical Tests," *Journal of Political Economy* 81, 607-637.

Fama, E.F. (1991), "Efficient Capital Markets: II," *Journal of Finance* 46, 1575-1617.

Feldman, B. and H. Till, (2006), "Backwardation and Commodity Futures Performance; Evidence Evolving Agricultural Markets," *Journal of Alternative Investments* 9(3), 24-39.

Hicks, J. R., (1939), *Value and Capital* , Cambridge: Oxford University Press.

Hirshleifer, D., (1988), "Residual Risk, Trading Costs, and Commodity Futures Risk Premia," *Review of Financial Studies* 1(2), 173-93.

Jegadeesh, N., and S. Titman, (1993), "Returns to Buying Winners and Selling Losers: Implications for Stock Market Efficiency,"*Journal of Finance* 48, 65-91.

Jegadeesh, N., and S. Titman,(2001), "Profitability of Momentum Strategies: An Evaluation of Alternative Explanations," *Journal of Finance* 56, 699-720.

Kaldor, N., (1939) "Speculation and Economic Stability," *Review of Economic Studies* 7(1), 1-27.

Keynes, J., (1930), "A Treatise on Money", Volume 2, Macmillan, London.

Lutkepohl, Helmut (1993), Introduction to Multiple Time Series Analysis, Second Edition (Springer-Verlag).

Miffre, J. and G. Rallis, (2007), "Momentum Strategies in Commodity Futures Markets," *Journal of Banking & Finance* 31, 1863-1886.

Moskowitz, T. J. and M. Grinblatt, (1999), "Do Industries Explain Momentum?" *Journal of Finance* 54, pp 1249-1290.

Samuelson, P. (1965), "Proof that properly anticipated prices fluctuate randomly," Industrial Management Review 6, 41-49.

Till, Hilary, (2007), "A Long Term Perspective on Commodity Futures Returns: Review of the Historical Literature," in H. Till and E. Joseph, Intelligent Commodity Investing: New Strategies and Practical Insights for Informed Decision Making London: Risk Books.

Working, H., (1948), "Theory of the Inverse Carrying Charge in Futures Markets," *Journal of Farm Economics* 30(1), 1-28.

Working, H., (1953), "Futures Trading and Hedging," *American Economic Review* 43, 341-43.

ON RELIGION AND LANGUAGE EVOLUTIONS SEEN THROUGH MATHEMATICAL AND AGENT BASED MODELS

M. AUSLOOS

GRAPES@SUPRATECS, ULG, B5a Sart-Tilman, B-4000 Liège, Euroland
E-mail: marcel.ausloos@ulg.ac.be

Religions and languages are social variables, like age, sex, wealth or political opinions, to be studied like any other organizational parameter. In fact, religiosity is one of the most important sociological aspects of populations. Languages are also obvious characteristics of the human species. Religions, languages appear though also disappear. All religions and languages evolve and survive when they adapt to the society developments. On the other hand, the number of adherents of a given religion, or the number of persons speaking a language is not fixed in time, - nor space. Several questions can be raised. E.g. from a macroscopic point of view : How many religions/languages exist at a given time? What is their distribution? What is their life time? How do they evolve? From a "microscopic" view point: can one invent agent based models to describe macroscopic aspects? Do simple evolution equations exist? How complicated must be a model? These aspects are considered in the present note. Basic evolution equations are outlined and critically, though briefly, discussed. Similarities and differences between religions and languages are summarized. Cases can be illustrated with historical facts and data. It is stressed that characteristic time scales are different. It is emphasized that "external fields" are historically very relevant in the case of religions, rending the study more "interesting" within a mechanistic approach based on parity and symmetry of clusters concepts. Yet the modern description of human societies through networks in reported simulations is still lacking some mandatory ingredients, i.e. the non scalar nature of the nodes, and the non binary aspects of nodes and links, though for the latter this is already often taken into account, including directions. From an analytical point of view one can consider a population independently of the others. It is intuitively accepted, but also found from the statistical analysis of the frequency distribution that an attachment process is the primary cause of the distribution evolution in the number of adepts: usually the initial religion/language is that of the mother. However later on, changes can occur either due to "heterogeneous agent interaction" processes or due to "external field" constraints, - or both. In so doing one has to consider competition-like processes, in a general environment with different rates of reproduction. More general equations are thus proposed for future work.

1. Introduction

Language and religion like sex, age, wealth, job, political affiliation, ... can be considered to characterize a human individual in its community[1,2] . Other characteristics may serve to identify a human being, like through its music interests[3,4] or its previous or present health conditions; see typical questions "who are you? " on visa application forms. These are all examples of some sort of "degrees of freedom" as so often found in statistical physics of critical phenomena[5] and dissipative structures[6] , self-organizing or not, in order to describe phase symmetries. Thus statistical physics ideas can find an interesting role outside physics[7,8] not only in econophysics but also in research pertaining to attitudes, behaviors, opinion formations, etc., as in socio-economic studies. In particular investigations on religion and language distributions, on one hand, and their evolution, on the other hand, can be studied as if such a degree of freedom is a variable defining some thermodynamic state, as a function of time and space[9] . These so called microscopic characteristics, can be studied through auto-correlations, or correlations with any other variable or "parameter" characterizing a population made of so called agents. The results can be next compared with available global data on macroscales.

Several pertinent questions can be raised, e.g. from a "macroscopic" point of view : (i) How many religions/languages exist at a given time? (ii) How are they spatially distributed ? ... From a "microscopic" view point: (iii) How many persons consider that they practice one religion/language? (iv) Does the number of adherents increase or not, (v) and how? and maybe why? (vi) Thus what are, if any, the basic equations ? (vii) Last but not least is there some possible modelization, through some agent based model?

However in this limited size contribution it is impossible to present a full state of the art concerning present studies by sociophysicists on languages and/or religions. Only a few topics can be tackled. The selection is admittedly biased and mainly due to a serious lack of competence of the author in front of huge research field activities. In Sec. 2 the differences and similarities between religions and languages are emphasized, with some stress put on the difference in time scales and the external field constraints. In Sec. 3, basic equations are written in to illustrate fundamental approaches to religion and/or language evolutions, without going into details. Many reviews exist elsewhere.

Sec. 4 serves to indicate that human networks are more complex than those studied by sociophysicists presently performing simulations of agent based models. In particular the notion of node intensity should appear to be relevant in view of the considerations in the previous sections. In Sec. 5 a more general set of equations is thereafter proposed and illustrated through an example involving the competitions between a few "degrees of freedom" of different types, like if there are describing struggles between ideologies, in a general sense. A conclusion and suggestions for future work follow in Sec. 6

2. Similarities and Differences

Through this section, it is hoped to outline a few "differences", also recognizing similarities, between languages and religions, from both an anthropological and a physics point of view. This is summarized through several items in Table 1 which are briefly commented upon[10] . This should serve as a perspective or input into modeling their sociological features in Sec. 3 and Sec. 5.

For example, the *origin* of languages and/or religions cannot be easily discussed as a simple statistical physics process. Basic anthropological studies, like those of refs.[11–16] , or refs.[17–20] are sometimes considered as hypotheses or theories. It is accepted that languages could not have originated as they are without a physiological modification of the human animal. Moreover, diversification is not only due to events like the Babel Tower construction, but is also rather due to different environmental, economic and sociological conditions. Similarly, religions could not originate from myths subsequent to the "scientifically unexplained" observation of natural phenomena. However the "final result", i.e. the appearance of a language or a religion[21] can be imagined by a physicist to correspond to a nucleation/growth process through homogeneous or heterogeneous fluctuations, *exordium* to competition between surface and volume free energy terms. It is relevant to point out to the attempt by Sznajd-Weron to demonstrate how to predict the initial concentration necessary for the fluctuation to grow to a viable size[22] in an opinion model of societies. Nevertheless it will take some time before one can connect such (microscopic) fluctuations (... of what ?) to the truly anthropological approaches. Always containing very profound and elaborate thinking.

First let it be recognized that the definition of a religion or an adherent (or adept) might not be accepted univocally, but the same can be said about languages. It is much debated between scholars on how to define a religion or a language. Like many, one can put on the same footing religions, philosophies, sects and rituals[20] . It is accepted, e.g., that a sect is a legally defined entity, - which in fact depends on the country where the *legal* definition is made. *Idem*, one could distinguish between adherents or adepts; there are also agnostics, atheists or "not concerned". Annoyingly there are various denominations for "closely related" religions which can impair data gathering and subsequent analysis. A similar set of considerations exists when discussing languages and dialects, slangs, etc.: e.g. there are three practical definitions of a language in[23] . The variety of religions and languages is surely enormous, to the point that they may even appear as fully individualistic ones, and only gathered into bins for defining communities.

It is then a habit of human beings to find roots and to connect to some important feature or on the contrary be singled out. Thus finding some genealogical-like trees for languages through their vocabulary or/and grammar, e.g. see[24–27] , or similarly for religions, e.g. see[28] can be done through various historical facts or measures. Such work on so called language trees is done like in many approaches about hierarchical systems. Yet it seems forbidden at this level and to-day to rank religions through any other indicators than materialistic ones, i.e. having led to some quantitative measure, - in contrast to written texts in some language, including translations, for which it is more common to discuss their "quality' or "values". One should be careful at another level: take the example pointed out by Nash[29] in the case of Christianity. He recalls that in searching for connections with other beliefs, scholars who first use a given terminology to describe e.g. pagan beliefs and practices, then marvel at the awesome parallels they think they have discovered. See also[30] . Nevertheless although it might be an audacious suggestion to consider the hierarchy of languages and religions when describing them through network schemes (see below) , it might be considered to use a measure, based on concepts, as the average overlap index (AOI) for some ranking[31–33] , as done for countries in econophysics studies. This has led to some original observation about communities.

It should be stressed that statistical physics needs data in order to produce some thought and answer a few questions. See the difficulty of obtaining and analyzing data in sociology studies. The data which is often used for describing the language distribution comes from[34,35] ; that for religion has been taken from[36,37] . However there are many other places where some data can be obtained, including detailed geographical ones[38,39] . Interestingly one can also connect to data obtained taking into account various economic aspects of religions[40] , maybe not so easily about the economics of languages.

Interestingly it is found, by survey takers, that there are about 6000 languages and 3000 religions. One may wonder why the order of magnitude is similar. The more so when it is absolutely clear that one can be a polyglot, but it is *very* difficult to be polyreligious, - even though the caveat pointed out here about holds much strongly here: religion being sometimes *very* individualistic. Therefore the log-normal plots of the *probability distribution function* (PDF) of the number of practitioners of a language[41,42] or a religion[43] are very strikingly similar, including on the value of the large tail, i.e. respectively -1.42[44] and - 1.67[45] . In both cases, the effect of the enhanced number of languages for very small sizes is nevertheless very perturbing, in particular since it announces the forever disappearance of rare cases.

A critical view of this data has to follow: in[43] a break was noticed in the PDF at 10^7 adherents, indicating an overestimation of adepts/adherents in or by) the most prominent religions, or a lack of distinctions between denominations, for these, - as can be easily understood either in terms of propaganda or politics, or because of the difficulty of surveying such cases precisely. Yet one paradoxical surprise stems in the apparent precision of the data. E.g., in several cases, the data in[36] seems to be precise up to the last digit i.e., in mid-2000 , there are 1057328093 and 38977 Roman Catholics and Mandeans respectively. In strong contrast there are 7000000 and 1650000 Wahhabites and black Muslims respectively, numbers which are quite well rounded. Thus a mere reading of the numbers warns about the difficulty of fully trusting the data. Nevertheless the analysis should be pursued bearing this *caveat* in mind. One can here point out an application of PDF studies on Jehowah witness activities, - as if they were economic ones[46] .

After having examined the global aspects, let us turn toward more microscopic considerations, i.e. the agents themselves, the role of opinion leaders and the external field effects on agents.

From this time point of view, and stressing the time scale, one can notice that a religion can seem to appear rather instantaneously, often as a so called sect, at the beginning, and its number of adherents can grow steadily (see the recent Mormon or Rastafarianism case) or not. New religions , necessarily called sects at their beginning, appeared after the second world war in Japan. Many disappeared. But in other cases, like the Black Muslims, Bahai, Mormons, Soka-Gakai, Universal Church of the Kingdom of God (founded in 1977, it has already more than 2 millions adepts), Jehowah witnesses, the Scientology church, etc. New languages can also appear, for very specific reasons: let Esperanto, FORTRAN and other computer "languages"", be mentioned here[47,48] . The notion of nucleation through homogeneous or heterogeneous fluctuations, in some imposed (thus external) field, could be thought of. A religion can also rather quickly disappear[49] , like the Antoinists in some coal mine regions of Western Europe, because its social role or its feeding ground disappears. Both cases though quite interesting are actually outside the realm of this paper. Yet the life time, or aging, of a religion can be studied, through the number of adherents, surely for modern times.

Notice the interesting fact when strong fluctuations arise: history is full of examples of individuals or entire groups of people changing their religion, - for various reasons: following the "leader", e.g. Constantinus, Clovis, or not changing at all under "external pressure" , leading to martyrdom, or like at inquisition time, or following a fatwah, but also cases of "internal pressure" (Khazars,[50] maybe) or so called adaptation under proselytism action, e.g. sun worshipping Incas in presence of catholic missionaries, Zoroastrians in Persia and Bogomils in Bosnia in presence of Muslim Arabs, etc. Such a competition through agent interactions or under "external field conditions" exist in many religion cases indeed. Thus, the number of adherents can evolve drastically due to such various conditions[51,52] , independently of the population growth size. Notice that it can be also a "legal field" which decides upon the use of a language. François I, King of France, decided in Villers-Coteret in 1529 that the true french was the one *spoken* in Isle of France, and nothing else. Back to religion history, a quarter century later, after the Peace of Augsburg treaty (1555), the ruling prince could decide in Germany whether his territory is protestant or catholic: *Cuius regio, eius religio.* In all cases, that does not mean that everybody became a Christian in the Roman Empire nor everybody spoke french in what was France, at that time, but it could lead to intolerance, wars of religions, and massacres. One can recall the case of "Matines Brugeoises" or "Brugse Metten" (1302) when Flemish "peasants" killed the french nobles, recognized as such, because they could not pronounce correctly "schild and vriend". Other cases are that of the red Khmers killing Vietnam educated intellectuals in Cambodia and that of Gileadites killing Ephraimites at a Jordan ford. Yet it is hard to find massacres due to "language heresy", though discrimination is known due to language conflicts, - see the problems of the French speaking community around Brussels, Belgium ! or the contempt of European Portuguese for the Brazilian Portuguese; Kurd and Armenian languages in Turkey; not too long ago, french in Canada, etc.

However religious (so called) heresy and differences has led to many more deaths: no need to mention Iran, China, Palestine, Balkan. This might be due to the fact that religions can serve as ground or excuses for political union: see Poland under the communist regime. Notice that attempts to suppress some "heresy" through harsh means does not always work: see the rise of Protestantism in Great Britain even through strong means[53] , though inquisition had some effect[51,52] .

See also a mapping of a sometimes called (religious or scientific) controversy on networks made of Intelligent Design Proponents (IDP) and Darwinian Evolution Defenders (DED) in order to analyze the local and global structural properties[54] of such strongly diverging opinion networks. See also the approach toward linguistics within the network concepts, not necessarily involving agents, in[55–57] .

Yet, it has been found in[43] that empirical laws can be deduced for the number of adherents, i.e. the *probability distribution function.* Two quite different statistical models were proposed, both reproducing well the data, with the same precision, one being a preferential attachment model[58] , like for heterogeneous interacting agents on evolving networks, e.g. it is more likely that one has the religion of one's mother or neighbor..... (leading to a log-normal distribution), another based on a "time of failure" argument (leading to a Weibull distribution function).

3. Basic Equations of Evolution

The most fundamentally relevant variables are thus accepted to be the number of practitioners of a language/religion[59] Only this number is treated here as the physics object/measure. Thus a language/religion is hereby considered as a (socially based) variable to be so studied like any other organizational parameter for defining a thermodynamic-like state. Since a religious state[60] is more individualistic than a linguistic state, one can better define the religious adherence of an agent than the linguistic one. Indeed one can hardly be multi-religious but one can be a polyglot. Of course one can switch easily, i.e. through "conversion", from one religious denomination to another; this is not so easy in language cases, - except through miracles. Thus the observation time of a religious state needs careful attention in surveys, but is less so drastic for language studies, as already hinted here above.

The dynamics of world's languages, especially their disappearance[61,62] has been recently considered through a Verhulst time t evolution equation[63] for the number density ρ_i of practitioners of some language i

Table 1. Comparing languages and religions through a few similarities and differences

	Languages			Religions		
	Refs.					Refs.
origin	11–13	physiology	"Babel Tower"	natural phenomena	myths	2,17–19
variety	14,23	huge:	dialects, slangs	huge:	denominations, sects	
semantics	16,23	grammar	vocabulary	images	rituals	20
agents	14,15	multilingual	frequent	polyreligious	rare	
impact factor		citations	libraries	"saints"	worship sites	
data set	34,35		WALS		WCT, WCE	36,37
number			more than 6000		more than 3000	
PDF	68		~ log-normal		~ log-normal	43
PDF tail	41,44		-. 1.42		- 1.67	43,45
genealogy	24,25,27		tree		tree	28
hierarchy			(AOI ?)		(AOI !)	
opinion leaders		authors	teachers	(high) priests	witches, shamans	
time scales	61,62	nucleation slow growth decay	slow François I	nucleation Contantine, Kazars decay	fast fast through avalanches	49
applied fields		Brugse Metten competition enforced civilization	discrimination media ... minorities	inquisition competition proselytism	Bogomils, Cathars economics often, strong decay, revival	40,52
network	55–57	small world			IDP-DED	54

$$\frac{\partial \rho_i}{\partial t} = r_i \, \rho_i \left(1 - \frac{\rho_i}{C_i}\right) \tag{1}$$

where C_i is the carrying capacity of the environment for the population speaking the i language and r_i is a positive or negative growth rate.

It was soon discussed that the evolution of one language must take into account some competition with other languages,[56,57,64–68] but also geographical constraints[69] and social structures[70] , sometimes extending the concepts toward hard evolutionist views[71,72] . Whence the Verhulst equation must be supplemented by a diffusion equation

$$\frac{\partial \rho_i}{\partial t} + \mathrm{div}\vec{j}_i = C_i \tag{2}$$

where $\vec{j}_i$ describes the net flux of the population speaking the language i into some area or in terms of a diffusion coefficient along some x axis

$$\frac{\partial \rho_i}{\partial t} = D_i \, \frac{\partial^2 \rho_i}{\partial x^2}. \tag{3}$$

There are several review papers like[66–68] discussing such cases.

A set of similar equations for religions can be written when attempting to quantify some religion dynamics[43,73–76] through the number of adepts. A population growth-death equation has been conjectured indeed to be a plausible modeling of such religion evolution dynamics, in a continuous time framework. The time evolution of several "main" religions was considered to be described, at a so called microscopic level, along the lines of the Avrami-Kolmogorov equation describing solid state formation in a continuous time framework[77] . The solution of which is usually written as

$$f(t) = 1 - e^{-Kt^n} \tag{4}$$

where $f(t)$ is the volume fraction being transformed from one phase to another; K and n are adjustable parameters.

A priori in analogy with crystal growth studies[77,78] , we have considered that a microscopic-like, continuous time differential equation can be written for the evolution of the number of adherents, in terms of the percentage with respect to the world population, of the world main religions, as for competing phase entities in Avrami sense

$$\frac{d}{dt}g(t) = \gamma t^{-h}[1 - g(t)]. \tag{5}$$

It can be solved easily giving the evolution equation for the fraction $g(t)$ of religion adherents in terms of a (positive for growth process) rate (or scaling) parameter to be determined, and a parameter h measuring the attachment-growth (or death) process to be deduced in each case, in a continuous time approximation. It should be emphasized that religions have appeared at some time t_0 which is somewhat unknown, or to say the least, loosely defined, due to a lack of historical facts but also due to the inherent process of the creation of a religion. In both, religion and language, cases, the initial condition (IC) which must be imposed in resolving the above equations is a major difficulty. Fortunately one might have some measure of ρ_i at some given time before integrating.

Thus to address some of these issues, classical scientific steps can be followed as in physics investigations[43,79,80] . *Accepting* as valid the number of adherents of religions from various surveys, "empirical" data can be subsequently analyzed. Examples of religions were found for which the number of adherents is increasing (e.g., Islam), decaying (e.g., Ethnoreligions) or rather stable (e.g., Christianity and Buddhism), - all giving interesting values for h. This has led to a more complete and more detailed analysis of the values of h and of its meaning for 58 "time series"[81] in[79,80] . Anomalous, or unexpected h values have been attributed to so called external field conditions, somewhat based on intuition.

However the main criticism of these equations is that they do not fully take into account social and historical conditions, - in some wide sense, the external/political fields and the environment, in a loose sense. For example in the case of the Antoinist which was founded by Louis Antoine at the end of the XIX-th century in some Liège suburb and had a number of adherents up to about 200 000, after 50 years of existence or so, and nowadays decays, it is found that the evolution equation is better represented through a rate equation which takes into account an external (sociological or not) condition through a parameter, c, i.e.

$$\frac{d\rho}{dt} = b\,\left(\frac{\rho}{t}\right) - c\,\rho \tag{6}$$

such that the evolution of ρ reads

$$\rho(t) = a \left(\frac{t}{t_1}\right)^b e^{-ct} \tag{7}$$

where a is an integration constant which can be adjusted. This equation states that the relative rate of change of the numbers of adherents $\frac{d\rho}{\rho}$ is proportional to $\frac{dt}{t}$ instead of to $\left(1 - \frac{\rho}{C}\right) dt$. and a scaling time t_1. For a proper display (Fig.1) a change of variables is made such that the previous equation reads

$$\rho(t) = A \left(\frac{t}{t_1}\right)^b e^{-\frac{t}{t_1}+\alpha} \tag{8}$$

Other formulations can be found that would present such a maximum after an appropriate time; they would need to be investigated and interpreted.

In concluding this section it is fair to point out that the spreading of languages and religions can be mapped into some approach like epidemics modeling[52] . At the same time, it can be thought that the spreading is similar to market sharing. Indeed the conclusions of Roach[52] when distinguishing between the behaviors of Cathar heresy in France and that of Bogomils in Bulgaria can be mapped either as an epidemic control or a market share battle depending on whether or not one looks at the overall result or at the cause of the result. Other spreading taking into account asymmetric grow and environmental conditions, can be found in anthropological research, see e.g. Paleoindian spreading in[82] . In most of these cases, the grow is analogous to crystal growth[83–88] conditions.

4. Networks

4.1. *Cluster Expansion*

In physics one should study the response of the system to intrinsic or extrinsic fields. In a mechanistical approach, the population of agents through a free energy F, Hamiltonian formalism or Langevin equation. In such a way, one would develop the quantity of interest as a series in terms of clusters, e.g., ordered along the increasing size of the cluster according to the number of spins in the cluster $< ... >$, as in

$$exp\left[-\frac{F-F_0}{kT}\right] = -\Sigma_i H_i S_i - \Sigma_{<ij>} J_{ij} S_i.S_j - \Sigma_{<ijk>} K_{ijk} S_i.S_j.S_k - \cdots \quad , \tag{9}$$

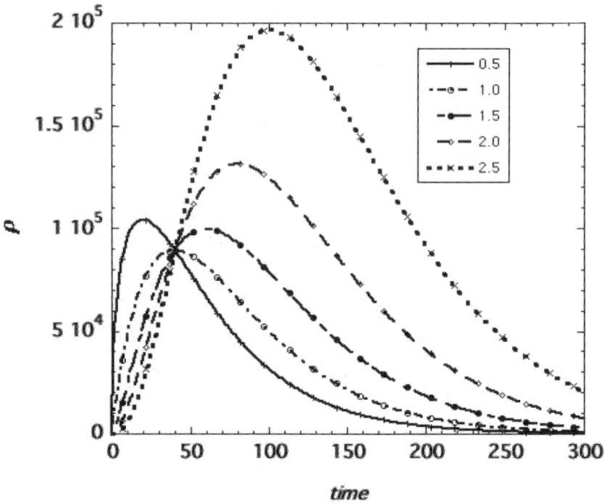

Fig. 1. Graph representing the evolution in the number of adepts under grow/decay conditions, as in Eq.(8), as for the Antoinist. The y- and x- axis values are arbitrary; for the display $t_1 = 40$ and $\alpha = 20$ values were chosen; the value of the b exponent is indicated in the inset

in obvious notations, i.e. each spin S_i representing an adept in an external field H_i and interacting with another adept through some interaction J_{ij}, etc., or similarly

$$\frac{\partial \Phi_i}{\partial t} = A_{ij}\Phi_j + B_{ijk}\Phi_j\Phi_k + C_{ijkl}\Phi_j\Phi_k\Phi_l + \cdots \quad , \tag{10}$$

for some information flux Φ_i. A vector generalization is immediately thought of.

4.2. *Node Intensity*

Although language is primarily *spoken*, most of the relevant research is about its *written* forms of communication, - which have significantly different sets of properties (known as registers) (e.g.,[89]) than spoken forms. It occurs that Web documents e.g. come from a wide range of styles, while blogs probably tend to be relatively informal, some being close to spoken language. In contrast, academic web sites contain large collections of very formal documents, such as research papers in e-journals, online copies of computer documentation, and university rules and regulations. Nevertheless, they also contain less formal genres such as personal home pages. Nevertheless one should clearly be precise on what is meant by "language investigations", since one could wonder whether someone pretending to be a language practitioner, indeed understands, reads, speaks, writes, ... the language. Moreover it seems relevant to question the quality level of the competence.

Similarly the time dependence of the *number* of adherents can be considered to be a very restrictive, and even wrong, way to "measure" the evolution of a religion. One should also "weigh" the level of adherence to a religion. For example, one could try as for languages to define a religion through its member intensity of participations in rituals, and in practicing the principles. Indeed, one could distinguish between *adepts* and *adherents*. An adept definition can be found in[90]

One can measure diverse quantities related to the religious effect as well. Moreover, one could consider religion dynamics from another ensemble of points of view or *indicators*.[91] As there are several definitions of a language,[23] similarly one could also define what a religion "is" in different ways.[20,92,93]

Thus it is clear that a religious/language "adherent" instead of being an analog of an up or down spin, is rather a vector for which each element can be a quantity measuring some *value* like one of those considered in sociology, i.e. a "quality". The measure of this quality being controversial depends on whether it is a self-assesment, on whatever scale, or a direct measure of some type. This make modelling very much complex.

No need to say that physicists are not the first ones to reflect on variability in religion distribution or adherence level. One may find already such considerations in books and papers by specialists of the history or sociology of religions, e.g.[20]

Therefore it is emphasized that one should describe society through agents having degrees of freedom which are both qualitative and quantitative in nature. Each node of the network becomes a quite complicated mathematical entity, surely not a scalar. Each node can be better considered as a vector node, each component being like a degree of freedom, represented by a tag, itself having an intensity, like on some energy level in a molecule. Interactions between the various degrees of freedom of different nodes can occur, through a matrix form; each link having a weight and a direction. That interaction matrix might not be symmetric. The more so if time delays are taken into account. Should one say, like molecules made of ions, or nuclei made of elementary particles interacting in complex ways? Should one at this stage recall Haven's[94,95] description of economy and economic agents along quantum mechanical-like concepts? Indeed quantitative and qualitative dynamical evolutions of agents and groups ("denominations") can find some theoretical source in many competition and organization physics models. Next, Potts vector or ferrroelectric type (Hamiltonian) models can be imagined for describing an ensemble of religious agent state or evolution.

Thereafter but *exordium* it seems natural to go back to the classical Verhulst ideas and its extensions for prey-predator problems in a Lotka-Volterra model by introducing some realistic conditions on the growth ratios and on the interaction coefficients between the agents in the populations.[96,97] - Indeed there is an obvious analogy with the problem of extinction of populations,[98] with those of religions[43,79,80] and languages,[62] and also to the very modern question of internet governance.[99]

5. Generalized Model

Leaving aside the networking description of the society, let us consider a set of Lotka-Volterra equations to describe the "struggles between ideologies", religions or languages, i.e. in a very general sense. It has been proposed to consider the following scheme.[100,101]

Let us consider a set ("area") with a population of N agents. Let the population be divided into $n+1$ factions or communities: each faction has a specific ideology, such that the number of members in the corresponding community is $N_1, N_2, \ldots, N_n$, but with a singled out fraction N_0 of people which are considered to be *not* concerned, not following an ideology, not speaking any language, at a given time. By definition at each time t,

$$N = N_0 + \sum_{i=1}^{n} N_i \tag{11}$$

The overall population $N(t)$ is assumed to evolve according to the generalized Verhulst law

$$\frac{dN}{dt} = r(t, N, N_\nu, p_\mu)N \times \left[1 - \frac{N}{C(t, N, N_\nu, p_\mu)}\right] \tag{12}$$

where N_ν represents the set $N_1, \ldots, N_n$ while $p_\mu = (p_1, \ldots, p_m)$ is a set of parameters describing the environment, and r is the overall population growth rate; which can be positive or negative.

The growth process is constantly disrupted by small extinction events, as in,[102] monitored through $r(t, N, N_\nu, p_\mu)$. As before $C(t, N, N_\nu, p_\mu)$ is the carrying capacity in the area.

5.1. *General Remarks*

In every community i one has to account for the following processes: deaths, dissatisfaction, unitary conversion, and binary conversion.

(1) First, the number of followers of an ideology can decrease through death or dissatisfaction with the ideology, i.e. through a term $r_i N_i$, where $r_i \leq 0$. In general $r_i = r_i(t, N, N_\nu, p_\mu)$.

(2) Conversion from one ideology to another can be made *without* direct contact between the followers of different ideologies. This type of conversion happens through the information environment of the population. In order to model the "unitary" conversion we assume that the number of people converted from ideology j to ideology i is proportional to the number N_j of the followers of the ideology j. A field f_{ij} coefficient characterizes the intensity with which this conversion from j to i occurs. Observe that f_{ij} is not necessarily a scalar. The corresponding modeling term is $f_{ij} N_j$; $f_{ii} = 0$. In general

$$f_{ij} = f_{ij}(t, N, N_\nu, p_\mu, C)$$

(3) A conversion to the i-th ideology can occur because of direct interaction between a member of this i ideology with a member of some other j-th ideology. It can be at first assumed that the intensity of the interpersonal contacts is proportional to the numbers N_j and N_i of the followers of the two ideologies. The coefficient characterizing the intensity of the binary conversion is b_{ij}. The larger is b_{ij}, the more people are converted to the i-th ideology. In general the binary conversion coefficients can be $b_{ij} = b_{ij}(t, N, N_\nu, p_\mu, C)$. Notice that if b_{ij} is finite, then $b_{ji} = 0$, since one converts from j to i. Of course $b_{ii} = b_{jj} = 0$: there is no self-conversion.

(4) Another conversion has also to be considered. It is possible that someone can convert to the ideology i simply by interacting with two members of ideology j and k. As above one can assume that the relevant term is proportional to the numbers N_j and N_k of the followers of the two interacting ideologies. The coefficient characterizing the intensity of this (still called) binary conversion is t_{ijk}. The larger is t_{ijk}, the more people are converted to the i-th ideology. In general the relevant binary conversion coefficients can be $t_{ijk} = t_{ijk}(t, N, N_\nu, p_\mu, C)$. Of course $t_{iii} = 0$: there is no self-conversion. Also t_{iik} does not exist, but t_{ikk} and t_{ijj} can be finite and are factors of N_k^2 and N_j^2 respectively. Other coefficients do not exist since one converts toward i.

In general one can have a co-evolution of the environment and the populations with some feedback, i.e. one might have $p_\mu = p_\mu(N, N_\nu, C, t)$, but this will not be discussed here.

In the simplest version of the model namely the case in which all the coefficients are time and p_μ independent, the model system becomes geared by

$$N = N_0 + \sum_{i=1}^{n} N_i \tag{13}$$

$$\frac{dN}{dt} = rN\left(1 - \frac{N}{C}\right) \tag{14}$$

$$\frac{dN_i}{dt} = r_i N_i + \sum_{\substack{j=0 \\ j \neq i}}^{n} f_{ij} N_j + \sum_{\substack{j=0 \\ j \neq i}}^{n} b_{ij} N_i N_j + \sum_{\substack{j=0 \\ j \neq i}}^{n} \sum_{\substack{k=0 \\ k \neq i}}^{n} t_{ijk} N_j N_k \tag{15}$$

A few remarks are in order.

Notice that arbitrary values are not allowed for the coefficients of the model. They must have values such that N, N_0, N_1, ..., N_n be nonnegative at each t.

Terms could apparently be observed to be similar when they refer to the same coupling. Indeed a $b_{ii}N_i^2$ term seems to be of the same nature as $t_{ijj}N_j^2$. One should not be misled, since the former does not exist, and the latter can never read $t_{jii}N_i^2$, since it does not pertain to the evolution equation of N_i. Nevertheless it is emphasized that the N_j^2 term, usually said of higher order, *must* be conserved in order to maintain correct symmetry conditions.

Moreover let us consider the i-th population and the binary conversion characterized by the coefficients t_{ijk} and t_{ikj} where j and k are different from i. One could at first think that $t_{ijk}N_jN_k$ and $t_{ikj}N_kN_j$ describe one and the same process in which the interaction between followers of the j-th and k-th ideology leads to a conversion of these to the i-th ideology. In general however one should not identify the two terms. In so doing in the general model we retain one additional degree of freedom, i.e., that which allows to distinguish between the ideology that is of the initiator of the interaction and the ideology of someone who is apparently simply a participant in the interaction.

Concerning the content of the f_{ij}, b_{ij}, and t_{ijk} parameters, one should be aware that they can arise from two ingredients. On the one hand the intended final state i maybe attained through imitation and the (positive or negative) influence of neighbours, - along usual majority/minority games, but on the other hand a *social value* can also be the cause of a change of ideology, i.e. a shift toward a *more rewarding social* ideology. The exact form of these coefficients should take such considerations into account in further studies.

Finally, these equations should be completed by two terms if one wants to describe real situations. The first one should be a constant (c) term, as introduced in Eq.(8) and a space dependent term of a diffusive type, $D_i \frac{\partial^2 N_i}{\partial x^2}$ as in Eq.(3). The former allows a decay due to say socio-historical circumstances, the latter allows to describe motion of populations in and out the area.

Below we shall consider the dynamics of populations of followers of the ideologies for the cases of presence of 2 ideologies in a given area, without diffusion. It will be shown that even without a constant forcing term some intricate situation nevertheless exists.

5.2. *Competition Between Two Ideologies*

In this section the competition for adepts that the presence of a second ideology introduces illustrates the above model. It is shown that an additional ideology leads to a measurable conflictual tension.

Let us consider the case of two ideologies with populations of followers N_0 and N_1. Let for simplicity all parameters be kept constant. One has

$$\frac{dN}{dt} = rN\left(1 - \frac{N}{C}\right) \tag{16}$$

$$\begin{aligned}\frac{dN_0}{dt} &= r_0 N_0 + f_{01} N_1 + b_{01} N_0 N_1 + \\ &\quad (t_{010} + t_{001}) N_0 N_1 + t_{011} N_1^2\end{aligned} \tag{17}$$

$$\begin{aligned}\frac{dN_1}{dt} &= r_1 N_1 + f_{10} N_0 + b_{10} N_0 N_1 + \\ &\quad (t_{110} + t_{101}) N_1 N_0 + t_{100} N_0^2\end{aligned} \tag{18}$$

$$N = N_0 + N_1 \tag{19}$$

The system has a fixed point when only an external field is applied for conversion

$$\breve{N}_1 = \frac{C\, f_{10}}{f_{10} - r_1} \tag{20}$$

When agent-agent interaction exists, one finds two fixed points, but only one satisfies $\breve{N}_1 \geq 0$. Its value depends on the initial condition, i.e. whether $N_1(0)/\breve{N}_1 \geq$ or ≤ 1. The number of converted agents is always $N_0 = N - \breve{N}_1$. As a numerical example let $C = 1$, $r_1 = r_0 = -0.01$ and $f_{10} = 0.02$. Then $\breve{N}_1/\hat{N}_1 = 0.6$, i.e. the number of followers of the ideology 0 decreases by 40%.

Fig. 2 shows a typical result obtained from the numerical investigation of Eqs.(16)-(19). A purely inertial growth and its decline are observed for the population of followers of ideology 1, simultaneously with the evolution of the population of followers of ideology 0. After a maximum of the latter, thus its decline, an inertial growth of the number of followers of ideology 1 takes place again. This can be usefully compared with the behavior discussed by[62] and that of Eq.(8), where in the latter case, no competition is assumed.

Even though the time is measured in Monte-Carlo time units, the coefficient is found to be of the reasonable order of magnitude. Indeed anthropologists[82] reports a measure of the birth rates of nomadic hunters and of agricultural settlers to be in the range 0.003-0.03/year: ethnographic evidence suggests than human population can expand at rates of up to 4% per year when colonizing new habitat (quotation from[82]). Thus in the present case, one has a reproduction rate of 0.005, - which seems to be a good order of magnitude. An equivalent order of magnitude, but with a negative sign can be taken for $r_1 = -0.005$; the value $f_{10} = 0.001$, chosen for the display, is somewhat more arbitrary at this time.

These results can be also compared to a model of telecommunication competition on a network made of nodes (customers),[103] some of them having no mobile phone at first and others belonging to one of several systems. The conclusion in[103] allows some interesting reflection by analogy with the cases of languages and religions. I quote from:[103] "schemes targeting local cliques within the network are more successful at gaining a larger share of the population than those that target users randomly at a global scale (e.g., television commercials, print ads, etc.). This suggests that success in the competition is dependent not only on the number of individuals in the population but also on how they are connected in the network.". It should

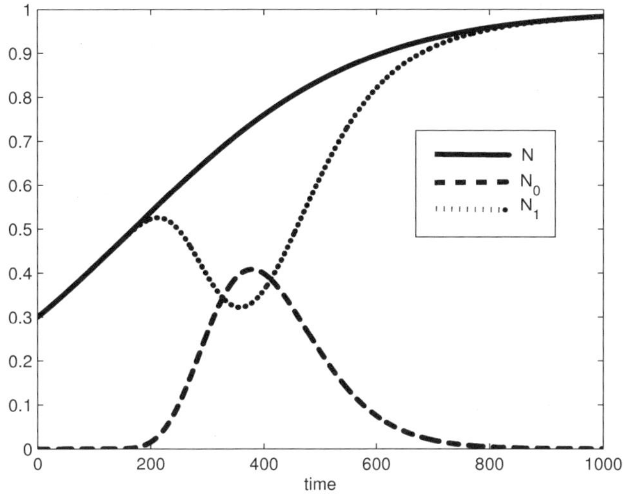

Fig. 2. Illustration of the case of two competing ideologies in populations N_0 and N_1, in an overall growing (N) population: with $r = 0.005$, $r_1 = -0.005$, and $f_{10} = 0.001$, - thus in the case of no b & no t influenced conversion

be added that the network in our above investigation is such that all agents are fully connected with each other, i.e. one considers a fully connected graph. This is equivalent to a mean field approximation study. Notice that the links are weighted through the f_{ij} and b_{ijk} coefficients.

5.3. *Tension Index*

Of course this causes some tension between the ideologies. A measure of this tension can be found through the index

$$T_{i;k} = 1 - \frac{N_i^{(k)}}{\hat{N}_i}, \tag{21}$$

where $N_i^{(k)}$ is the population of the followers of the i-th ideology when the k-th ideology is present in the area. If the ideology is alone then $N_1^{(1)} = \hat{N}_1$ and the tension index is $T_{1;1} = 0$. If N_1 decreases because of the competition with the second ideology, then the tension between the ideologies characterized by the tension index $T_{i;k}$ increases. The above definition for the tension holds even if N_1 follows some time dependent trajectory.

Notice that the tension index can be generalized for the case of an arbitrary number of ideologies in the area.[100,101] Let m ideologies be present.. The tension on the i-th ideology in presence of two other ideologies, k and

l can be defined as

$$T_{i;k,l}(t) = 1 - \frac{N_i^{(k,l)}(t)}{\hat{N}_i} \tag{22}$$

where $N_i^{(k,l)}$ is the population of followers of the i-th ideology when the ideologies k and l operate in the area. Similarly, the tension on the i-th ideology in presence of three other ideologies, j, k, l can be measured through

$$T_{i;j,k,l}(t) = 1 - \frac{N_i^{(j,k,l)}(t)}{\hat{N}_i} \tag{23}$$

where $N_i^{(j,k,l)}$ is the number of followers of the i-th ideology in presence of ideologies j, k, l. In such a way one can define a series of indices for the quantification of the tensions among ideologies competing for followers.

6. Discussion and Conclusions

In considering the present studies of physicists on language and/or religion dynamics, one is not only interested in their historical course and anthropological features, but also in attempting to unify the description of natural and social phenomena, for the good of humanity progress.

Here above one has touched upon several aspects, alas neglecting important ones. Mentioned aspects are the size and fraction distribution of denominations in the world. The fluxes are not much studied though. One has insisted on the number of adherents/practitioners, and the role of reproduction rates in the populations, defined though ideologies. The external field influences have been stressed. The conversion through media influence, direct proselytism and indirect, by disgust, conversion have been underlined. Similarities and differences between religion and languages are numerous. For a physicist, the various time scales are of interest. Same for the basic nucleation-growth problem. General differential equations can be written in the spirit of prey-predator systems for describing competitions. The difficulty to introduce quality aspects of language competence and religious practice in simulation through agent based models is an actual challenge. It implies a rethinking of the structure of complex networks.

Many neglected aspects seem to be due to shyness. As the direct comparison of selected word sets of different languages can be used to estimate their historical distance (i.e. the measurement of Levenshtein distances between two languages gives an idea of the time of their first common ancestor), one could define and evaluate religion distances, through various historical indicators. Notice that an interesting phase transition-like problem is the evolution from polytheism to monotheism. Finally, the economics of religions has already much attracted researchers,[40,46,49] but it could receive more attention from econophysicists.

In conclusion, mathematical studies and agent based models on religion and languages are fascinating challenges for physicists in the future. Combining research in different fields, like anthropology, economy, sociology and so called hard science will remove barriers between fields and thus will allow not only scientific progress but also provide bases for human better understanding of its life in communities.

Thanks to P. Clippe, F. Petroni, S. Pirotte, G. Rotundo, A. Scharnhorst, D. Stauffer, J.Steele, N. Vitanov for various critical comments and sometimes joint work on this subject. Moreover this paper would not have its form nor content without the persuading influence of E. Haven. Thanks also to the CHESS organizers for some financial support having allowed some presentation of these ideas in Saskatoon in Aug. 2009.

References

1. N. Chomsky, *Reflections on Language* (Pantheon Books, New York, 1975);N. Chomsky, *Aspects of the theory of syntax*, (Cambridge, MA: MIT Press, 1965); N. Chomsky, *Language and Mind.* San Diego, Harcourt Brace Jovanovich (1972).
2. E. Durkheim, *Les Formes élémentaires de la vie religieuse* ; transl: *The Elementary Forms of Religious Life* (1912).
3. R. Lambiotte and M. Ausloos, Uncovering collective listening habits and music genres in bipartite networks, *Phys. Rev. E* **72** (2005) 066107 (11 pages)
4. R. Lambiotte and M. Ausloos, On the genre-fication of Music: a percolation approach, *Eur. Phys. J. B* **50** (2006) 183-188.
5. H.E. Stanley, *Phase transitions and critical phenomena* (Oxford Univ. Press, Oxford, 1971)
6. P. Glansdorff and I. Prigogine, *Structure, Stabilité et Fluctuations*, (Masson, Paris, 1971).
7. D. Stauffer, Introduction to statistical physics outside physics, *Physica A* **336** (2004) 1-5.
8. A. Carbone, G. Kaniadakis, and A. M. Scarfone, Tails and Ties Topical Issue on Physics in Society, *Eur. Phys. J. B* **57** (2007) 121-125
9. J. Mimkes, A Thermodynamic Formulation of Social Science in *Econophysics and Sociophysics*, B. K. Chakrabarti, A. Chakraborti, and A. Chatterjee, Eds. (Wiley-VCH, Berlin, 2006). pp. 279-310.

10. The order of items in the Table is somewhat arbitrary but is intended to go from a general, anthropological-like point of view, to a more statistical physics one. Moreover the list of references is very biased, and does not recognize the enormous amount of works on the subject; surely many of which the author is unaware of, - what he deeply regrets.
11. M. Ruhlen, *A guide to the world's languages*, Stanford University Press, Stanford (1987).
12. S. Johansson, *Origins of Language - Constraints on Hypotheses*, see *http : //www.arthist.lu.se/kultsem/pro/SverkerJohanssonsem.pdf*.
13. J. Diamond, *The Third Chimpanzee: The Evolution and Future of the Human Animal* New York: Harper Perennial, (1992, 2006) 141-167.
14. L. Campbell, *Historical Linguistics: An Introduction*, MIT Press, Cambridge, 1999.
15. J. Nichols, Diversity and stability in languages, in: B. D. Joseph, R. D. Janda (Eds.), *The Handbook of Historical Linguistics*, Blackwell Publishing, Malden/Oxford/Melbourne/Berlin, 2003, pp. 283-310.
16. J. H. Greenberg, Some universals of grammar with particular reference to the order of meaningful elements, in: J. H. Greenberg (Ed.), *Universals of Language*, MIT Press, London, 1963, pp. 73-113.
17. M. Eliade, *A History of Religious Ideas, vol. I, From the Stone Age to the Eleusinian Mysteries*, Chicago, IL: University of Chicago Press (1978).
18. G. Swanson, *The Birth of The Gods: The Origin of Primitive Beliefs* (University of Michigan Press, Ann Arbor:,1960)
19. D. Dubuisson, *The Western Construction of Religion: Myths, Knowledge, and Ideology* (2003)
20. D. C. Dennett, *Breaking the Spell: Religion as a Natural Phenomenon*, Penguin Group, (2006).
21. M. Afzal Upal, Simulating the Emergence of New Religious Movements, *Journal of Artificial Societies and Social Simulation* **8** (2005) 1; *http : //jasss.soc.surrey.ac.uk/8/1/6.html*
22. K. Sznajd-Weron, Instabilities in population dynamics, *Eur. J. Phys. B* **16** (2000) 183-187.
23. J.M. Klinkenberg, *Des langues romanes* (Duculot, Louvain-la-Neuve, 1994)
24. M. Serva and F. Petroni, Indo-European languages tree by Levenshtein distance, *Europhys. Lett.* **81** (2008) 68005
25. F. Petroni and M. Serva, Language distance and tree reconstruction, *J. Stat. Mech.* P08012 (2008).
26. D. Zanette, Self-similarity in the taxonomic classication of human languages, *Advances in Complex Systems* **4** (2001) 281-286.
27. S. Wichmann, P. M. C. de Oliveira, V. Velupillai, A. Muller, D. Bakker, and A. P. Grant, A universal law of language taxonomics, preprint (2008)
28. There is to my knowledge at this time no "minimum spanning tree" presented for religions as there are for languages. However an interesting and artistic one can be found concerning a genealogical aspect of the cistercian monasteries; see a reproduction of the original drawing in Constance H. Berman, *The Cistercian Evolution.* The Invention of a Religious Order in

Twelfth-Century Europe (University of Pennsylvania Press, Philadelphia, 2000) figure 25, page 105, "Engraving of an eighteenth-century genealogical tree for monasteries for men of the Cistercian Order. (Dijon, Bibl. Mun. 90-098). A religious martyr tree can also be seen in.[36]

29. R. H. Nash, *The Gospel and the Greeks: Did the New Testament borrow from Pagan Thought?* (Dallas: Word Publishing, 1992), p.126.
30. Yeon-Mu Choi and Hyun-Joo Kim, *Physica A* **382** (2007) 665-671.
31. M. Gligor and M. Ausloos, Cluster structure of EU-15 countries derived from the correlation matrix analysis of macroeconomic index fluctuations,*Eur. Phys. J B* **57** (2007) 139-146
32. M. Gligor and M. Ausloos, Convergence and cluster structures in EU area according to fluctuations in macroeconomic indices, *Journal of Economic Integration* **23** (2008) 297-330
33. M. Gligor and M. Ausloos, Clusters in weighted macroeconomic networks : the EU case. Introducing the overlapping index of GDP/capita fluctuation correlations, *Eur. Phys. J. B* **63** (2008) 533-539
34. M. Haspelmath, M. Dryer, D. Gil, and B. Comrie (Eds.), *The World Atlas of Language Structures* (Oxford University Press, Oxford, 2005).
35. B.F. Grimes, *Ethnologue: languages of the world* (14th edn. 2000). Dallas, TX: Summer Institute of Linguistics; www.sil.org.
36. D. Barrett and T. Johnson, *World Christian Trends AD 30 - AD 2200. Interpreting the annual Christian megacensus.* William Carey Library (2001).
37. D. Barrett, G. Kurian, and T. Johnson, *World Christian Encyclopedia (2nd edition)*. New York: Oxford University Press (2001).
38. *http : //www.census.gov/prod/www/religion.htm; http : //www.census.gov/compendia/statab/*
39. *http : //www.census.gov/compendia/statab/cats/population.html.*
40. L. Iannaccone, Introduction to Economics of Religion, *J. Econ. Lit.* **36** (1998) 1465-1496.
41. S. Wichmann, On the power law distribution of language family sizes, *J. Linguist.* **41** (2005) 117-131.
42. P. M. C. de Oliveira, D. Stauffer, F. S.W. Lima, A. O. Sousa, C. Schulze, and S. Moss de Oliveira, *Physica A* **376** (2007) 609-616.
43. M. Ausloos and F. Petroni, Statistical dynamics of religions and adherents, *Europhys. Lett.* **77** (2007) 38002 (4pp)
44. P. M. C. de Oliveira, D. Stauffer, S. Wichmann, and S. M. de Oliveira, A computer simulation of language families, *J. Linguist.* **44** (2008) 659-675 .
45. D.H. Zanette and S.C. Manrubia, Vertical tranmission of culture and the distribution of family names, *Physica A* **295** (2001) 1-8.
46. S. Picoli Jr. and R.S. Mendes, Universal features in the growth dynamics of religious activities, *Phys. Rev. E* **77** (2008) 036105.
47. M. Boulton, *Zamenhof, Creator of Esperanto* (Routledge, Kegan & Paul, London, 1960).
48. M. Zemankova and C.M. Eastman, Comparative lexical analysis of FORTRAN code, code comments and English text, ACM Southeast Regional Conference Proceedings of the 18th annual Southeast regional conference,

Tallahassee, Florida, 193 - 197 (1980).
49. M. Ausloos, unpublished
50. C. Zuckerman, On the date of the Khazars'conversion to judaism and the chronology of the kings of the Rus Oleg and Igor, *Revue des études byzantines* **53** (1995) 237-270.
51. P. Ormerod and A.P. Roach, The medieval inquisition: scale-free networks and the suppression of heresy, *Physica A* **339** (2004) 645-652.
52. A.P. Roach, unpublished
53. P. Ormerod and A. P. Roach, Social Networks and Ideological Movements in History: Burning and the Rise of English Protestantism, ... preprint.
54. A. García Cantú and M. Ausloos, Organizational and dynamical aspects of a small network with two distinct communities: Neo creationists vs. Evolution Defenders, *Scientometrics* **80** (2009) 457-472.
55. R. Ferrer i Cancho and R.V. Solé, The small-world of human language, *Proc. R. Soc. B* **268** (2001) 2261-2265.
56. S.N. Dorogovtsev and J.F.F. Mendes, Language as an evolving Word Web, *Proc. R. Soc. London B* **268** (2001) 2603-2606;
57. Jinyun Ke, Tao Gong, and W. S-Y Wang, Language Change and Social Networks, Social networks play an important role in determining the dynamics and outcome of language change. ; J. Ke, T. Gong, W. S.-Y. Wang, talk at 5th EVOLANG conference, Leipzig, Germany, March 31, 2004; *Comm. Comput. Phys.* **3** (2008) 935-949.
58. A.L. Barabási and R. Albert, Emergence of scaling in random networks, *Science* **286** (1999) 509-512.
59. It is sometimes hard to know or to be sure whether an adherent, a disciple, ... is truly a member of a religious denomination or church. However this caveat pertains to usual problems encountered in sociological investigations.
60. ... admitting indifference, atheism, agnosticism, ... as a sort of religion, from our point of view.
61. M. Nowak and D. Krakauer, The evolution of language, *Proc. Natl. Acad. Sci. USA* **96** (1999) 8028-8033.
62. D.M. Abrams and S.H.Strogatz, Modelling the dynamics of language death, *Nature* **424** (2003) 900.
63. P.F. Verhulst. *Recherches mathématiques sur la loi d'accroissement de la population.* Nouveaux Mémoires de l'Académie Royale des Sciences et Belles-Lettres de Bruxelles, **18**, 1-45 (1845).
64. V.M. de Oliveira, M.A.F. Gomes, and I.R. Tsang, Theoretical model for the evolution of the linguistic diversity, *Physica A* **361** (2006) 361-370.
65. M. Thelwall and L. Price: , Language Evolution and the Spread of Ideas on the Web: A Procedure for Identifying Emergent Hybrid Word Family Members, Journal of the American Society for Information Science and Technology (JASIST), **57** (2006) 1326-1337
66. C. Schulze and D. Stauffer, Microscopic and macroscopic simulation of competition between languages, *Phys. Life Rev.* **2** (2005) 89-116
67. C. Schulze and D. Stauffer, Computer simulation of language competition by physicists, in *Econophysics and Sociophysics: Trends and Perspectives*,

B.K. Chakrabarti, A. Chakraborti and A. Chatterjee, Eds. (Wiley-VCH, Weinheim, 2006) pp. 311-318
68. C. Schulze, D. Stauffer, and S. Wichmann, Birth, survival and death of languages by Monte Carlo simulation, *Comm. Comput. Phys.,* **3** (2008) 271-294
69. C. Schulze and D. Stauffer, Competition of languages in presence of a barrier, *Physica A.* **379** (2007) 661-664.
70. J. W. Minett, and W. S. Y. Wang, Modelling endangered languages: The effects of bilingualism and social structure, *Lingua* **118** (2007) 19-45 ; W. S. Y. Wang and J. W. Minett, Vertical and horizontal transmission in language evolution, *Trans. Philological Soc.* **103** (2005) 121-146, *http : //www.isrl.uiuc.edu/amag/langev/* ; W. S. Y. Wang, and J. W. Minett, The invasion of language: emergence, change and death, *Trends in Ecology and Evolution* **20** (2005) 263-296.
71. F. Hashemi, An evolutionary model of the size distribution of sect/congregations, *J. Evol. Econ.* **10** (2000) 507-521.
72. C. Buckley and J. Steele, Evolutionary ecology of spoken language: co-evolutionary hypotheses are testable, *World Archaeology* **34** (2002) 26-46.
73. T.M. Johnson and D. Barrett, Quantifying alternate futures of religion and religions, *Futures* **36** (2004) 947-960.
74. J. Hayward, Mathematical Modeling of Church Growth, *J. Math. Sociol.* **23** (1999) 255-292.
75. J. Hayward, A General Model of Church Growth and Decline, *J. Math. Sociol.* **29** (2005) 177-207.
76. O. Shy, Dynamic models of religious conformity and conversion: Theory and calibrations, *Eur. Econ. Rev.* **51** (2007) 1127-1153.
77. R. Cloots, N. Vandewalle, and M. Ausloos, Simulations of the kinetic growth of $YBa_2Cu_3O_{7-d}$ grains, *J. Cryst. Growth* **166** (1996) 816-819.
78. A. Gadomski, Stretched exponential kinetics of the pressure induced hydration of model lipid membranes. A possible scenario *J. Phys. II France* **6** (1996) 1537-1546.
79. M. Ausloos and F. Petroni, On World Religion Adherence Distribution Evolution, *http : //arXiv.org/e − print/* : 0801.1010
80. M. Ausloos and F. Petroni, Statistical Dynamics of Religions, *Physica A* **388** (2009) 4438-4444.
81. More precisely there are 56 data sets for specific religions, in the World Christian Encyclopedia (WCE) and World Christian Trends (WCT) reference books,[36,37] most of them being in the *main denomination* bracket, i.e. in the upper part of the PDF as obtained from the surveys taken between 1900 and 2000. Some data refers to 3000 religions which are put together, as "Other Religions" in the WCT tables. Moreover some data refers to *Atheists* and *Nonreligious* persons,[36] Thereafter for conciseness, those three sets are called "religions".
82. J.Steele, J. Adams, and T. Sluckin, Modelling Paleoindian dispersals, *World Archaeology* **30** (1998) 286-305
83. *http : //physicsweb.org/articles/news/*11/1/1/1

84. J. Candia and E. V. Albano, *Int. J. Mod. Phys. C* **19** (2008) 1617-1634.
85. N. Vandewalle and M. Ausloos, Inner Patterns and Front Propagation of a Dynamic Random Impurity Model, *Phys. Rev. Lett.* **77** (1996) 510-513.
86. M. Ausloos, N. Vandewalle, and R. Cloots, Magnetic Kinetic Growth Models, *J. Magnet. Magnet. Mater.* **140-144** (1995) 2185-2186.
87. N. Vandewalle and M. Ausloos, Lack of universality in two-dimensional multicomponent spreading phenomena, *Phys. Rev. E* **52** (1995) 3447-3454.
88. M. Ausloos and N. Vandewalle, Growth models with internal competition, *Acta Phys. Pol. B* **27** (1996) 737-746.
89. D. Biber, Variation among University spoken and written registers: A new multi-dimensional analysis. In P. Leistyna & C. F. Meyer (Eds.), *Corpus Analysis: Language Structure and Language Use* Amsterdam: Rodopi (2003) (pp. 47-70).
90. *http : //en.wikipedia.org/wiki/Adept*
91. C. Herteliu, Statistical Indicators System regarding Religious Phenomena, *J. Study of Religions and Ideologies* **16** (2007) 111-127.
92. *http : //www.religioustolerance.org/rel_defn.htm*
93. *http : //www.religioustolerance.org/var_rel.htm*
94. A. Khrennikov and E. Haven, Quantum mechanics and violations of the sure-thing principle: the use of probability interference and other concepts, *J. Math. Psychol.*, **53** (2009) 378-388.
95. E. Haven, Private information and the 'information function': a survey of possible uses. *Theory and Decision* **64** (2008) 193-228.
96. Z. I. Dimitrova and N. K. Vitanov, Influence of adaptation on the nonlinear dynamics of a system of competing populations, *Phys. Lett. A* **272**, 368-380 (2000).
97. N. K. Vitanov, Z. I. Dimitrova, and H. Kantz. On the trap of extinction and its elimination, *Phys. Lett. A* **349**, 350-355 (2006).
98. A. Pekalski and M. Ausloos, Risk of population extinction from periodic and abrupt changes of environment, *Physica A* **387** (2008) 2526-2534.
99. L. N. Olivera, A. N. Proto, and M. Ausloos, Modeling the information society as a complex system, (to be published) (2009).
100. N. K. Vitanov, Z. I. Dimitrova, and M. Ausloos, Verhulst-Lotka-Volterra (VLV) model of ideological struggle, *Physica A,* in press.
101. N. K. Vitanov, Z. I. Dimitrova, and M. Ausloos, On the dynamics of populations which members can change their properties. arxiv...
102. C. Wilke and T. Martinetz, Simple model of evolution with variable system size, *Phys. Rev. E* **56** (1997) 7128-7131.
103. E. F. Legara, A. Longjas, and R. Batac, Competition in a social structure, *Int. J. Mod. Phys. C* **20**, (2009) 1-7.

TOWARDS THE ESTABLISHMENT OF CRITERIA FOR DISCUSSING MUSIC IN TERMS OF SYMMETRY AND ASYMMETRY (AN OUTLINE)

GYULA CSAPÓ
Department of Music, University of Saskatchewan
Saskatoon, SK, S7N 0X1, Canada
E-mail: gyula.csapo@usask.ca

Symmetry and *asymmetry* are universally present in music: our subject matter is vast, systematic applications abound. One cannot hope to do justice to this subject in an essay (only perhaps in a multi-volume book), let alone in a comprehensive manner that would include even the most important examples. The *theoretical problems* alone, concerning even the seemingly obvious cases, are considerable. Many of these arise from the *complex,* often *interactive, simultaneous* and *temporal, i.e. multidimensional* manner in which *musical parameters*[1] behave. Thus, a rigorously systematic approach is necessary. This essay will attempt no more than to create scaffolding to facilitate future access to this subject. The bibliographical source references following the article will guide the readers to the editions of the musical examples. The terms *symmetry* and *asymmetry* (S/A for a while from here on) will be used to describe ephemeral and transitory, definitive, referential and background phenomena alike; and everything in-between. Further, S and A will be treated as a matter of *degree*, not as absolutes. In many instances, S/A will be thought of metaphorically or symbolically. With this caveat, many of the theoretical problems will no longer appear insurmountable. As stated before, examples of (systematic, conscious uses of) S and A in Western Art Music are universally present, all the way from Gregorian Chant to the music of Ockeghem; all the way from J.S. Bach through the Classical and Romantic eras to the increasingly conscious and mathematically ambitious applications from the Twentieth and Twenty-first Centuries (e.g. in the music of Bartók, Stravinsky, Schönberg, Webern, Boulez, Stockhausen or Xenakis). Therefore, I attempt to treat the subject by *outlining the different models of how the question maybe thought of* in musical terms. Specific examples will be provided to each section after the article. Although I consider these examples instructive, they by no means represent more than an arbitrary assortment of *possible* cases – and this fact alone is not without a pleasurable harvest of peculiar and specific conclusions in the end.

---

[1] "Musical parameters" even in a most restricted sense must at least comprise *duration, pitch, loudness* and *timbre (tone colour)*. Please also see next footnote.

1. Examples of Some Theoretical Issues

1.1 Concerning Messiaen's Irreversible Rhythms

Music (put in simplified terms) deals with two aspects of sound: the *horizontal* and the *vertical.* Horizontal may involve **A**: *temporal*, **B**: *linear* (i.e. melodic) principles of construction. A famous case of **A** is Olivier Messiaen's so-called "irreversible" rhythm (**Ex. 1**).

Symmetry is conceived here in terms of *durations* yet executed in terms of *attack points*. Allegedly around a point in time, the preceding and subsequent rhythmic values are supposed to form symmetry. In reality, this happens around the middle of a *duration* where no musical attack/event occurs at all. When we hear these attacks in succession, we are expected, in retrospect, to somehow halve the central duration in order to fully come to terms with the symmetrical structure. In the absence of any *periodicity* prior to that point to induce us to do so, it is unlikely to happen. In musical performance, we process this rhythmic series in time[2], receiving temporal information attack by attack. What we should be sensing is a dynamical change at attack points in correlation with duration: the longer the duration until the next attack, the higher the dynamical level. As our processing occurs in time and is dependent on cumulative memory, the temporal trajectory of this process is barely going to exhibit the simple temporal symmetry intended. Yet the reader should be aware of how Messiaen thought of this structure. He wrote, explaining the title *The Charm of Impossibilities and the Relation of the Different Subject Matters* of his first chapter in *The Technique of My Musical Language*: "This charm, at once voluptuous and contemplative, resides particularly in certain mathematical impossibilities of the modal and rhythmic domains. Modes which cannot be transposed beyond a certain number of transpositions, because one always falls again into the same notes; *rhythms which cannot be used in retrograde, because in such a case one finds the same order of values again – these are two striking impossibilities.*"[3] The difference between "non-retrogradable" and "symmetrical" is not merely rhetorical: it is conceptual[4]. Messiaen demonstrates an awareness of the problem above, and it is truly "charming" how he makes this seeming contradiction his – and our – best friend in his music.

---

[2] For a thorough treatment of this subject, the reader is referred to two articles by Karlheinz Stockhausen: *Structure and Experiential Time*, In: Die Reihe Nr.2, pp. 64-74, 1958, and *...how time passes...(... wie die Zeit verging...)*, Die Reihe Nr.3, pp. 10-40, 1959.

[3] Olivier Messiaen: *The Technique of My Musical Language*. Paris: Alphonse Leduc, 1956, Chapter I, page 13. Italics are mine G.C.

[4] Although the differentiation is blurred when one reads this: "the symmetry of the rhythmic groups is a retrograde symmetry" (ibid., Chapter V, page 21).

1.2 Symmetry or Equilibrium? A Case for Generative Principles Awakening Latent Symmetries

Example 2: the Gregorian hymn *Ave Maris Stella.* Gregorian chant is one of the most intriguing and rewarding of all subject matters in music from our viewpoint. Here, the surface seems devoid of outright symmetry, yet the generative principles of this music ceaselessly act towards equilibrium. There are two phases in interaction: the *impulse* and *resolution* phases of any given melody[5]. An *impulse* is a melodic upsurge, an increase of energy invoking higher frequencies, leaps or moves *upwards.* Each *impulse* achieves this by an undulating movement with small checks, brakes and even reversals built into each upsurge. It is the eventual connection of all "rogue" notes into scalar congruity that defines the *resolution* phase. In *Ave Maris Stella*, there are four segments (A,B,C,D). Segment A is an *impulse*, section B is its *resolution.* Segment C and D are in the same relationship. As Segment C is a lesser *impulse* than Segment A, it follows that Segment D is a shorter *resolution* moving within a smaller ambit than Segment B. The larger the ambit, the higher the number of note-attacks within each segment. In *Ave Maris Stella*, notes 1 –2 leap a perfect fifth up. Note 3 adds another step. Notes e, f, and g are left out, in a gap. Note 4 brings the g in (a move towards temporary balance), but the upward motion continues. Note 8 fills in the upper c (like note 4 did the g). By the end of the phrase, notes 1, 2 and the peak-note 7 emerge as a solid frame, and notes 4 and 11 (the g) as lower neighbours to the a (i.e., notes 2, 5, 10 and 12). This g is later responsible for note 30, c, the lower neighbour of – and only resolved by – the final d. Notes e and f (still left dangling after the entire Segment A) resolve in Segment B (notes 16,18 and 19). The large leap of notes 1-2 is resolved by its literal (symmetrical) reversal by notes 14-15. This is reinforced in Segment C by notes 24-25. When notes 1-2-3 are considered, their *mirror image* is created by notes 24-25 and 30. The entire six-note segment comprised of notes 5-10 is repeated one perfect fifth lower by notes 15-20. The last five notes of Segment B are echoed in Segment D. Notes 2-3-4 are *mirrored* by 21-22-23, *retrograded* by 6-7-8 and *mirror-retrograded* by 32-33-34. Notes 3-4-5 are *mirrored* by 6-7-8 and 16-17-18, *retrograded* by 21-22-23. The motive 10-11-12 is found *mirrored* as 26-27-28 and 33-34-35, while 14-15-16-17 is *mirror-retrograded* as 34-35-36-37.

---

[5] Markand Thakar describes so much more thoroughly how these concepts work in Pre-Baroque counterpoint, in Markand Thakar: *Counterpoint: Fundamentals of Music Making.* New Haven, CT: Yale University Press, c1990.

I called this material most rewarding and intriguing at the outset. It tells us that asymmetry here is always "informed" in a generative way by symmetry, much as in communicating vessels in physics. In analysis we should always look for discrepancies in "symmetrical" materials and for correspondences in "asymmetrical" materials. It also informs us that one must *shift the theoretical ground from descriptive to generative theory* (as per the concepts of *melodic impulse and resolution*) to readily find symmetry – if not symmetry, than symmetry-in-the-making – in seemingly utterly asymmetrical material. Of course, the word symmetry here is used as symbolic.

1.3 Haydn's "Menuet al Rovescio"

A wondrous example of temporal symmetry is represented by this "retrograded menuet" from *Piano Sonata* Hob XVI: 26 where it constitutes the Second Movement. Please note that the Trio follows the exact same scheme. The Menuet part is given in **Ex. 3** A comment will follow shortly.

Ex. 4 demonstrates temporal symmetry in Webern's *Symphony Op. 21*, Second Movement. Note that the tone-row is folded in two and is outwardly symmetrical from the central tritone interval all the way to the tritone between the first and last notes.

Temporal symmetry in music always sets up a *process* by which we *bypass* symmetry. We are either approaching it or we are leaving it. And what exactly happens *at the centers* in Haydn's *Menuet al Rovescio*? Precisely: nothing.

1.4 Vertical (i.e., harmonic-registral-spatial) Symmetry

This occurs in music when there is a singular point serving as a center of the registral space (of a certain section or the whole of a work). A famous precursor of this is the second piece in Bartók's *Fourteen Bagatelles*. A thorough study of seminal works by Edgard Varèse is of particular interest here[6]. Webern's *Symphony Op. 21*. also offers us a rich harvest.[7] In **Ex. 5** below, Webern takes

---

[6] The reader may wish to consult, for example, Jonathan W. Bernard: *The Music of Edgard Varèse*. New Haven, CT: Yale University Press, c1987. Varèse also thinks in terms of balance between sound masses and so did Beethoven.

[7] Its title is using the traditional genre name *symphony* in a non-traditional allusion to "sym-", a common root with the word *sym*metry. Webern's seven years old daughter Christine may have just been playing with building blocks, relying on symmetry to ensure stability, to prompt Webern's dedication of this work to her... Webern in his turn explored primarily spatial (vertical) symmetry in the First Movement, and primarily temporal (horizontal) symmetry in the Second (and of course both, in both movements).

great care to present his vertical row as *two sets* of perfect fourths, each comprising six notes, one from below (d – g – c –f – b-flat – **e-flat**) and one from above (e – b – f-sharp – c-sharp – a-flat – **e-flat**): this is why two e-flats meet at an octave. As the e-flat is a tritone to the central a (the single missing tone from the two sets), this arrangement demonstrates the insight Webern possessed in conceiving these overriding relationships. These features also show just what a simple twelve-tone row can conceal (or reveal).

These (and other) substantial works of the Western musical canon are based on our pitch system of equal temperament.[8] Our perceptual system is non-linear. As we proceed on a chromatic scale upwards, intervallic distances presented by the notation system as "equal" semitones are in fact gradually increasing in terms of the *differences* between those semitones measured in Hz. The higher the semitones in register, the larger the Hz difference between the tones. So, the "symmetry" conceived on such a ladder is not a symmetry in reality; however, it maybe a symmetry in perception.[9] What one must remember with equal force is that this fact alone does not diminish the beauty and validity of these examples as musical statements. These phenomena *are* indeed (perceived as) symmetries in the virtual, musical sense of the word.

2.1 S/A as Temporal Equilibrium. Phrase Structure (e.g., Classical Periods)

A simple and frequent case of symmetry and equilibrium is the *period* in the temporal domain of music[10]. One of the basic formal units of Classical music, the *period* is a typical product of this tendency.[11] **Ex. 6** is a period. Its symmetry is recursive, as it is manifest in *circular time*. On paper, it appears simply as a unit of 4+4 measures of equal length, made of closely related materials and answering the D function of m. 4 (see ?) by the T of m. 8 (see !). The harmonic

---

[8] For a good treatment of how this is achieved the reader is referred to Llewelyn Southworth Lloyd: *Intervals, Scales, Temperament.* London: Macdonald & Jane's, 1978.

[9] The reader is referred here to a discussion of this phenomenon in Allen Strange: *Electronic Music: Systems, Techniques, And Controls.* Dubuque, Iowa: W.C. Brown Co., 1983., pp. 9-10.

[10] One must be aware here of how the music of each period and each distinct culture may construct time differently. For instance, here we must constantly remind ourselves of the circular nature of musical time in the Classical era, effected by mostly harmonic means. It is in conjunction with such cycles that periods are playing their role.

[11] By Classical music I specifically mean the particular historical period that occurred in Vienna roughly between the years 1750 through 1827, and whose primary composers were Joseph Haydn, Wolfgang Amadeus Mozart and Ludwig van Beethoven. The period preceding this is the Baroque era; what follows it is the Romantic period and afterwards we talk about Modern music (Twentieth century), then about the New Music of the 1950-es up to the 80-es, then of the Post-modern period or Contemporary Music, although these latter definitions are subject to debate and all of these are categories of convenience.

cycle, however, can wreak havoc on physical time in Classical music, as the next section will demonstrate.

2.2 S/A as a Harmonic Phenomenon in Classical Music

The essential underlining functional cycle T-SD-D-T is based on what I could term symmetry by analogy, for the T to the SD is what the D is to T. This simple relation creates a gap between SD and D. The T to SD move is a fall from an overtone to a fundamental. Likewise, D to T. All *prolongations* constitute *asymmetry*. Symmetry is restored when the harmonic circle is closed, extemporaneously. For instance, the "resolution" of the cadential 6/4 chord after the *cadenza* (as in a Concerto) makes the actual physical time the *cadenza* takes immaterial. This is well in line with what will be seen about "drones" in Section 3 ahead.

Symmetry in music emerges as time *recovered*, as an innate *perceptual etalon* against which deviations from it – asymmetry – is measured. Although it is often delegated from one parameter (time) to another (harmony), it tends to be asserted.

2.3 About Asymmetrical Attack Numbers within Symmetrical Temporal Units

A curious phenomenon often "meets the ear": that of irregular attack numbers within phrases of equal lengths. Apparently, composers like to inject *asymmetry* to counter prevalent symmetrical tendencies. Let us consider **Ex. 6** again (the number of "melodic attack points" for each unit under the brackets is given). There are intricacies within this seemingly simple period. A comparison of m.1 and m. 2 reveals important asymmetries concerning *repeated* versus *non-repeated* note-pairs. The upbeat/downbeat is a *note repetition* in m. 1 but ceases to be one in m. 2, where there is now a semitone move up to the downbeat. This change calls attention to the fact that the note repetition itself is used *to link* the two motives, just on account of the repeated melodic b in m 1. This is not the case in the response, for the next upbeat to bar 3 repeats the note of the *first downbeat* of m. 2 instead of repeating the note immediately preceding it. This reverses the initial "upbeat–downbeat-by-note-repetition" rhythm, starting on the second quarter of m. 1 and structurally linking it by note repetition to the upbeat. This device ties together the two initial rhythmic motives, counteracting rigidity.

M. 1 presents a *chaconne*-like rhythm that places a fuller, opened-up position of the G-major chord on the second quarter. This is the typical emphasis on the second quarter beat of a *chaconne*. The asterisks above notes in the example indicate shifting emphases within the phrase. This first asterisk is in parenthesis expressing its lesser significance in comparison with the next one (bar 2, high a), which, by virtue of its *melodic accent*, increases this tendency, reshaping the very phrase in which it participates. This is why the sudden *piano* appears in m. 3, helping the phrase to congeal. The bracket underneath unifies the descending line of five attack points. A kind of syncopation occurs, where the melodic leap (up a major sixth) counteracts rhythmic regularity. In m. 3, the double asterisk signals an *added* attack at a place of the measure where none occurred before. This is another device to emphasize the arrival of the first SD function. Indeed, the next downbeat bears a *sforzato* sign: the arrival of the mid-point of the period, sung out by the violin ornament. Again, the *piano* in m. 3 is crucial in facilitating this.

Here one must backtrack, because by now a *dynamical reversal* had also happened. While the first two bars represent *dynamical increase* from their first quarters to the second, in m. 4 there is now a *decrease* for the first time.

The rest of the example should be self-explanatory. Note, that the attack-point numbers strongly emphasize asymmetry within the essentially symmetrical structure. Please observe how the repeated upward major sixth melodic leaps (m. 2 and 5) are countered by the scale descending a major sixth from e all the way down to g (mm. 6 – 7). The c-sharp in m. 3 is essentially *recalled* by the c-sharp on the third quarter of bar 7. The c-sharp of m. 3 is highlighted by being placed on the third beat (of any bar) for the first time in the work. The sharp *appoggiatura* in m. 7 unifies the c-natural (of the dominant-seventh chord) and *that* c-sharp. The entire chromatic scale flows out of this pungent "syzygy". The full functional cycle of mm. 3-4 (by virtue of the faint signal of the T on the last sixteenth of m. 4) is projected and prolonged, quite clandestinely, in mm. 6-7-8.

The example shows how forces of asymmetry are fuelling this music, subtly asserting their dynamic presence over no matter how symmetrical a frame. This important dialectics advertises symmetry on the surface and uses asymmetry behind the scenes. Please note that *Bar-form* (Stollen-Stollen-Abgesang) or *anapaestic form-rhythm* is also present in this (and many other) Haydn period(s), precisely for its dynamic quality and counter-symmetrical tendency.

In an analogous case, please witness how Mozart reaches back for the g of m. 1 in m. 2, stealthily putting a dose of asymmetry into the soup in **Ex. 7:** Mozart: Piano Concerto K.V. 467 in C-major, First movement main theme.

3. S/A in cyclic time: a scientific metaphor from Magnetic Resonance Imaging[12]

1.1. Pure time –
1.2 Pure time, listening presence: think again of Peter Brook: *The Empty Space*: his definition of theatre. Please note: John Cage's *4'33"* deals with 1.2.

Pure time in musical terms is total randomness, unknown outcome[13]. When coupled with a listening presence in the sense Peter Brook defines theatre in *The Empty Space* (when one crosses an empty space and someone is watching it, theatre is born)[14], this listening presence introduces *memory*. Chaos cannot be remembered. As the work of John Cage highlighted to us, we *read patterns into randomness and chaos as we listen; we shape it* – as he shaped it by his choices as a composer: choices of procedure, choices of material that he subjected to "chance operations". These *musical* choices made his music entirely Cage, made it *sound* Cage. It is these *choices* that shaped Cage's music, it is the musical insight that shaped all of these choices, resulting in the lightness, playfulness of what we listen to when we hear this magnificent music. The proportions of sounds to silences, the space, the kinds of sounds (even when they are not "defined" in any orthodox way) – these are what we hear and *remember*. Now, remember, the key word is *remember*! Remembering implies something cyclic. A computer – with very large but *never infinite* memory – can only approximate randomness. The presence of memory implies the element of repetition. Of *certain* sounds, the sounds of an instrument. (think of *Etudes Australes* or of the individual personality of each note in *Sonatas and Interludes for Prepared Piano*). When there is repetition, there is "recycling". We thereby take hold of time. Once we take hold of time, we end up with multiple times. Why? Because, when "time" appears (as remembered, cyclic), then it becomes self-contained in terms of *itself* as per its characteristics, recycled from the moment we first encountered and recognized it as such, i.e., *as time* (musical time, listening time, inwardly speaking). So, it is obvious that each recycled "shape" (say, timbre, wave-form) is a *drone*: self-contained, exuding its own "time(less)ness", its own eternity. Sounds – multiple waveform morphologies –are as many living, pulsating *drones*, and we remember each for its own specific *flavour of eternity*, among the multitudes of these eternities. Much in the manner Stockhausen describes "how time passes", we need *change* as well as *recycling* for memory

[12] Periodicity/aperiodicity will be used here as in an extended sense as a conceptual framework as well. I invite the reader to share with me a bold perspective and a novel *context*.

[13] In line with the concept of time expounded in John Cage: *Silence: Lectures and Writings*. Middletown, CT: Wesleyan University Press, 1961.

[14] Peter Brook: *The Empty Space*, p. 3. London: MacGibbon & Kee, 1969, c1968.

to be at work. By way of a *scientific metaphor*, let us look into the case of the *spin echo* experiment:

"In nuclear magnetic resonance, *spin echo* refers to the refocusing of precessing nuclear spin magnetisation by a *180° pulse* of resonant radiofrequency. The NMR signal observed following an initial excitation pulse [...] decays with time due to both spin-spin relaxation and any *inhomogeneous* effects which cause different spins to process at different rates e.g. a distribution of chemical shifts or magnetic field gradients. Relaxation leads to an irreversible loss of magnetisation (decoherence), but the inhomogeneous dephasing can be reversed by applying a 180° or *inversion* pulse [...] that inverts the magnetisation vectors. ***If the inversion pulse is applied after a period T of dephasing, the inhomogeneous evolution will rephase to form an echo at time 2T.*** The intensity of the echo relative to the initial signal is given by exp(– 2*T* / *T*2) where *T*2 is the time constant for spin-spin relaxation. Another method for generating spin echoes is to apply three successive 90° pulses. After the first 90° pulse, the magnetization vector is exchanging energy through dipole-dipole interactions and in a time τ, forms what is often referred to as a "pancake" in the x'-y' plane. A further 90° pulse is then applied such that our "pancake" is now in the x'-z' plane. When considering the two types of relaxation, spin – lattice and spin – spin (T1 and T2) we assume the first to take an infinite amount of time as such allowing the spin vectors to precess about the z axis. Now, the angle each spin makes with the z' axis is equal to the angle it previously made about the y' axis. At this point any change in angle that now takes place will require a change in energy thus implying a spin – lattice interaction is necessary. ***This implies a permanent memory of the state of the system as it was at time τ.***[15] After a further time τ2 a third pulse is applied and our Magnetization vector is back in the x' – y' plane and will lie in the same direction as for a (90 – τ – 180) spin echo sequence formally discussed. Then after final delay of τ we see what is commonly referred to as a stimulated echo [...]" [16] The above *metaphor* prompts me to consider the following:

1.3 Periodicity/aperiodicity. In a system with high entropy, energy dissipates, only to find *creative response-related* input. Musical sound (pitch) is periodic waveform/vibration, noise is aperiodic, yet noise *remembered* becomes periodic.
1.4 I call *drone* or *hyper-periodic* the vibration with observed qualifications of timbre attached. I thus formulate:

---

[15] ***Bold italics*** within the quote are mine. G.C.
[16] Cited from Wikipedia, January 17, 2010: http://en.wikipedia.org/wiki/Spin_echo.

as we scan simultaneous but distinct musical events *in time* (yes, in extremely fast succession but *never* simultaneously), it follows that no *harmonic spectrum* identifiably exists "at once". Harmonic spectra, or components of such, are processed in time as we comb through their different regions in perception.

HYPOTHESES:

A) is it, then, true that *sound-waves of identical morphology* could be picked up "sooner" than those of divergent morphology? "Sooner" here means that we actually "revert" to a previously experienced "time" and pick such signals up as a "spin echo" of a particular energy flow, distinct from spectra to spectra.

B) the experience of continuum, then, exists only within sounds of identical morphology or within *patterns* (if multiple morphologies are involved). Within cyclic phenomena "time" (its drone quality) appears to us *as* it (physical time *as change*) disappears.

C) it must also be true that memory is conditioned by the considerations above.

It is from this line of thinking that we understand why Cage negates the notion of "harmony" as he focuses his inquiry on *time* in music.[17]

1.5 One can thus formulate a theory of timbre/time/perception as mutually interdependent.

1.6 This in turn leads to the formulation of what I call *synthetism* (tracking of "origins" by timbral segregation/reiteration) by the memory-recognition of cyclicities of timbral signatures etched on all pitches, regarding their life-spans eternal).

All of the above may then be regarded as a question of S/A in terms of the periodicity/aperiodicity principle. Symmetry is a balance – a loss of physical time and a gain of *time as an image*[18]. Asymmetry characterizes processes to regain what is perceived "lost". Any impulse synchronized with a cycle,

---

[17] Cf. also my *Krapp's Last Tape:* a) sine-wave (total continuum) b) live violin, muted (subdued time) c) distorted violin sound via contact microphone (dissected time). Gyula Csapó: *Krapp's Last Tape (after Samuel Beckett)*. Budapest: Editio Musica Budapest, 1982. Also on Budapest Music Center CD 13.

[18] This term has been coined by Morton Feldman, see his: *Essays*. Walter Zimmermann, ed. Kerpen: Beginner Press, c1985.

(however long ago elements of that cycle have been present) reinforces, actualizes that cycle, triggers the presence of that drone *as an (re)acknowledgement of a continuum.* Non-cyclic impulses counteract this, establishing independent cycles, or not.

4. Projecting Pitch Collections (Satie: *Gymnopédie No. 1*, circle-of-fifths representations)

Ex. 8: Erik Satie's *Gymnopédie No. 1*: the arrival of "new" notes in circle-of-fifths representation. The entire piece is folded into two: a symmetry of two times 39 measures, a total of 78 measures. Asymmetry within this is caused by the modal change in bar 72 (where the corresponding f-sharp of m. 33 now becomes an f natural). Thus, from bars 72 through 76 and in 78, we have f naturals *and* c naturals as opposed to the f-sharps and c-sharps in the corresponding section in mm. 33 – 39 (with the exception of bars 38 and 77, which remain identical). We gain more insight into this structure by representing it on a ladder of the circle of fifths. The *pc contents* of the accompaniment pattern in bars 1 – 16 (*in circle-of-fifths order*, bottom up, g–d–a–*gap*–b–f-sharp–c-sharp) and that of the melody in the right hand (m. 5 –18) is identical.[19] The oscillating bass line changes in m. 17 *for the first time*. It continues in m. 18 and leads to a meeting of bass line *and* melody on pc e in m. 19. Curiously, this note is *repeated* in m. 20 in the bass (the first such event in the piece), and the same pc is *extended* (until the end of m. 21) in the top line. This e is a remarkable *long-withheld complementary note*, filling the *gap* in the circle-of-fifth. The arrival of this note stretches this period to nine bars instead of the previous eight, denying the expected symmetry. Bar 21 gives us a new note with urgency (the f-natural), and the next bar again fills in the *gap* by giving us its circle-of-fifths neighbour, the c-natural. At this point we have obtained a contiguous chain of nine fifths (f – c – g – d – a – e – b – f-sharp – c-sharp): the total pc contents of *Gymnopédie No. 1*.[20]

In summary: the sharing of pc contents between "accompaniment" and "melody" in mm. 1-18 of *No. 1* is a symmetry (pc contents projection). This shared "perspectivelessness" sets up the want to fill the gap – the lack of pc e – by m. 19, which in turn triggers the advent of an expansion (mm. 21 and 22) of

---

[19] Causing the impression of the famous "perspectiveless harmony" of Erik Satie. See also Alan M. Gillmor: *Erik Satie*. New York, London: W.W. Norton and Co., 1992; c1988 by G. K. Hall & Co.

[20] *Gymnopédie No. 2* will add b-flat, e-flat and a-flat, thus closing the entire circle (the addition of the d-flat would merely duplicate the c-sharp); *No. 3* uses the diatonic notes of the C-major scale plus the b-flat, no more.

the pc contents further. The conclusion emerges that the ending of the first half *reinstates* the f-sharp, *retracts* the f-natural but *keeps* the c natural (see the modal cadence, m. 39), whereas the ending retains both (f natural, c-natural) in bars 77 and 78.

5.1 Registral Distribution (Virtual Musical Space: Examples from Mozart, Bartók, Webern)

Let us now look at Mozart's *Piano Concerto in C-major, K.V. 467*, First Movement, mm. 68-70, oboe, in comparison with m. 74, the first entry of the solo piano, **Ex. 9.** One cannot fail to notice that the latter piano figure is a *mirror image* of the oboe line, projected onto the diatonic G-major scale.[21] Looking at a larger picture, this spatial symmetry is but a reflection of the spatial symmetry of the entire section. Its origin is the main theme of the movement, which already exhibits spatial symmetry in and by itself. The g – c – f boundaries of the initial theme (bars 1 –3) are projected onto the large canvas of the entire section m. 1– 28 (at which point the entry of the second theme occurs). The registral boundaries of this entire section are defined by the low g in the double-bass (as in bar 1 already; and repeated but *never exceeded downwards* in this section) and the high f (m. 24, first violins and flute), which acts as a "ceiling". The music repeatedly "bumps its head" into this highest boundary before bouncing back from it via the chromaticised scale starting in m. 26. The center of the g – f distance thus created is the middle c (top, first beat, m. 1): the entire section is spatially symmetrical around this starting point.[22] As these events are precursors of the mirror symmetry in mm. 68 – 70 and 74, the latter is no surprise. When one adds to this the fact that – although differently harmonized by Mozart in bars 128 – 131 – the solo piano material there is *actually harmonizable* (when transposed to C-major) by putting the main theme underneath it, then a sort of concealed "derivative symmetry" emerges between the two (compare mm. 1– 4 to 128 – 131 transposed into C-major).[23]

In Béla Bartók's *From the Island of Bali* [24] it is apparent, that the f-sharp and a-flat, sounded simultaneously in the penultimate bar are spatially

---

[21] Of course, this projection immediately introduces asymmetry into this, but the symmetrical morphology and its impact are undeniable.

[22] With the caveat of *non-linear perception* as discussed in Section 1.4.. Just how strong a role g-c-f as a structure plays here can further witnessed, e.g., in m. 66, where this structure is *reversed downwards* in the main notes f-c-g.

[23] This relationship was discovered by the Hungarian pianist Gábor Csalog.

[24] Béla Bartók: *Mikrokosmos*, volume IV. number 109. London: Boosey and Hawkes, New Definitive Edition, 1987.

symmetrical to the middle g, a pitch at the center of the previous "musical summary" of mm. 40 – 41 of the two main collections of which the entire piece is conceived, namely, the b – c – f – g-flat (right hand, bottom up) and the g-sharp – a – d – e-flat collection (left hand, bottom up). Oddly, the g – the "tacit" center of the whole piece – is sounded only once (curiously spelled as *a-double flat*, m. 15) before its repeated appearances in m. 34 (spelled as g for the first time). Put together, the pc-collections of mm. 1 and 2 (which of course mirror one another), yield the g-sharp – a – b – c – d – e-flat – f – g-flat *octatonic scale*. When this is stretched by a semitone on both ends, an octave g would result.

5.2 Registral Counterpoint, Spatial Symmetry

The First Movement of Bartók's *Music for Strings, Percussion and Celeste* is a fugue, where the entry points of all voices follow symmetrical circle-of-fifth patterns. The fugal nature of the polyphony here also creates *counterpoint* in a sense of projected spatial relationships, *both simultaneously and over time*. When the celeste first enters, the theme and its mirror image look at each other "in the mirror" of the celeste texture-pedal (mm. 78 – 81).[25] And in the end of the movement, original and mirror *simultaneously* converge on the central note a. The entry points of the fugue and their timings, (measured in eight-notes) are provided in **Ex. 10.**

5.3 Physical Space: Placement of Instruments, from Stereophony to Total Use of Space

Bartók prescribes a stereophonic seating arrangement for his *Music for Strings, Percussion and Celesta.* The strings are divided into two entities, left and right. The music uses stereophony as a part of its essential structure. *Sonata for Two Pianos and Percussion* uses the joint resonance of the two pianos as an acoustical amplifier of the percussion instruments, unifying them within their *shared resonant space*. These procedures demonstrate a unique awareness of musical thought conceived in terms of acoustical symmetries.

In Cage's *Thirty Pieces for Five Orchestras* (1981), the groups surround the audience and exchange Morse-like repeated sounds, evocative of uncoordinated, beautiful signals that meet in the ear of the listener across the audience space. Xenakis composed a large number of architectural pieces where instruments follow statistical distributions of sound as they are dispersed within

[25] Citation from Ernő Lendvai's apt description of this unforgettable passage in: Lendvai Ernő: *Bartók Stílusa.* Budapest: Editio Musica Budapest, 1955. English approximation mine. G.C.

the audience, as in his Terretektorh (1966) and *Nomos Gamma* (1968). His *Polytope* (1967) is a musical projection of an architectural design. Back in the past, one must mention Giovanni Gabrieli (1554(57) – 1612) for his symmetrical use of space in Venice's San Marco church[26], and of course Berlioz's *Requiem*, with its four brass orchestras and eight pairs of timpani placed in the four cardinal directions within the performance space (originally the St. Eustache Church in Paris).

In my work-in progress since 1980, sound spectrum and *sound source location in the total auditory space are linked*. The so-called stage has seven instruments that can sustain and an equal number that fade. They play within a "closed-circuit" weighed locrian scale. When the circuit is broken ("overfilled"), extension instruments (oboes, flutes and violas) play glissandi, *flattertongues* and non-related pitches from within the audience. Suddenly, the full spectrum of white noise from the back of the audience replaces these sounds from a pair of loudspeakers. When the sound shifts further to another pair of loudspeakers positioned to the sides (center, left and right) of the listeners, the spectrum changes to a full chromatic, and when it further shifts (to the front left/right side speakers), to a full diatonic spectrum, as it gives its way back to the fourteen instruments plus Japanese *rin* on the stage.

My *The Great Initial* (2005-2009) unites the gradual passage of all imaginable age groups (from the oldest towards the ever younger) by inscribing their long-held spoken vowel sounds onto a single channel (monaural) recording. Here, *a chorus that never met*, was *made to sound as a single sound source* in order to emphasize the centrality and oneness of all human voices and to keep the listener's attention focused and *away from* dispersal in space, at a singular point of gravity.

5.4 Dynamics: Imaginary Listening Positions

One can imagine a listening position between *forte-piano* juxtapositions, indeed an imaginary seat from where they are heard "symmetrically", i.e., in an equilibrium (against which they are calculated). However far-fetched this idea may seem, it would be possible to design locations from the vantage points of which *piano* and *forte* would be heard *as equalized*. Since the respective *timbres would be altered* by how their dynamics actually are *played* (as opposed to the levels at which they would be *heard*), composers should put this idea to interesting uses.

---

[26] For further reading please refer to http://en.wikipedia.org/wiki/Giovanni_Gabrieli (cited on January 20, 2010)

The human ear is capable of distinguishing thousands of *dynamical levels.* Yet our notation system offers only a few Italian initials. In computer music these limitations and limitations regarding timbral and spatial control are superseded. If one considers the natural ratio between the noise of the bow on a violin versus the sound produced as "symmetry", then all sorts of shifts of this ratio may be digitally produced and finely controlled, resulting, for instance, in the highly "asymmetrical" case of hearing the bow-noises of fifty violins but the sound of only one.

6.1 S/A Phenomena as a Fundamental Question of Tonality and Atonality

In order to achieve tonality, one must have *uneven distribution* of notes. When *intervallically equidistant* scales occur, they are tonally neutral. This will have to induce us to revisit **Ex. 5**. (Webern: *Symphony, Op. 21*). This is obviously *tonal music,* not only in that it has a central note (around which the entire registral space is organized symmetrically, and this organization is *not equidistant*), but also because of the shared notes among the different forms of the row and how Webern positions these in time. Is Webern, then, a "tonal" composer, in this extended sense of the word? Absolutely. (Other instances like the Second Movement of *Variations for Piano, Op. 27.* corroborate this point magnificently).

6.2 S/A as a Force Relevant to the Temporal Propagation in Music

The opening tone row theme of Schönberg's *Fourth String Quartet* is curious in that it presents a veritable *teleological vortex* pulling it towards its last note. Its rhythmic phrasing highlights the notes printed in bold: **d** – c-sharp – a – b-flat – **f** – e-flat – e – c – **a-flat** – g – f-sharp – **b.** Schönberg's procedure (regarding his invention, the twelve-tone row and its inversion, retrograde and retrograde inversion) has its true origins deep in the history of music, for instance in Ockeghem and in J.S.Bach's fugues. So does the "cadential" perfect fourth leap up to this "last note". (The rhythmic structure is pure Beethoven).

7. Symmetry and Asymmetry in Musical Form (Bartók's bridge-forms vs. memory forms, Schubert's Aufenthalt (Schwanengesang, D957, No. 5.))

An outgrowth of the simpler and most beloved A-B-A "memory" form, the bridge-form is nonetheless very different in Bartók's musical universe. His *Cantata Profana, String Quartets No. 4, 5; First* and *Second Piano Concerti* are all highly relevant examples here. The most powerful example to define this difference, however, is the moment of recapitulation in *Cantata Profana*. This recapitulation is *simultaneous with our realization of the impossibility of return*. Its *simultaneity*, its "neverness" (to defy recapitulation, to cause us grasp the irrevocable fact that the stags will *never* return home) sends shivers across the spine of any sensitive listener, rendering this perhaps the most memorable "recapitulations" in the history of Western music. What we witness is how the *metamorphosis* bridge forms in Bartók's music: that a *transformation* occurs by a *rite of passage* through the mid-point to the point of no return, despite everything symmetrically returning.

Schubert's *Aufenthalt* (*Schwanengesang*, No.5) exhibits a seven-part symmetrical formal structure: A–B–C–D–C–B–A. Curiously, although it is superficially similar to many of Bartók's *bridge-forms*, musically and conceptually it is much closer to the drone-continuity idea[27] than to the *transformational* concept of Bartók.

8.1 S/A in Timbral Semiotic Space

I suggest two fundamental ways in which S/A relationships may be conceived in a **timbral semiotic space**. The first would be along a *linear timbral scale* based on perceptual timbral distances among the instruments used in a given composition. This could be an intuitively conceived scale or one that relies on measured overtone structures (based on sonograms). The scale would present a gradation, starting from one extreme, gradually reaching another, in terms of timbre. The mid-point of this scale then maybe thought of as a sort of axis of symmetry, around which structures would unfold in observation of the above relationships.

The second way would require a more intricate operation, where *three-dimensional* representations of timbral relationships would serve as a basis,[28] and S/A would then be conceived accordingly in a semiotic/spatial system of co-ordinates.

[27] As discussed in Section 3 by way of the MRI metaphor.

[28] This would follow the model worked out in John M. Grey: *Multidimensional Perceptual Scaling of Musical Timbres*. In: *Journal of the Acoustical Society of America*, Vol. 61, No. 5, May 1977, pp. 1270 – 77. Other published work of interest and importance here is David Wessel: *Timbre Space as a Musical Control Structure*. *Rapport IRCAM* 12/78, 1978 – Centre Georges Pompidou. Also published in vol. 3 No.2 of the *Computer Music Journal*, June 1979, M.I.T. Press, Cambridge, Mass.

8.2 Timbral Analogies as Cases of S/A

This category would focus on the *lines* drawn across the aforementioned three-dimensional timbral distance/vicinity structures, where equations and symmetries would be conceived not so much on the basis of the individual entities themselves, but in terms of *movement from one to another* and their multiple equivalent movements in such a system of co-ordinates. In my own *Concerto for Viola and a Changing Environment*, I conceived the solo viola to operate in a constantly changing timbral space where nonetheless recurring *analogous relationships* emerge. Here, timbral transition-lines may be used to create analogous counterparts to them, linking other timbral entities, thus providing parallel symmetries on account of these analogies.

9. Conclusion

Symmetry in music emerges as a universal frame of reference. When conceived as such, the concept has the capacity to *unify musical parameters* that are often treated separately by limitations of language and those of the rigidity of analytic systems. All relationships, correspondences ever dealt with and detected in analysis refer to it, even if unwittingly or unknowingly. I do not readily recall any attempt in musical analysis where the theorist would deal only with incongruities, chaos or disconnect, zealously unearthing these (they leave this task to critics). Yes, we all tend to see connections instead, connections everywhere. Yet it would be fruitful to balance these two extremes of approach. What more can be hoped for than that this essay may encourage *that* to start happening. Symmetry is gravity. We can move only by resisting it. Our resistance is called asymmetry. Symmetry is what we violently resist and strive for at precisely the same time and our resistance and striving is asymmetry. Symmetry is the Invisible Attractor that just isn't. It is a Black Hole. Let's now remove the /: the *oneness* of SA is evident.

References

Bartók, Béla
Music for Strings, Percussion and Celeste. Budapest: Editio Musica Budapest, Z. 9001, Philharmonia scores. Reprint of c1937 by Universal Edition, Vienna.

Bartók, Béla
Mikrocosmos. Volumes I – VI. London: Boosey and Hawkes, New Definitive Edition, 1987.

Bartók, Béla
Fourteen Bagatelles for Piano, Op. 6. Budapest: Editio Musica, c1950.

Bernard, Jonathan W.
The Music of Edgard Varèse. New Haven, CT: Yale University Press, c1987.

Brook, Peter
The Empty Space. London: MacGibbon & Kee, 1969, c1968.

Boulez, Pierre
On Music Today. Cambridge, Mass., Harvard University Press, 1971.

Cage, John
Etudes Australes. New York: C.F. Peters, 1977.

Cage, John
Sonatas and Interludes for Prepared Piano. New York: C.F. Peters, 1960.

Csapó, Gyula
Krapp's Last Tape (after Samuel Beckett). Budapest: Editio Musica Budapest, 1982. Also on Budapest Music Center CD 13.

Feldman, Morton
Essays. Ed. by Walter Zimmermann. Kerpen: Beginner Press, c1985.

Gillmor, Alan M.
Erik Satie. New York, London: W.W. Norton and Co., 1992; c1988 by G. K. Hall & Co.

Grey, J. M. (1975)
Exploration of Musical Timbre. Stanford Univ. Dept. of Music Tech. Rep. STAN-M-2.

Grey, J. M. (1977)
Multidimensional Perceptual Scaling of Musical Timbre. Journal of the Acoustical Society of America, Vol. 61, pp. 1270-1277, 1977.

Haydn, Joseph
Symphony No. 88 in G-major. München: G. Henle, 1958 -

Lendvai, Ernő
Bartók Stílusa. Budapest: Editio Musica Budapest, 1955. In Hungarian.

Lloyd, Llewelyn Southworth
Intervals, Scales, Temperament. London: Macdonald & Jane's, 1978.

Mozart, Wolfgang Amadeus
Concerto for Piano and Orchestra in C-major, K.V. 467. Kassel, Basel, London, New York: Bärenreiter, 1961. Neue Ausgabe sämtlicher Werke, Serie V, Werkgruppe 15, Band 6.

Schönberg, Arnold
Fourth String Quartet, Op. 37. New York: Schirmer, c1939.

Strange, Allen (1983)
Electronic Music: Systems, Techniques, And Controls. Dubuque, Iowa: W.C. Brown Co., 1983.
Stockhausen, Karlheinz
...how time passes...(... wie die Zeit verging...), Die Reihe Nr.3, pp. 10-40, 1959.

Stockhausen, Karlheinz
Structure and Experiential Time, In: Die Reihe Nr.2, pp. 64-74, 1958.

Thakar, Markand
Counterpoint: Fundamentals of Music Making. New Haven, CT: Yale University Press, c1990.

Webern, Anton
Symphonie Op. 21. Vienna: Universal Edition UE 12 198, 1929.

Wessel, David L. (1978, 1999)
Timbre Space as a Musical Control Structure. Rapport IRCAM 12/78, 1978 – Centre Georges Pompidou. Also published in vol. 3 No.2 of the *Computer Music Journal*, June 1979, M.I.T. Press, Cambridge, Mass.

WHAT IS QUANTUM THEORY TELLING US ABOUT HOW NATURE WORKS?

P. GOYAL

Perimeter Institute for Theoretical Physics,
Waterloo, ON N2L 2Y5, Canada
**E-mail: pgoyal@perimeterinstitute.ca*
www.philipgoyal.org

Quantum theory is an extraordinarily successful physical theory. But what does it mean? What implications does it have for the mechanical conception of Nature that underlies classical physics? Remarkably, some eighty years after the creation of quantum theory, we still lack clear answers to these questions. In this paper, we discuss the nature of the obstacles that stand in our way, and describe recent work to overcome them by attempting to *reconstruct* the mathematics of quantum theory from a small number of simple physical ideas.

Keywords: Quantum theory

1. Introduction

Quantum theory is perhaps the most empirically successful theory in the history of physics. In the eighty or so years since its creation, it has provided us with precise mathematical models that account for a vast range of physical phenomena ranging from the principles of chemical bonding, the nuclear reactions that fuel the stars, exotic phenomena like superconductivity and superfluidity, and many others besides. In short, quantum theory has been the faithful companion of physicists for eighty years, and has consistently met the challenge of providing mathematical models of new physical phenomena.

Quantum theory also underlies much of the modern technology that fuels our lives. The transistor, the basis for the modern computer that underlies all information processing technology, requires quantum theory for its design and modelling. The same holds true for the laser and the light-emitting diode, which jointly provide the basis for optical communication networks (the backbone of the telecommunications industry) and optical data storage (in the form of CDs and DVDs, for example), and many other technologies. Yet more quantum technology (such as quantum computers and quantum cryptography) is waiting in the wings, promising to transform our lives still further.

The empirical successes of quantum theory leave little doubt that the mathematical rules of quantum theory—the *quantum formalism*—accurately capture fundamental features of the workings of the physical world. Since their formulation over 80 years ago, there have been very strong indications that these rules describe a physical reality that cannot be encompassed within the view of reality that underpins classical physics. However, the precise nature of the implications of the quantum formalism for our understanding of the material world remains obscure.

2. Classical Physics and the Nature of a Physical Theory

To understand better the nature of the problem and its importance, it is helpful to begin by considering classical physics. Underlying classical physics is a definite conception of the nature of reality, which traces back to the pioneering work of such figures as Descartes, Galileo, and Newton in the seventeenth century. The essential idea is that the totality of all that exists in the phenomenal world (namely that aspect of the world registered by our senses or instrumental extensions thereof such as microscopes or telescopes) is matter moving on the fixed stage of space in step with a universal time in precise accord with mathematical, deterministic laws of motion. And that, in principle, one can probe this matter as precisely as one wishes without disturbing its nature or its motion, and, through such probings, one can in principle discern the underlying laws of motion and thereby aspire to a theoretical model of physical reality that allows arbitrarily precise prediction of the behaviour of physical systems.

When mathematised in a fairly direct way, this conception leads to the *classical mathematical framework*, which forms the mathematical basis not only for Newtonian mechanics but for all of the theories of classical physics—chiefly electromagnetism (describing the phenomena of electricity and magnetism), thermodynamics and classical statistical mechanics (both describing the phenomena of heat and temperature)—that were built up during the two centuries after Newton.

Hence, classical physics has a clear bipartite structure: (i) a definite, clearly statable conception of reality (the mechanical view of material reality) that we are able to readily grasp without the aid of mathematics, and (ii) a mathematical framework which one can regard as a natural translation of this conception into the language of mathematics.

That classical physics has this bipartite structure has major consequences. Firstly, insofar as the development of classical physics is concerned, this bipartite structure imparted *flexibility:* if the mathematics in which Newton clothed the mechanical conception proved too restrictive, his intellectual descendants were able to refer back to the much broader mechanical conception for guidance on how to change the mathematics[a].

Second, this 'clockwork universe' view of the phenomenal world was also *readily graspable* by those who had no knowledge of the mathematics. Consequently, this view was not confined to the physics community, but propagated to other areas of science (such as chemistry, biology, psychology, economics, and sociology) which were able to adapt this view to their own domains of enquiry and often also able to successfully adapt the mathematical techniques developed by physicists. More broadly, this view also propagated beyond the sciences to areas such as philosophy and theology, where it had a profound impact which, in the more than 300 years since Newton announced his mechanics, has played a vital role in the radical transformation that has occurred in our understanding of our place in the world.

3. Quantum Physics

In contrast to classical physics, whereas quantum physics possesses a mathematical framework that can be used to model physical systems, it does *not* presently possess a clear conception of physical reality which, when suitably mathematised, leads to this mathematical framework. In order to understand how this situation arose, it is helpful to consider some history.

---

[a]Fundamental changes were indeed made on several occasions. First, whereas Newton believed that all that exists is atoms in motion (with necessarily finitely [or countably infinitely] many degrees of freedom), Faraday and Maxwell introduced the idea of dynamical *fields* with continuously many degrees of freedom. Second, whereas Newton believed that everything that exists changes in step with a universal time and moves on the fixed stage we call space, Einstein (in his theory of general relativity) introduced the view that spacetime is itself a dynamical entity. In spite of these developments, the essential features of the mechanical view of reality (such as the deterministic and reversible nature of change, and the in-principle complete accessibility of the state of a physical system by means of suitable measurements) remained largely stable during the course of the development of classical physics. Perhaps the singular exception was the threat posed by the assertion of the second law of thermodynamics that certain physical processes are irreversible, and assertion that conflicted with Newton's assertion in the fundamental reversibility of all dynamical processes.

3.1. *Historical Development of Quantum Theory*

By the 1890s, classical physics was in a highly developed state. Newtonian mechanics precisely accounted for both the motion of bodies on the Earth and detailed astronomical observations of the orbits of the planets, moons, and comets; Maxwell's electromagnetism provided a comprehensive understanding of electric and magnetic phenomena (including a quantitative explanation of radiation phenomena), and thermodynamics provided a detailed understanding of the phenomena of heat and the process by which heat is converted into useful mechanical motion ('work'). However, there were some clouds on the horizon—curious experimental facts which stubbornly resisted explanation within the framework of these theories. One of these facts concerned the frequencies of radiation given off by heated bodies—the classical theories were unable to account even qualitatively for the experimental data. The tentative explanations of these stubborn facts (initiated by Max Planck) would, over the next 30 years, bring about the development of quantum theory—an entirely new theoretical foundation for physical theories that would replace the mathematical framework of classical physics.

Quantum theory did not come into existence in a straightforward and principled manner. On the contrary, the path by which it came into being is a tortuous, which is difficult to adequately summarise. In essence, the creation of the quantum formalism was guided by heuristic ideas (such as de Broglie's idea that each particle has an associated 'guiding wave'), and by ingenious mathematical guesswork about what aspects of the existing mathematical structures of classical physics needed to be changed and what aspects could be retained. Remarkably, even though the conceptual basis was patchy, and did not amount to a coherent picture of reality, it was somehow enough (in the hands of Heisenberg and Schroedinger) to give rise to an empirically-valid mathematical formalism.

To illustrate the *ad hoc* nature of the building up of the quantum formalism, I shall give one example. One of the striking features of the quantum formalism is the assertion that, in contrast to the situation in classical physics, when a measurement is performed on a quantum system, the measurement outcome is not wholly determined by the state of the system. Rather, the *probability* of the outcome is determined by the state of the system, but which outcome will actually be obtained is essentially undetermined. That is, the measurement outcome that is obtained cannot be wholly accounted for by reference to the state of the system immediately prior to the measurement. Now, given the importance of this feature, it is remarkable that, in the building up of the quantum formalism, there was no real inkling that the formalism had to possess this probabilistic feature. In fact, when Schroedinger, one of the founders of the theory, wrote down his famous quantum mechanical equation for the electron, he believed that it could be interpreted as simply a new kind of classical field equation that is fully deterministic. It was only afterwards that it was realised that such an interpretation could not possibly stand, and that Schroedinger's equation had to be interpreted in a probabilistic manner[b].

The process, then, by which quantum theory came into being led to a situation where, by 1925–6, although a well-defined empirically-adequate mathematical framework had been constructed, the physical origin and meaning of many of its features was obscure.

3.2. *Immediate Implications of the Quantum Formalism*

Once the quantum formalism was created, many implications were quickly drawn from it which suggested that the world of quantum phenomena it described violated fundamental tenets of the classical mechanical conception of reality. For example, as previously mentioned, one of these classical tenets is that it is possible, in principle, to perform a measurement on a system that reveals everything about its physical state without causing any disturbance to the system. In contrast, the quantum formalism implied that this is not the case—a measurement can, in principle, only give partial information about the state of a system, and necessarily causes disturbance to the system. Moreover, as already mentioned above, whereas the mechanical conception implies that all measurement outcomes are determined by the state of the system, the quantum formalism implied that measurement outcomes are *not* wholly determined by the state of the system, but have an inherently indeterminate element. These implications seemed to radically undermine key tenets of the mechanical view of classical physics.

Many of the founders of quantum theory, such as Bohr, Heisenberg and Pauli, believed that the above-mentioned implications of the formalism had to be taken seriously, and that the classical mechanical view needed to be replaced by something fundamentally new. In order to illuminate these non-classical features of the quantum formalism, some of these physicists

[b]More precisely, in Schroedinger's equation, the wavefunction $\psi(\mathbf{r}, t)$, was initially interpreted by Schroedinger to be a field much like the fields of classical physics. Later, it was realised, principally by Born, that this interpretation could not stand, and that $\psi(\mathbf{r}, t)$ actually encoded the probability that a measurement of position would yield a particular result.

attempted to identify concepts in the existing philosophical literature or to develop new concepts of which these features could be regarded as particular instantiations, thereby placing these non-classical features in a broader philosophical light. For example, Bohr developed the concept of complementarity, which he expressed as meaning that the process of coming to know anything about some aspect of reality unavoidably had the effect of bringing into existence some aspect of it at the expense of simultaneously rendering inaccessible some other aspect of it.[1] Bohr believed that this concept had general validity beyond the field of physics itself, in areas such as psychology and sociology, for instance.[2] From his point of view, the impossibility in quantum theory of performing a measurement that provides complete knowledge about a physics system (but only one property of it at the expense of another) was simply a special case of the general principle of complementarity. Similarly, Heisenberg and Pauli both suggested that the Aristotelian notions of potentiality and actuality could be useful in understanding the relationship between the state of a system and the outcomes of measurements performed upon the system, and in understanding why the classical mode of thought led to inconsistencies when applied to such simple experimental situations as the well-known double slit experiment.[3,4]

3.3. *Development of New Conception of Nature*

The ideas explored and developed by Bohr, Heisenberg, Pauli and others were important first steps in developing a new, non-mechanical conception of reality to underpin quantum theory. However, after their pioneering efforts, the development of a non-mechanical conception along the lines they envisaged was not continued in any kind of sustained way by the mainstream physics community.

As a result, today, more than 80 years after quantum theory was created, physicists possess a well-defined mathematical formalism in which (following its successful application to a vast number of physical systems) they have an enormous confidence, but lack a clear underpinning conception of reality with which to make sense of it.

The lack of such an underpinning conception impacts the development of physics in several ways. First, if one wishes to discover new quantum phenomena or to harness quantum physics in new technology, then, given the counter-intuitive nature of its predictions, one needs some guidance as to what interesting and useful quantum phenomena might follow from the quantum formalism. A clear conception of an underlying reality has the potential to provide such guidance; conversely, the lack of such a conception impedes the rate at which new phenomena are discovered and the rate at which quantum physics is harnessed.

Second, when attempting to apply the quantum formalism to new physical domains, a clear conception of reality becomes especially important to guide one's intuition. For example, an outstanding problem in theoretical physics is the construction of a theory of quantum gravity—a theory that somehow unites the insights of quantum theory and general relativity, the two great pillars of modern physics. In carrying out such a project, it is important to know what assumptions about the physical world lie implicit in the quantum formalism. For example, is the formalism implicitly dependent upon the structure of spacetime? If so, what is the nature of that dependency? More broadly, how does the quantum formalism depend upon the fundamental concepts of classical physics such as matter, energy, momentum, and locality? At present, since we do not have a clear conception of reality that underpins the quantum formalism, we do not know what features of the physical world the quantum formalism depends upon. So, we do not have clear answers to any of these questions.

More broadly, from an interdisciplinary standpoint, the lack of a clear conception of reality underpinning the quantum formalism means that the core essence of quantum theory (whatever that may be) cannot properly disseminate to other areas of science and, more broadly, into other areas of academia and, beyond that, into the general intellectual discourse of society.

3.4. *Deciphering Quantum Theory*

Historically, various attempts have been made to decipher the physical content of quantum theory. The primary methods of doing so which have borne fruit have been to interpret it (often by first reformulating the quantum formalism in some way) or to prove that it possesses particular physical features.

As an example of the latter, it has been shown that quantum theory possesses such non-classical features as non-locality,[5] contextuality,[6] and no-cloning,[7] each of which strongly suggests that quantum physics is at odds with the mechanical view in fundamental ways. For example, non-locality is the feature that implies that, if a pair of quantum systems are allowed to interact and are then separated by an arbitrarily great distance, the outcomes of measurements performed on each of them separately can be correlated in a way that cannot be accounted for on any so-called 'local hidden variable model' of the supposed underlying reality. In short, one is driven to the idea that there is in fact *some kind of connection* between physical systems that have interacted in the past which endures irrespective of their distance and which enforces subtle but very real correlations between them. If so, this connection is of a type never seen in classical physics: unlike the forces of gravity, electricity or magnetism, the connection is unattenuated by distance, and is specific to the systems which previously interacted. If true, non-locality fundamentally alters the status of space, and with it another key element of the mechanical view: space is no longer the entity which mediates all interactions between material bodies; instead, there are also an ever-changing web of connection between systems that are inherently non-spatial.

Although these methods have yielded important insights, their principal limitation lies in the fact that they take the quantum formalism as a *given*. Recall that, in classical physics, we possess a clear conception of reality which, when mathematised in a natural way, yields the mathematical framework of classical physics. However, the quantum formalism has a great deal of mathematical structure which is not present in the mathematical framework that underlies classical physics. For example, states of systems are represented by complex vectors that live in a complex vector space, and temporal evolution of states are represented by unitary transformations of these vectors. Presumably, there are *physical* reasons why this is so, but what are these reasons? By themselves, the above-mentioned efforts to decipher quantum theory do not answer this question.

3.4.1. *Reconstructing Quantum Theory*

In the last three decades, there has been a growing realisation that, if we are to make significant further progress, we must try to discover what *physical* ideas are implicit in the quantum formalism. That is, rather than taking the formalism as a given, we ought to try to formulate a set of simple physical principles from which the quantum formalism can be derived. Such a *reconstruction* of the quantum formalism would essentially distill the *physical content* of the quantum formalism into a set of clear statements expressed in natural (everyday) language, thereby rendering the entire content of the quantum formalism available to the kind of conceptual analysis that is necessary to develop an underpinning conception of reality.

Over the last thirty years, significant efforts have been made to reconstruct quantum theory (see Ref. [8] for details). As a result, we are in the extraordinary situation where there now exist many detailed proposed reconstructions of quantum theory based on a wide variety of different starting points. However, thus far, none of the existing reconstructions has been able to obtain the quantum formalism in its standard form without making abstract mathematical assumptions of key importance.

4. Quantum Theory from Complementarity

In Ref. [9], I have recently shown how it is possible to reconstruct the core of the quantum formalism, namely Feynman's rules of quantum theory, directly from a single postulate:

> ***Pair Postulate***: each sequence of measurement outcomes obtained in a given experiment is represented by a *pair* of real numbers, and the probability associated with this sequence (which is all that one can learn about in the experiment) is a continuous, non-trivial function of both components of this real number pair.

By making use of elementary symmetry and consistency conditions, and without assuming that these real number pairs have any other algebraic structure, one finds that these pairs must be manipulated according to the rules of complex arithmetic. Furthermore, one finds that these complex numbers combine according to Feynman's sum and product rules, with the modulus-squared yielding the probability of a sequence of outcomes. The reader is referred to Refs. [9,10] for details.

The pair postulate, which is at the heart of the approach, expresses the simple idea that a measurement performed on a physical system is only able to yield information about *one-half* of the degrees of freedom of the system. This is a simple formalisation of the principle of *complementarity* originated by Bohr.[1]

In classical mechanics, one assumes that one can simultaneously attribute exact values of position and velocity to a particle. However, the postulates of quantum theory imply that one can attribute either an exact value of position, or of velocity, but not both simultaneously. In order to understand this quantum restriction on a fundamental classical notion, Bohr developed the Principle of Complementarity.[11] According to this principle, any description of a physical phenomenon is indissolubly bound up with a particular experimental arrangement, and such an arrangement provides information about certain aspects of the phenomenon to the exclusion of other aspects that, in classical thinking, are equally necessary for a complete description of the phenomenon. This principle, which was partly inspired by the dialectical aspect of Kierkegaard's philosophical thought,[2] served to underpin the quantum restriction by treating it as a special case of a general philosophical principle, namely that our apprehension of reality involves bringing particular aspects of reality into consciousness by choosing to attend to reality in a particular way and, in so doing, providing a perspective on reality to us that is inescapably incomplete.

By way of exploring the generality of the principle, Bohr and others pointed to analogous phenomena in the mental and organic realms. For example, Pauli developed an analogy between physical complementarity and the relationship of the unconscious and conscious aspects of mental experience:

> On the one hand, modern psychology demonstrates a largely objective reality of the unconscious psyche; on the other hand, ever bringing into consciousness, i.e. observation, constitutes an interference with the unconscious contents that is in principle uncontrollable; this limits the objective character of the reality of the unconscious and invests reality with a certain subjectivity.[4]

Bohr regarded complementarity as one of the most fundamental lessons of quantum phenomena for our physical world-view.[1,11,12] What we have shown is that it is possible to construct a route that leads directly from the fundamental concept of complementarity to the formal structure of quantum theory (in the form of Feynman's rules).

5. Implications for a Conception of Nature

The above reconstruction makes two key assumptions. First, the operational assumption of indeterminacy, namely that measurements are probabilistic in nature. Second, the physical assumption that complementarity holds true, namely that, in a given experiment, one can only access one half of the degrees of freedom needed to theoretically describe a system. Both of these assumptions are at odds with fundamental assumptions in the mechanical view of classical physics, namely determinacy (a measurement outcome is determined by the state of the system) and transparency (in principle, there exists a measurement which is capable of accessing all of the degrees of freedom of the state of a system).

If we accept indeterminacy and complementarity as a given, what conception of nature do they suggest? What are they telling us about how Nature works? A reconstruction of quantum theory, as the one sketched above, is an important stepping-stone to a new conception of Nature in that it focusses our attention on just a small number of non-classical features. But it is still only a stepping-stone. So, how does one proceed?

One of the important jobs of a conception is to provide a unified understanding of the separate assumptions that underlie a reconstruction. In the above reconstruction, indeterminacy and complementarity have been postulated for essentially different reasons. So, in this case, we would like to find an overarching understanding of why Nature has *both* these features of indeterminacy *and* complementarity, and not just one or the other.

We would also like to understand indeterminacy and complementarity in themselves. For example, it is one thing to accept that measurement outcomes are indeterministic as an operational principle (that is, as a summary of what we find in our experiments), but quite another to accept at a philosophical level that things happen without any cause whatsoever. How are we to get a philosophical handle on such an idea? In the case of complementarity, why is it that a measurement can access *one-half* of the degrees of freedom of the state of a system, and not some other fraction, for instance?

Currently, all of these questions remain unanswered, and some are the focus of current research. More broadly, it is still an open question whether one actually needs a new conception of nature in order to understand the indeterminacy and complementarity, or whether these can still be underpinned by a classical mechanical conception of reality.

My personal view is that the most fruitful way forward is the one which Bohr and Pauli long ago asserted, namely to accept complementarity and indeterminacy as fundamental features of reality, and to attempt to evolve our understanding so as to grasp their nature. That there is a deep psychological resistance to such ideas amongst physicists is clear—it could hardly be otherwise if one has grown up with classical physics and has had the mechanical conception of reality repeatedly reflected back to oneself within the attitudes and culture of the society in which one has grown up. However, there exist many philosophical resources that may be able to help us loosen the grip of these cultural conditionings, such as Whitehead's process philosophy.[13] Several physicists, such as Bohm[14] and Wheeler[15] have, in recent time, attempted to develop novel conceptions of reality which take some of the non-classical features of quantum theory as fundamental.

The view of reality which one can dimly make out is one which is fundamentally *participatory*: agents have some irreducible freedom to choose what questions to ask of Nature through their measurements (the non-triviality of this choice been assured by complementarity), and Nature responds with an answer which is only partly determined by variables under experimental control. The answer in turn brings into a state of definiteness some aspect of reality at the expense of some other feature, which is rendered into a state of indefiniteness. Hence, the temporal trajectory of reality is bound up inextricably with the choices that we make. In this conception, what exists is not in a state of complete definiteness, but is rather somewhere along a spectrum between complete definiteness and complete indefiniteness. As Heisenberg pointed out some time ago,[3] there seems to be close connection here with Aristotle's concepts of potentiality and actuality, which he developed in response to Parmenides paradox of change.[16]

One startling feature of the above reconstruction is that no mention is made of space or its properties. This stands in sharp contrast to the manner in which the quantum formalism was obtained by Schroedinger and Heisenberg, where the classical notion of space was presumed at the outset. That is, the above reconstruction strongly supports the view that the quantum formalism is *independent* of space and its properties; that is, it is intrinsically *non-spatial*. This suggests that the classical mechanical view that what exists fundamentally is matter *in space* is incorrect, and that what exists is, in fact, *not necessarily tied to space*. This point of view is already suggested by the existence of Bell non-locality; the above reconstruction provides even clearer support of this view. This insight is of particular relevance to the programme of quantum gravity, and suggests that a theory should be sought in which spacetime is emergent rather than being posited at the outset.

References

1. N. Bohr, *Philosophy of Science* **4**, 289 (1937).
2. A. Pais, *Niels Bohr's Times* (OUP, 1991).
3. W. Heisenberg, *Physics and Beyond* 1971. Translated from the German original.
4. W. Pauli, *Writings on Physics and Philosophy* (Springer, 1994).
5. J. S. Bell, *Physics* **1**, 195 (1964).
6. S. Kochen and E. P. Specker, *J. Math. and Mech.* **17**, p. 59 (1967).
7. W. H. Zurek and W. K. Wootters, *Nature* **299**, 802 (1982).
8. P. Goyal, *Phys. Rev. A* **78**, p. 052120 (2008).
9. P. Goyal, K. H. Knuth and J. Skilling (2009), arXiv:0907.0909.
10. P. Goyal, K. H. Knuth and J. Skilling, The origin of complex quantum amplitudes, in *Bayesian Inference and Maximum Entropy Methods*, ed. P. Goggans (A.I.P., 2009).
11. N. Bohr, *Nature* **121**, 580 (1928).
12. N. Bohr, *Essays 1958-1962 on Atomic Physics and Human Knowledge* (Interscience Publishers, 1963), ch. 1.
13. A. N. Whitehead, *Process and Reality* 1929.
14. D. Bohm, *Wholeness and the Implicate Order* 1980.

15. J. A. Wheeler, Information, physics, quantum: The search for links, in *Complexity, Entropy, and the Physics of Information*, ed. W. H. Zurek (Addison-Wesley, 1990).
16. T. Lear, *Aristotle: The Desire to Understand* (Cambridge University Press, 1988).

BELL'S INEQUALITY: REVOLUTION IN QUANTUM PHYSICS OR JUST AN INADEQUATE MATHEMATICAL MODEL?

ANDREI KHRENNIKOV

International Center for Mathematical Modeling in Physics and Cognitive Sciences, Linnaeus University, Sweden
E-mail: andrei.khrennikov@lnu.se

The main aim of this review is to stress the role of mathematical models in physics. The Bell inequality (BI) is often called the "most famous inequality of the 20th century." It is commonly accepted that its violation in corresponding experiments induced a revolution in quantum physics. Unlike "old quantum mechanics" (of Einstein, Schrödinger Bohr, Heisenberg, Pauli, Landau, Fock), "modern quantum mechanics" (of Bell, Aspect, Zeilinger, Shimony, Greenberger, Gisin, Mermin) takes seriously so called quantum non-locality. We will show that the conclusion that one has to give up the realism (i.e., a possibility to assign results of measurements to physical systems) or the locality (i.e., to assume action at a distance) is heavily based on one special mathematical model. This model was invented by A. N. Kolmogorov in 1933. One should pay serious attention to the role of mathematical models in physics. The problems of the realism and locality induced by Bell's argument can be solved by using non-Kolmogorovian probabilistic models. We compare this situation with non-Euclidean geometric models in relativity theory.

Keywords: Bell, Boole, Kolmogorov, Euclid, Lobachevsky, completeness of quantum mechanics, "death of realism", quantum nonlocality

1. Introduction

We start with a remark that the creation of *mathematical models* is the basic tool of modern science. One cannot "understand" the reality, but only create its models. Then any model should be tested by experiments. Unfortunately, it happens (and quite often) that people forget this basic principle of scientific methodology. A mathematical model is sometimes identified with reality as it is. We can mention the *Newtonian model*. During a few hundred years its mathematical counterparts, e.g., continuous space-time and continuous evolution, were considered as features of physical reality.

We also emphasize the role which was played by *Euclidean geometry* during practically two thousands years. Geometry of physical space was firmly identified with this special mathematical model (not only in physics, but even in philosophy, cfr. Kant). It took a lot of time and effort to understand the correct place of those mathematical models (Newtonian mechanics and Euclidean geometry) in physics.

Quantum mechanics (QM) is strongly coupled to probability theory. In fact, QM is a special probabilistic theory. It cannot predict motion of an individual quantum system, e.g., an electron. QM can only provide probabilities of results of measurements performed on huge ensembles of quantum systems. One might dream of the creation of a better model of processes in the micro-world. However, nowadays even such a dream is forbidden. By the *Copenhagen interpretation* QM is a *complete theory.* It is in principle impossible to find a new mathematical model which would provide a finer picture of processes in the micro-world.

Completeness of QM was postulated by Bohr. On the basis of general philosophic and methodological arguments it is not easy to accept this postulate. For a "normal brain", it is not easy to start to think in the Copenhagen way: "mankind had approached the final description of the micro-world." Never in the future, a finer model (which will be adequate to the experimental situation) will be created. Bohr considered completeness of QM as a consequence of the *principle of complementarity.* The latter he "derived" from the *Heisenberg uncertainty relation.* We remind the reader of the historical basis of this important philosophic principle. It is well known that Bohr formulated his principle of complementarity as a personal reaction on the impossibility to create a classical-like model describing individual micro systems*. His physical condition was not especially good at that time. He understood that he should either recognize that he and his collaborators were not able to create a satisfactory model of microscopic processes or claim that creation of such a model is impossible.

Completeness of QM and, in fact, the principle of complementarity were questioned in the famous article of Einstein, Podolsky, and Rosen[1] (EPR). Bohr rejected their argument. However, his answer was written in such a form that not so many physicists, if any, could understood his argument. Nevertheless, Bohr's position was supported by various *NO-GO theorems* – statements on the impossibility of creation of new mathematical models

*He attempted to do this during many years preceding the formulation of the principle of complementarity.

which would provide a finer picture of micro-processes. The first one was presented by von Neumann.

Nowadays the most popular NO-GO theorem is *Bell's theorem.*[2] This theorem induced revolutionary changes of views on QM. Although formally Bell's theorem was directed against local realism (i.e., a possibility to go beyond QM and at the same time respect locality), it finally (as a grimace of scientific psychology) generated the idea of *quantum non-locality.* This idea, that QM is based on action at a distance, became very popular in the quantum community, especially among people doing quantum information.

My main aim is to show that this idea of quantum non-locality, as well the general argument against local realism, is a perverse consequence of the misuse of the standard scientific methodology. In manipulations related to Bell's inequality (BI) a majority of physicists (not all!) completely forgot the role which is played by mathematical models in physics and science in general. J. Bell did not formulate explicitly what model of probability theory was used. None of his later adherents to his arguments did so either. I (as well as a few other scientists educated in quantum probability) found that he implicitly assumed the use of the Kolmogorovian model of probability. This model was invented by Andrei Nikolaevich Kolmogorov[3] in 1933. Its essence indicates a possibility of the representation of observables (of any origin) by so called *random variables.* These are functions, $\lambda \to a(\lambda)$ which are defined on the space Λ endowed with a probability measure P. We remark that the symbol λ is commonly used in quantum physics to denote so called hidden variables (HV). These are "finer variables" or to say "prequantum variables" which determine the values of all quantum observables, compatible as well as incompatible. Adherents of HV say that at the moment our measurement devices are too rough (too badly designed) to detect these HV. However, in the future, e.g., in a few hundred years, the precision of measurements will be improved so much that HV will be approached. However, as a consequence of Bell's argument, it is commonly assumed that such HV (if they exist at all) will be non-local.

My main point is that the assumption of Kolmogorovianity, i.e., a possibility to use a single Kolmogorov probability space, was not justified in the experiment demonstrating the violation of BI.

2. EPR-Bohm-Bell Experiment

This experiment was designed on the basis of the experiment proposed by EPR.[1] However, it differs essentially from the original EPR. The main difference is that the EPR-Bohm-Bell experiment is on correlations between two subsystems of a composite system. These correlations correspond to *simultaneous* measurements on two subsystems. The EPR experiment was designed for conditional measurements: first the state preparation by measurement on one subsystem and then, under this condition, measurement on another subsystem. There is another important difference which we discuss now.

The original *EPR experiment* was proposed for *continuous quantum observables,* position and momentum. In fact, it has never been done! Typically, in a performance of the original EPR experimental difficulties are emphasized. It is not widely known that, for this experiment, a local realistic model can be easily constructed, e.g., Khrennikov and Volovich.[4] Those who know this fact try to ignore it by pushing it to subconsciousness.

In any event, Bohm modified the original EPR-experiment by considering discrete observables, instead of continuous observables[†]. He proposed to work with *spin.* We recall that the operator of the spin projection to any axis has the discrete spectrum – it represents (in the mathematical model of QM) a physical observable taking a discrete set of values. This version of the EPR experiment is often called the *EPR-Bohm experiment.*

Although this experiment was originally designed for electrons, it has been never performed for them. It was executed (first by Aspect[5]) for photons – for measurement of *polarization* of light. In QM the operator of the polarization projection to any axis has the discrete spectrum. For non-physicists, we remark that the classical illustration of spin and polarization is given by vectors which can be assigned to a moving particle. However, this illustrative picture is purely heuristic. By the Copenhagen interpretation it is forbidden to consider not only such vectors, but even the trajectory of a particle. At the same time such graphical illustrations are widely used in all textbooks on QM.

In a few words, the EPR-Bohm experiment can be described in the following way. A laser sends a pulse to a special crystal. The latter is engaged in a *nonlinear* interaction with this pulse. It emits two pulses which can be sent with the aid of two glass-fibers to two labs. They can be located on a large distance from each other, a few hundred kilometers in recent experiments. In each of these labs light passes through a polarization beam splitter which is oriented along some axis. These axises can be chosen in different ways in two labs. In principle, it is possible to work in plane ge-

---

[†]However, he did not stress the fundamental difference between the original EPR and his experiment, namely, the existence of a local realistic model for the former and the absence for the latter.

ometry. Thus axes can be encoded by θ_1 and θ_2, where $\theta_i, i = 1, 2$, are the angles between the axes of orientations of polarization beam splitters and, e.g., the x-coordinate axis. Each of these splitters splits the corresponding incoming signal with respect to polarization and sends it to two detectors, say $D_{i,+}$ and $D_{i,-}$, where $i = 1, 2$ indexes labs. If $D_{1,+}$ clicks then the value $+1$ is assigned to the observable of θ_1-polarization, say $a_{\theta_1} = +1$. If $D_{1,-}$ clicks then $a_{\theta_1} = -1$. In the same way we introduce another observable, say b_{θ_2}, corresponding to clicks of detectors $D_{2,\pm}$ in the second lab.

To define each of the observables properly, we should exclude *simultaneous clicks of two detectors in the same lab*, e.g., the situation when at the same instance of time both detectors in lab 1, $D_{1,\pm}$, click. This is a crucial point. Here the quantum description of light plays the fundamental role. In QM, light is represented not by a wave (electromagnetic field, continuously propagating in space), but by *photons*. By QM the nonlinear crystal emits not two beams of classical electromagnetic radiation, but pairs of photons.

These are so called *entangled* quantum systems. They "remember" on the nonlinear crystal which emitted them. So to say, they remember their "common house." However, such a picture of entanglement may be considered as old-fashioned. It is merely in the spirit of the EPR paper.[1] A modern quantum guy would not be fine with the reduction of entanglement to the memory of preparation. He will speak of non-local entanglement, i.e., that each of the photons in a pair of entangled systems feels permanently what happens with his partner. This is the essence of quantum non-locality. However, to come to such a picture of entanglement one uses violation of BI.

We leave the sandy ground of entanglement and turn to the experiment under consideration. Thus by QM in each of the labs a photon will pass the corresponding polarization beam splitter. It cannot split a photon, unlike a classical electromagnetic pulse. The latter (the classical one) is split into two beams. Intensities of these beams are described by *Malus law.* In QM we have the same Malus law as we do in classical electrodynamics. But, instead of intensities, we speak of probabilities for an incoming photon to go to respective detectors.

Thus, in this experiment with quantum particles, photons, we escape double clicks in the same lab, since only one photon should pass the polarization beam splitter in each lab (at least theoretically). It is possible to match clicks in different labs and find corresponding probabilities – the frequencies of matched clicks. For example, $P(++)$ is the probability that detectors $D_{1,+}$ and $D_{2,+}$ click simultaneously. The frequency definition of probability is used. Suppose that the source (nonlinear crystal) produced N pairs of entangled photons and suppose that the result $++$ was obtained in N_{++} cases then $P(++) = \frac{N_{++}}{N}$. Of course, this probability depends on the orientations of polarization beam splitters, i.e., $P(++) \equiv P_{\theta_1,\theta_2}(++)$.

BI is an inequality which can be proved for a signed linear combination of such probabilities. The form of this inequality is not important for us. We are interested in assumptions for its derivation.

The crucial point is that probabilities involved in BI cannot be defined in a single experimental framework. One should select four different pairs of angles: for $\theta_i, i = 1, 2, 3$, take $(\theta_1, \theta_2), (\theta_2, \theta_3), (\theta_3, \theta_4), (\theta_4, \theta_1)$, and perform EPR-Bohm experiments for four pairs of settings. We emphasize that experiments for different settings are incompatible. On the same photon it is impossible to perform measurements corresponding to two different selections of orientations of polarization beam splitters. A detector kills the photon. The click of a detector implies that this photon is dead and it is impossible to send it to another polarization beam splitter and then to repeat detection.

Such a combination of a few EPR-Bohm experiments, for different pairs of orientations of polarization beam splitters, we call the *EPR-Bohm-Bell experiment.* From the statistical viewpoint it differs crucially from the original EPR-Bohm experiment. In the latter, one did not try to combine data collected for incompatible settings of measurement devices.

3. Implicit Assumption of Bell

To derive BI, Bell assumed that mathematically probabilities obtained for different settings (different pairs of angles) can be described by a common probability space. In other words, he assumed that all observables a_{θ_i}, b_{θ_i}, $i = 1, ..., 4$, can be realized as random variables on the same probability space. We call this condition *probabilistic compatibility of observables* (PC),[6].[7] Thus, Bell claimed (implicitly) that the Kolmogorovian model of probability theory can be used in the EPR-Bohm-Bell experiment. Unfortunately, this assumption was not explicitly formulated in his papers[‡]. Nevertheless, it is present in his considerations and it can be easily identified through the use (by Bell) of a single probability which was denoted by the symbol $d\rho(\lambda)$.

[‡]I am even not sure whether Bell did know about the Kolmogorov model.

In my books,[8,7] I questioned justification of appealing to PC for incompatible observables. It is easy to see from my works,[8,7,6] that violation of PC induces violation of BI. Thus BI can be violated without appealing to such revolutionary assumptions as non-locality or death of realism – "just" as a consequence of violation of PC.

It also should be remarked that the EPR-Bohm-Bell experiment was presented in the idealized form. We did not mention that photonic detectors have very low efficiency. Thus each detector produces a cutoff of the original ensemble of pairs of entangled photons. Since it is unknown which photons are more preferable to be detected, one cannot exclude that by varying orientations of the polarization beam splitter we change the ensemble of detected photons. In the BI studies it is commonly assumed that sampling of detected photons is *fair*, i.e., its statistical properties do not depend on orientations, see[9] for an extended analysis of this problem. However, the assumption of *fair sampling* has never been checked experimentally. Moreover, there is a rather common opinion that it cannot be checked even in principle. I and my graduate student Guillaume Adenier elaborated a test[9] of the hypothesis of fair sampling. However, it has never been done. Experimenters, who continue to perform thousands of tests of BI and who are fine by obtaining its violation, rejected our proposal to check this important assumption. A few laboratories showed interest, but finally nothing has been done. It can be easily explained, since each experiment demonstrating violation of BI is considered as a step towards the confirmation of quantum mysteries. Suppose one performs an experiment to check the fair sampling hypothesis. If this hypothesis can be confirmed, such a result will not provide profits to this experimenter. The common reaction will be that this result could be expected. Even if somebody obtains data supporting unfair sampling, then this result will not have a great impact. The quantum majority firmly believes in quantum non-locality or at least the impossibility to use the realistic description. People will either simply ignore this experiment or express doubts in its correctness.

I performed a detailed study of BI-type inequalities under the condition that the assumption of Kolmogorovianity, PC, does not hold.[8] It was shown that, instead of the standard BI-inequality, one obtains generalizations of BI, so called Bell-Khrennikov inequalities, see Smolyanov and Truman.[10] They match well with experimental data obtained in the EPR-Bohm-Bell experiment.

4. Possible Sources of Non-Kolmogorovianity (Probabilistic Incompatibility)

In principle, one could not exclude that, e.g., non-locality is the real source of violation of PC. Thus my argument does not imply that Bell was definitely wrong. However, by understanding that BI can be easily violated as a consequence of violation of a very general condition – PC, we should not restrict our mind to Bell's ideas on possible sources of violation of BI. Other possible sources should be taken into account, none less seriously:

a). Unfair sampling: it can be induced by the variation of orientations of polarization beam splitters in combination with the functioning of detectors;

b). Rejection of Einstein's hypothesis on light quanta, photon: by coming back to the classical wave model of light we definitely have a departure from the canonical description of the EPR-Bohm-Bell experiment (clicks of detectors could not be associated to photons, a click may be created as a collective effect of a few pulses, see[11]).

5. Corresponding Experimental Tests in Geometry

My viewpoint is that the experimental test of violation of BI is simply a test of validity of the *Kolmogorovian model of probability theory.* Therefore, it is useful to compare these tests with a corresponding test which were proposed by *Lobachevski and Gauss* to test the limits of application of the Euclidean model.

Lobachevski is known as the first mathematician who invented non-Euclidean geometries, namely, *hyperbolic geometry.* It is less known that he also proposed the first reasonable test of space geometry. The corresponding story about Gauss is known much better. Therefore, we start with it here. We follow (with some modifications) the electronic book:[12]

One of the most famous stories about Gauss depicts him measuring the angles of the great triangle formed by the mountain peaks of Hohenhagen, Inselberg, and Brocken for evidence that the geometry of space is non-Euclidean. He tested the inequality:

$$\alpha_{12} + \alpha_{23} + \alpha_{13} < 2\pi, \tag{1}$$

where α_{ij} is the angle between corresponding sides of triangle. Gauss understood how the intrinsic curvature of the Earth's surface would theoretically result in slight discrepancies when fitting the smaller triangles inside the larger triangles, although in practice this effect is negligible, because the Earth's curvature is so slight relative to even the largest triangles that can be visually measured on the surface. Still, Gauss computed the magnitude of this effect for the large test triangles because, as he wrote, "the honor of science demands that one understand the nature of this inequality clearly".

On the other hand, if the curvature of space was actually great enough to be observed in optical triangles of this size, then presumably Gauss would have noticed it, so we may still credit him with having performed an empirical observation of geometry, but in this same sense every person who ever lived has made such observations. The first person to publicly propose an actual test of the geometry of space was apparently Lobachevski, who suggested that one might "investigate a stellar triangle for an experimental resolution of the question." The stellar triangle he proposed was the star Sirius and two different positions of the Earth at 6-month intervals. This was used by Lobachevski as an example to show how we could place limits on the deviation from flatness of actual space, based on the fact that, in a hyperbolic space of constant curvature, there is a limit to how small a star's parallax can be, even for the most distant star. The first definite measurement of parallax for a fixed star was performed by Friedrich Bessel (a close friend of Gauss) in 1838, on the star 61 Cygni. Shortly thereafter he measured Sirius (and discovered its binary nature). Lobachevski's first paper on the new geometry was presented as a lecture at Kasan in 1826, followed by publications in 1829, 1835, 1840, and 1855 (a year before his death). He presented his lower bound on "characteristic length" of a hyperbolic space in the later editions based on the still fairly recent experimental results of stellar parallax measurements.

Suppose at the moment that somebody before Lobachevski and Gauss (or nowadays somebody badly educated in geometry) made measurement of angles in a light triangle and see that his data violate the"Lobachevski-Gauss inequality" (1). He might speculate that he found a trace of non-locality: measurement of, e.g., the angle α_{12} has influence on measurement of α_{13} and α_{33}. Another possibility is to speculate on the death of realism. Angles are not properties of a triangle. They are created by our measurement devices.

6. Conclusion

The role of BI in modern QM was extremely overestimated. Of course, its violation can be considered as a sign of non-locality or impossibility to use the realistic description in QM. However, this is only a sign. It cannot be considered as a crucial argument. In the derivation of BI an explicit assumption, Kolmogorovianity of data (PC of observables), was used. This assumption can be easily violated without appealing to Bell's revolutionary ideas on non-locality and death of realism. The main problem is the study of the functioning of photodetectors, cf.[11] It is surprising how little can one say on the coupling of violation of BI and the functioning of detectors. Another problem is to try to perform the EPR-Bohm-Bell experiment for heavy particles. Such experiments, when properly done, would clarify a lot of the sources of violation of PC.

References

1. Einstein, A., Podolsky, B., Rosen, N. Phys. Rev. 47, 777–780 (1935)
2. J. S. Bell, *Speakable and unspeakable in quantum mechanics.* Cambridge: Cambridge Univ. Press, 1987.
3. Kolmogorov, A. N., *Grundbegriffe der Wahrscheinlichkeitsrechnung,* Springer Verlag, Berlin (1933); reprinted : *Foundations of the Probability Theory,* Chelsea Publ. Comp., New York (1956).
4. A. Yu. Khrennikov and I. V. Volovich, *Soft Computing* **10**, 521 - 529 (2005).
5. A. Aspect, J. Dalibard, G. Roger, *Phys. Rev. Lett.,* vol. 49, pp. 1804-1807, 1982.
6. A. Yu. Khrennikov, *Journal of Russian Laser Research,* **28**, 244-254 (2007).
7. Khrennikov, A., *Contextual Approach to Quantum Formalism,* Springer, Berlin (2009).
8. Khrennikov, A., *Interpretations of Probability,* VSP Int. Sc. Publishers, Utrecht/Tokyo (1999).
9. Adenier, G., and Khrennikov, A., *J. Phys. B: Atomic, Molecular and Optical Physics* **40**, 131-141 (2007).
10. Smolyanov, O.G., and Truman, A., *Doklady Mathematics* **66**, 440-445 (2002).
11. Adenier, G., arXiv:0906.1539 (2009).
12. Reflections on relativity. http://www.mathpages.com/rr/rrtoc.htm

STRUCTURES AND SYMMETRIES IN PHYSICS

CHARY RANGACHARYULU

Department of Physics and Engineering Physics, University of Saskatchewan, Saskatoon, SK, S7N5E2 Canada

Symmetries play a very significant role in describing the dynamics of physical structures and phenomena. While preserved symmetries enable physicists to establish systematics and predict regularities, broken symmetries open up new avenues of research as they admit new degrees of freedom. Quite often, physicists resort to mathematical symmetries to define the patterns and use metaphors to convey meaning. A caution is warranted not to take the symbolisms too literally and to be aware of limitations while borrowing physics language into other disciplines.

1. Introduction

Broadly, we may state that research in physics is the investigation of dynamics among inanimate beings and interactions in the physical universe. The vast range of dimensions that these studies encompass is awe-inspiring. Physics at microscopic scale working with subnuclear degrees of freedom may concern with objects smaller than 10^{-15} meters in size, weighing less than 10^{-36} kilograms and interacting in time scales shorter than 10^{-30} seconds. At the other end, macroscopic physics at the cosmological scale may deal with objects larger than 10^{21} meters weighing as much as 10^{41} kilograms, and time scales of several billion years. A physicist employs some model prescriptions to deal with phenomena involving only a few bodies, or a many-body system treated with effective values for physical parameters.

Regardless of the dimensions and complexity of the topic at hand, a physics problem is subdivided between the structures of the objects and the interactions among them. In this regard, we consider the identities of objects to be independent of the nature of interactions to which they are subject, and that the interactions have their own characteristics independent of the objects they act upon. This is a basic unverifiable assumption at the heart and soul of all physics theories and models. It is due to the fact that structures are deduced only through interactions, while the interactions are inferred from what they do to structures. At least in some ways, it is the physics counterpart of Kurt Gödel's theorem of incompleteness of mathematics [1]. The effective field theories or structural arguments are purely mathematical simplifications which do not deny the fundamental identities of entities and interactions.

An object's structures are defined by its spatial extensions, and interactions occur over space and time. It is therefore obvious that all physical descriptions rely heavily on space-time models. It is well known that Einstein revised the Newtonian concepts of three-dimensional space and one-dimensional time to a four-dimensional space and time continuum. He arrived at this hypothesis to provide a conceptual foundation for the mathematical relations which he and other notables such as Lorentz and Poincaré had derived, as they sought to describe the empirical evidences stemming from diverse physical phenomena [2]. The important point is that a quest to provide a coherent description of experimental data necessitated the revision of established concepts; it is not a desire to change the concepts that drives thinking in physics. In all these attempts to describe physical phenomena, symmetries (both preserved and broken ones) are very powerful tools in the hands of physicists. It is to be noted that Newtonian and Galilean relativity is asymmetric in space versus time, while the Einstein-Lorentz transformations bring in space-time symmetry.

Physicists deal with very complex systems involving many interacting partners, which poses the serious challenge of having to solve a many-body problem. It is somewhat amusing that a few-body problem of one physicist becomes a many-body problem of another. For example, an atomic or molecular physicist, dealing with processes of interest to a physical chemist or a material scientist, would treat an atomic nucleus as a point-object which simply provides a center of force to the external world. This force can be sufficiently described by the nucleus' mass and atomic number. In contrast, a nuclear physicist working with classical nuclear models will deal with the neutrons and protons inside those atomic nuclei as point objects with electric charge of zero and +e (elementary electron charge), respectively. For a particle physicist, however, a proton or neutron are very complex objects, each made up of three valence quarks[1] and an infinite number of "sea quarks" as well as a swarm of gluons mediating the interactions.

---

[1] In the present-day particle physics, proton and neutron are not elementary particles. But, they are made up of "quarks", which are confined to the interior of the strongly interacting objects. See, for example ref. 3

Thus, it is obvious that in their quest to describe natural phenomena, physicists resort to methods of approximation that aim to be conceptually simple and amenable to mathematical manipulation. There is a vast amount of empirical evidence that many diverse physical properties (eg. electrical conductivities or magnetic properties) are due to the dynamical behavior of a few constituents of matter while most others are passive participants adding to the bulk properties. Thus, a major enterprise in physics is a study of systematics where we measure the change or constancy of a physical observable as we vary parameters in a controlled manner. There are also other observable properties which simply relate to the overall mass or electric charge of the system. A very useful example is the "giant resonance" of excitations in atomic nuclei [4]. At any accelerator laboratory there is a concern with radiation arising from the flux of fast neutrons emitted from this resonance in materials intercepting a stray beams of particles. Their energies are very predictable, making the system design rather straightforward. This feature is not only of simple academic interest, but also has important practical applications. In physics and chemistry, these systematic studies have had a long history with impressive results [5].

2. Symmetries in Physics

As in other human endeavors, symmetries play an essential role in physics. Their impact was explained very eloquently by Hermann Weyl in his book, *Symmetry* as follows: "Symmetry, as wide or as narrow as you may define its meaning, is one idea by which man through the ages has tried to comprehend and create order, beauty and perfection."[6] Hermann Weyl wrote this book in the 1930s. I would venture to suggest that had he written it in the 1950s or later, he would have mentioned that in physics it is not only preserved symmetries which are important but broken symmetries which also play an essential role in advancing the discipline[2].

In physics, symmetries are manifested as, a) laws of conservation where a physical observable remains constant (or invariant); b) selection rules dictating whether a process can occur through emission or absorption of specific quanta of particles or radiation; or c) the occurrence of certain excitation properties or the existence of particles. As an example, the mathematical treatment describing the empirical fact that the speed of light is independent of an observer's frame of

[2] Seemingly, not all disciplines share this enthusiasm for broken symmetries. Sociologists point out that broken symmetries in societal contracts produce disastrous consequences such as ill-health, lack of self esteem, emotional disregard, etc. [7], and they must be avoided like the plague.

reference led to Einstein's space-time symmetry argument. This is a vivid example of introducing a symmetry to account for an invariance principle.

A crowning achievement of 1950s particle physics was the identification of breakdown of the parity operation as a fundamental symmetry. Parity symmetry is an exotic operation wherein we ask what happens to a system if we change the signs of all spatial coordinates (i.e. $x \rightarrow -x$, $y \rightarrow -y$, $z \rightarrow -z$). Will it remain unchanged or will it become a totally different object? If the system remains unchanged, it is said to have "good parity". If it ends up being a different system, the parity is not conserved and the parity symmetry does not hold. During the first half of the 20th century, parity was assumed to be a "good" attribute of all particles. In the late 1940s, particle physicists were puzzled by the decay of a meson[3] showing two different parity modes: one with positive sign and the other with negative sign, indicating that the parent particle may not have a fixed parity. The resolution for this puzzle came from a pair of theoretical physicists Lee and Yang [8] as they questioned the validity of parity as a good symmetry. They provided a mathematical formulation and explained the physics implications for an experiment that one can carry out in the laboratory. In physics, there is little interest in mathematical reasonings with no observables to test the model in the laboratory.

An important development in this problem was defining the observables which are accessible to experimental verification. The principle is straightforward: construct a measurement operator which may be a product of several other operators. The net operator must be "odd" under the symmetry operation in question, i.e. if "A" is the symmetry operator, construct an "M", the measurement operator, such that $A \times M = -M$, which is the original operator with a sign change. An experimental measurement of M and -M on the system will give identical results if A is a good symmetry of the system. Any deviation from the equality in the experimental result will be a measure of breakdown of the symmetry A[4].

---

[3] Mesons are a family of unstable particles. They are produced in nuclear and particle interactions. In the early descriptions of particle physics, mesons were treated as elementary particles. In present day treatments, a meson is a composite object made up of a quark and an anti-quark pair.

[4] It is conceivable that we can extend this principle to test symmetries in other disciplines as well. It is, however, essential that the outcome of the application of each operator is predictable without any ambiguity. Say, in political studies, we want to study the voting patterns of constituents. Even if we can ascertain that all poor people vote in a way opposite to that of rich people and that all Tories are opposed to Labor, it would still leave the system ambiguous since some people belonging to a party may be rich while some others of the same party affiliation may be poor. Great caution is thus warranted.

In the case of Lee and Yang's work, an experimental group [9] performed the measurement which confirmed their theoretical expectations. It is important to note the methodology here. The theory was guided by a problem that needed a solution, *viz.*, why is it that from among the same species some particles decay to a negative parity final state while others decay to a positive parity state? The physicists examined whether the existing theoretical framework could account for the experimental observations and found this was not possible due to the symmetries built into the mathematics. They then asked if there was room to extend the mathematical formulations to account for the experimental data regarding parity violation in certain decay processes, *without contradicting other experimental observations*. After identifying a recipe to do this, they predicted further physical consequences which could be verified in the laboratory. The experimental group took the lead from that suggestion and much to the excitement of the particle physics community, the theoretical prediction was confirmed. This has become a textbook example. It may be safe to say that particle physics has not been the same since. This parity breaking opened new avenues to test the goodness of other symmetries. It also provided a prescription of how to go about testing symmetry breakdowns in physics.

The above discussion refers to a physical symmetry with a distinct observable in the laboratory as we manipulate external forces. The scenario also involved scientists questioning a symmetry which had become a dogma in the physics community: parity was considered to hold good at the microscopic level even though there are violations at the macroscopic level. For example, chirality (handedness) is seen at the (macroscopic) molecular level in optical rotation properties of glucose or anisotropic crystal structures. It is bemusing to note that the parity violation in particle physics is still limited to one of the four fundamental interactions, the so-called weak interaction, while the macroscopic phenomena at atomic or molecular or even larger scale are mainly electromagnetic processes[5].

In passing we may note that Einstein was invoking a new space-time symmetry to account for the constancy of the speed of light, while Lee and Yang introduced reflection asymmetry to account for parity breakdown.

There are also algebraic symmetries built into hypothetical Hilbert spaces[6]. We can label the members of a group belonging to an algebraic symmetry as

---

[5] Gravitational forces are not considered here. Their influences are minutely small for the processes discussed here.

[6] Hilbert space is a fancy term commonly used by physicists where we consider the whole family of basis states and composites as belonging to a world of their own. Once we define the basis states,

components of a vector representing that specific group. There is no physical force parameter which can measure the component or the vector in the laboratory[7]. However, this concept proves to be very powerful again in the systematics and selection rules of levels and transitions, etc. Heisenberg was among the first to introduce such a symmetry in physics. Soon after the neutron was discovered in 1932, he observed that the principal difference between a proton and a neutron is that the proton carries electric charge while the neutron does not. With the quantum mechanical formulations of particle spin and its projections still fresh in his mind, he figured it would be a mathematical convenience to consider the neutron and proton as two states of different "charge projection" of a single hypothetical particle. This new degree of freedom, defined in analogy with spin angular momentum, is called "isospin". With this assignment, one can blindly and almost mechanically apply all of the mathematics developed for spin (e.g. vector addition, angular momentum algebra, etc.,) to the treatment of neutron-proton symmetries and asymmetries. This simple idea resulted in enormous mathematical activity, physics implications, and experimental investigations not limited to these two particles, but also extending to several other families of particles both in nuclear and particle physics[8].

---

then we can build any composite state as a combination or a set of combinations of basis states. For example, we can define the economy of a state as Hilbert space, which comprises of natural resources, technological infrastructure, agriculture etc.. as basis states. We can subdivide any of these states as further sub-Hilbert spaces. Say, we can consider the agriculture space to be made up of rice, wheat, pulses etc. and then build a complex structure of the country's or world's economy.

[7] Abstract symmetries are not unique to physics. One finds them in music, poetry and other fine arts.

[8] One may easily see that any discipline which deals with a finite number 'n' of characteristics distinguishing each of its constituents, may employ a model of vectors with members having projections s_i= -s, -s+1, -s+2,....s-1, s, where $s=\frac{(n-1)}{2}$. As an example, if a sociologist needs to analyze data on the basis of gender, the system can be viewed as a set of vectors where the members are each of s=1/2 with, say, projection -1/2 and +1/2 for male and female members respectively. Similarly, a political scientist studying the characteristics of members in a two-party system (say, Democrats and Republicans or Tories and Grits) can make use of the same labeling. Caution must be exercised when more than one parameter plays a role in the system analysis. To illustrate this point, a sociologist interested in analyzing data on the gender basis in separate age groups must build two sets of vectors of appropriate dimensions and use the correct tensor algebra depending on whether the different vectors span the same or separate Hilbert spaces. The details of these formulations can be found in standard tensor algebra books [7]. This logic can be extended to an arbitrary number of variables, each treated on an individual basis with the result being a tensor product of all terms.

Another example of interest is found in nuclear physics: certain isotopes[9] are more stable against breakup and exhibit simpler excitation structures in comparison to their neighbors. The number of neutrons or protons exhibiting these features were called magic numbers and were found to be 2, 8, 20, 28, 50, 82 and 126. Initial attempts at building a model for this phenomenon employed an analogy with an elastic three-dimensional spring oscillator. This assumed the isotropy of space, and allowed a simple calculation of the model's results, which were the magic numbers 2, 8, 20, 40, 70, ... There is a clear disagreement with observations. It was thought that the model could not be completely wrong since it did reproduce the first three magic numbers. An anecdote says that Fermi asked his research assistant, Maria Goppert Mayer, to introduce a spin-orbit term[10] in the potential and calculate the results [11]. The story had a happy ending in that Mayer was one of the recipients of the Nobel Prize for resolving the magic number puzzle. This is an example where no new physics was brought in, but existing ideas of the spin-orbit interaction were extended from atomic and molecular spectroscopy to nuclear structure.

There was, however, a consequence of Mayer's work in that the particles in the system have a memory of the relative orientation of spin vector with respect to orbital angular momentum. A system retaining this memory breaks a symmetry and in this case the degeneracy of energy levels is lifted. Once the spin-orbit term was introduced, there arose a natural question about its implication for level structures, transition rates, probabilities, etc. and, how to verify these predictions. This is how physics functions, in iterative ways. A theory or model makes a prediction that needs testing, or an observation is made that needs to be clarified mathematically. The experimentalists set themselves the task of designing, performing and interpreting measurements, which will

[9] In nature, a chemical element may consist of atomic nuclei with fixed positive electric charge but different masses. The atom corresponding to each atomic mass is called an isotope of the element. For example, hydrogen occurs in nature with masses A=1, 2 and 3. It is therefore said to have three isotopes. Ordinary water contains hydrogen of A=1, while heavy water has deuterium (heavy hydrogen of A=2). The A=3 hydrogen isotope is a toxic, radioactive substance.

[10] In the mid-1920s, it was firmly established that one must introduce a quantum-mechanic specific internal angular momentum to particles, called spin. Once we specify a particle, its spin (σ) is also fixed and the projection (m), has a finite set of values with $m = -\sigma, -\sigma+1, \ldots, \sigma-1, \sigma$. While the original atomic and molecular structure models were unaware of spin and limited calculations to the orbital angular momentum particle motions, they needed to introduce this extra degree of freedom to describe the systematics of atomic and molecular spectra. In the course of development, it was also recognized that the product of spin and orbital angular momentum terms known as the spin-orbit term was important. This is one example where vectors in two different spaces (one an intrinsic property and the other a kinematical property) are coupled to construct a vector which spans both spaces.

either confirm or disprove the theory. Theoreticians will be engaged in presenting model frameworks to describe the observations and make further predictions. Thus, they feed into each other's activities.

While preserved symmetries are of interest in their own ways, broken symmetries are even more fascinating for physicists. The preceding examples of parity breakdown and the attempt to account for magic numbers in nuclear physics are cases in point. The first involves identifying the direction of motion of a particle as breaking symmetry, while the second is the alignment of intrinsic spin and orbital angular momentum (the angular momentum with respect to a center of force). Because a symmetry is a measure of indistinguishability, broken symmetries or asymmetries are identifying marks[11]. In the case of parity breakdown, discussed above, the intrinsic properties of some particles are not decoupled with their spatial orientation or the direction of their motion. Physicists are quite ambiguous if this is characteristic of some particles or if it is due to the nature of weak interaction which mediates the phenomena. Nevertheless, these asymmetries leave distinguishable marks on the particles and phenomena in the laboratory, while the preserved symmetries render them indistinguishable.

3. Nöther's theorem and consequences

A system under investigation, be it physical or otherwise, can be described by an equation of state. How does one know by looking at the system's equation of state whether or not a symmetry is broken, and how does one build an equation of state by knowing the broken and preserved symmetries?

Emmy Nöther [13] derived a simple, elegant theorem addressing this point. In physics, we write a Hamiltonian operator, H, which represents the total energy content of the system. We also define an operator, A, and a corresponding observable, say, a. To test whether the observable a is a conserved quantity for the system, we only need to verify,

---

[11] It is of interest to note some similarities of these observations with other disciplines. In psychology, symmetry is seen as the absence of process memory, while asymmetry is the memory that processes leave on objects [12]. Compare it with our conclusions in physics. Very striking parallelism exists. Also, in social exchanges, symmetry is defined as the equivalence of return between give and take. Clearly, the two parties are not distinguishable when the "give" and "take" are equal. Equivalently, symmetry corresponds to "lack of knowledge" while the asymmetry is gain of knowledge.
It is amusing to note that the pre-19th century architecture was mostly based on bilateral symmetry as a guiding principle with the assumption that God's image of man is such a symmetry. Needless to say, the human anatomy with the heart on the left and the liver on the right is a manifestation of broken symmetry.

i) $HA = AH$ and
ii) $\partial A/\partial t = 0$

If A spans the space of H, and if the two conditions are satisfied [i.e. i) H and A commute and ii) A does not have an explicit time dependence], the observable corresponding to A is conserved or it is a constant. Indeed, this reasoning can be extended to more than one observable. If there are two operators A and B, each of which satisfy the conditions that $HA=AH$ and $HB=BH$, then the observables corresponding to both A and B will be simultaneously conserved.

The above conditions are necessary but not sufficient for the existence of simultaneously conserved observables. There is a spectacular example in quantum mechanics where this feature poses a seemingly perplexing puzzle.

3.1 Angular Momentum Conservation

Spatial isotropy imposes rotational invariance and as a result the angular momentum of a system is conserved. Following the general convention, we define J as the operator of angular momentum, and J_x, J_y, and J_z as its three Cartesian components. We can construct Hamiltonians such that J^2 and the three angular momentum components each separately commute with the Hamiltonian. One might then anticipate that the total angular momentum and the three projections corresponding to the three components of the angular momentum operator will be conserved. However, it turns out that the angular momentum and *at most one* of its three components can be *simultaneously* conserved. This property is explained when one recognizes that the products of the pairs of angular momentum component operators do not commute. It is interesting to note that, contrary to the common classroom discussions, this result is not a quantum mechanical artifact but is also manifested in classical physics reasoning. One reconciles this apparent dilemma with the understanding that at most one component can serve the purpose of a measurement operator yielding an observable quantity. The other two components will then be transformation operators, rotating the physical state to a different orientation [14], and affect the magnitudes of the observable corresponding to the measurement operator[12].

---

[12] It is tempting to bring in an analogy from psychology or sociology. Let us define a prosperity operator as a vector operator P, with three components P_i, P_m, and P_p corresponding to intellectual, mental and physical prosperity, respectively. From the quantum mechanical analogy, we can write $P^2 = P_i^2 + P_m^2 + P_p^2$. Clearly, the components are not commutative, since changing one of them may affect the other two. For example, making a society more intellectual for a fixed P might make it richer or poorer in material or physical prosperity. Seeking an eigenstate of the P^2 and P_i operators, we can define the operators which increase or decrease the P_i component by one unit as

Thus, for observables corresponding to a set of operators to be conserved, each operator must commute with the Hamiltonian and the product of each pair of operators must also be commutative. Still this statement constitutes a necessary condition but not a sufficient one. A trivial example for this condition to be insufficient would be the situation where we define a Hamiltonian that does not span the coordinate space of the operator corresponding to the observable of interest.

4. Algebraic Symmetries

We have already introduced an algebraic symmetry, which defined isospin as a hypothetical observable. Along similar lines, particle physicists define several other symmetries which depict the system in a quantitative way and introduce conservation principles. Noteworthy among these are the baryon and lepton[13] number conservations. While in recent decades a few other number counting schemes have been introduced, we will mention the strangeness number. This was introduced in the late 1940s in reference to the discovery of the parity-violating mesons mentioned above. Baryon number, strangeness and isospin together allowed physicists to categorize the particle families as baryon octets, decuplets, meson nonets, etc. The most impressive result of this exercise was to enable the prediction of a very exotic Omega baryon and articulate its properties, production and decay mechanisms simply from these symmetry arguments [15].

The most expensive, time- and manpower-consuming task that the particle physics community has ever undertaken is the Higgs Boson search [16]. This search arose because of the postulate that a spontaneous local symmetry breaking can account for origin of mass. As we write this article, the researchers at the Large Hadron Collider (LHC) CERN, Europe laboratory are gearing up for these experiments[17]. Regardless of the final physics result, it is truly remarkable that particle physics communities from several countries around the world have joined forces investing their resources to settle this question.

$P_i^{\pm} = P_m \pm iP_p$ or $P_m = (P^+ + P^-)/2$ and $P_i = (P^+ - P^-)/2i$. In this construction, P_m and P_p do not have well defined, unique values. If one devises a system where a component is changed without affecting the other components, then all the three operators are commutative, P is no longer constant, and this model does not apply.

[13] Baryons participate in strong interactions while leptons do not. Strong interactions are responsible for the binding of atomic nuclei and they act over very short distances.

5. General Remarks

While physics can claim several stellar achievements in furthering knowledge of the physical universe, it is inherently incomplete and fails to reconcile several internal contradictions. Here, we note a few issues in no particular order:

- The overarching principle of microscopic physics based on fundamental building blocks relies on the reversibility of time, while the macroscopic world as we know it acknowledges the forward arrow of time. Andrei Sakarov [18], in his description of origins of the Universe, required that this reversibility was absent at the beginning.
- The microscopic world as described by quantum mechanics is non-deterministic and acausal while determinism and causality are the guiding themes of macroscopic phenomena. In classical physics, the chaoticity of a system arises out of our ignorance, although the dynamics are described by the deterministic formulations. On the other hand, in quantum mechanical or field theoretical descriptions, the chaos is inherent in the microscopic system itself: the *probabilities* of interactions and outcomes govern the phenomena.
- It is somewhat puzzling to examine the implications of entropy and symmetry. As is well known, a system tends toward a state of maximum entropy, that is, a state of maximum disorder. If we consider a system of identical particles, the spins are randomly oriented in all directions and we say the system is isotropic with no preferred axis of reference. Thus, while an external observer sees an isotropic system of perfect symmetry, it is inherently one of maximum disorder.
- Physicists thrive on simplistic phenomenological models and concepts to provide operational prescriptions of dynamics and kinematics. For example, models which imagine spring-like forces and billiard ball-like collisions between subatomic particles may be appealing, but are not realistic pictures of the microscopic world. As physicists address various phenomena at different levels of microscopic or macroscopic manifestations, they resort to varying levels of approximation to keep the problem tractable with the corresponding mathematical manipulations and model interpretations of the entities and interactions. It is remarkable that they can get away with impunity as they deal with inanimate universe.

The cognate disciplines, when they use physics metaphors or employ these techniques in their fields of study, are best advised to bear in mind the inadequacies of these concepts. They must be aware of the limitations, internal contradictions and inherent ambiguities in physics language and concepts. A particle physicist can get away with some assumptions of particles and a few variables as parameters to describe the intrinsic properties and external surroundings of them. Emotions, feelings and changing values do not play a role in the inanimate world. However, a sociologist or a psychologist cannot afford to ignore the individuals and treat people as a statistic for all purposes. It will be even more worrisome if a health scientist or an economist employs these techniques without carefully scrutinizing the implications of such formulations to draw conclusions about societal health or economic well-being.

Lastly, it must be remarked that the Principle of Reductionism, which has been the guiding principle of physics for several centuries, may soon lose its prestigious position as physicists look for "God particle" and local-symmetry breakings as if to rescue this principle. The imminent discoveries might, though unintended, question the validity and practicality of the principle of discernable elementary building blocks of matter and fundamental interactions.

I benefited a lot from Mr. Chris Polachic's careful reading of the manuscript and very thoughtful comments and questions. I am very thankful to him. The contents, for their merit or folly, are my own responsibility.

References:

1. K. Gödel, *On Formally Undecidable Propositions Of Principia Mathematica and Related Systems,* translated by B.Meltzer, Pub: Oliver&Boyd, (1962).
2. See for example, Ray Skinner, *Relativity*, Pub: Blaisdell Publishing Company, (1969).
3. See for example, Particle Data Group, "Review of Particle Physics", *Physics Letters* **B667**, 1(2008).
4. S. S. Hanna in *Giant Multipole Resonances*, topical conferences, Oak Ridge, Tennessee Ed. Fred Bertrand, Pub: Harwood Academic, (1979).
5. See for example J. P. Elliott and P.G. Dawber, *Symmetry in Physics*, Pub: Macmillan Press Ltd, (1979).
6. Hermann Weyl, *Symmetry*, 1952 edition, Pub: Princeton University Press
7. Johannes Siegrist, *European Review*, **13,** 145 (2005).
8. T.D. Lee and C.N. Yang, *Physical Review* **104**,1956 (1956).
9. C. S. Wu, E. Ambler, R.W. Hayward, D. D. Hopper, and R.P. Hudson, *Physical Review* **105,** 1413, (1957).
10. See for example, M.E. Rose, *Elementary Theory of Angular Momentum*, (1957) Pub: John- Wiley; D.M. Brink and G.R. Satchler, *Angular Momentum*, Pub: Oxford University Press, (1993).
11. M. G. Mayer and J. H.D. Jensen, *Elementary Theory of Nuclear Shell Structure*, Pub: John Wiley and Sons (1955).
12. M. Leyton, *Symmetry, Causality and Mind*, Pub: The MIT press, (1992).
13. J.J. Sakurai, *Modern Quantum Mechanics*, Pub: Addison-Wesley, (1994).

14. H. Goldstein, *Classical Mechanics*, Pub: Addison-Wesley, (1971).
15. V.E.Barnes et al, *Physical Review Letters*, **12**, 204, (1964).
16. see for example, D. Lincoln, *Understanding the Universe: from quarks to the cosmos*, Pub: World Scientific, (2004).
17. Visit, *http://lhc.cern.ch/lhc*
18. A. Sakharov, *Journal of Experimental and Theoretical Physics*, **5**, 32 (1967).

INTEGRATING DIFFERENT MODES OF INQUIRY FOR PRE-SERVICE TEACHERS

TANYA AUGSBURG* and J. ACACIO DE BARROS
Liberal Studies Program, San Francisco State University, 1600 Holloway Ave., San Francisco, CA 94132-1722
**E-mail: tanya@sfsu.edu*

At San Francisco State, about half of the students enrolled in Liberal Studies (LS) plan to be elementary school teachers. The LS faculty consists of an anthropologist, a physicist, a geographer, a performance scholar, and a writing composition specialist. Between 2007-2009 the five faculty members developed a curriculum that integrated our areas of expertise and attempted to make students aware of different disciplinary approaches. In this paper, the authors will discuss the changes of curriculum at the Liberal Studies program at San Francisco State University, how we integrated the disciplines, and how we addressed the needs of our students planning to be elementary school teachers.

1. Introduction

In a 1989 paper, Jacobs introduces us to a 4th grader, Mike, who thought that mathematics is "something you do in the mornings." Though sadly funny, Mike's view is not surprising. It contains a certain element of truth. In school, classes are divided in single-subject time slots. Math is taught separately from Science; Science disconnected from History; History unrelated to English; and so on. This shouldn't be the case. The elementary school classroom is ideal for interdisciplinary approaches. A glimpse of the California Frameworks (Ong and Lundin, 2002) reveals that a typical implementation of the science curriculum can be integrated into other subjects. However, curricular segmentation, with its consequences to students' fragmented epistemological views, is the norm, not the exception. This problem only increases at advanced levels, where specialists teach single-subject courses. It is not hard to see why Mike sees math the way he does.

By emphasizing standardized tests on Math and English, the No Child Left Behind Act (NCLB) may be exacerbating curricular segmentation. It sends the message that what matters is context-free Math and English. NCLB is arguably the reason why other subjects are being pushed aside, with 80% of the San Francisco Bay Area elementary school teachers spending less than one hour per week in science (Dorph et al., 2007). This is striking, given San Francisco Bay Area's contributions to high-tech and science.

If elementary school teachers should draw connections between disciplines, then they need to be trained to be interdisciplinary. Teachers with this training will be able to create integrative lesson plans. For example, a teacher discussing the industrial revolution may look at it from historic, scientific-technological, or economic points of view (just to mention a few). But a teacher trained in thinking about problems interdisciplinarily may be able to integrate those aspects in his/her lesson plan, enabling a more comprehensive view of the industrial revolution. Thus, we believe that teacher preparation should include interdisciplinary preparation (Koirala & Bowman, 2003; Jacobs, 1989).

In this paper, we describe the interdisciplinary collaboration of the Liberal Studies (LS) faculty at San Francisco State University (SFSU). We focus on the process of changing the LS program to increase its interdisciplinarity, and in particular on the issues of pre-service teacher preparation. We start by situating our program in the context of teacher preparation in the State of California, and then describe the process of transition going from a multidisciplinary program to an interdisciplinary program with a teacher track (for definitions of 'interdisciplinary' and 'multidisciplinary', see Augsburg, 2006 or Repko, 2008). We will recount how a faculty from diverse disciplines and backgrounds worked together to create the curricular changes. Finally, we will discuss some of the challenges and successes we had working as part of a very heterogeneous and diverse faculty, including some compromises we had to make as a group. As we believe, our experiences demonstrate interdisciplinary teamwork within an academic setting that illustrates that disciplinary silos can be bridged for the betterment of addressing complex societal problems, such as improving teaching preparation.

2. Teacher Preparation

In California, teacher preparation begins at the undergraduate level and is completed at the postgraduate level. After graduating from college, prospective elementary school teachers are required to obtain a multiple subject teaching credential. The Teacher Preparation and Licensing Law of 1970 (Ryan Act) emphasized educational breadth in a range of courses that were distributed among four areas:

1. English, including Grammar, Literature, Composition, and Speech;

2. Mathematics and the Physical or Life Sciences;
3. Social Sciences, other than education and education methodology;
4. Humanities and the Fine Arts, including foreign languages. (Lewenstein, 1972).

Such educational breadth was designed to reflect the needs of elementary school teachers, as they are required to be proficient in a range of subject matter. This is because the multiple subject credential allows its holder to teach multiple subjects from grades K-8. However, starting with the sixth grade (the gray area of middle school), teaching positions exist that require single subject credentials.

During their credential candidacy, a pre-service teacher's proficiency in the above areas is currently established by a series of standardized tests, the California Subject Examinations for Teachers (CSET). A credential candidate needs to score satisfactorily in the corresponding CSET exams to obtain his/her credentials. For Multiple Subject credentials (K-8), pre-service teachers take the Multiple Subject part of CSET, corresponding to three subtests: (I) Reading, Language, and Literature; History and Social Sciences; (II) Science; Mathematics; (III) Physical Education; Human Development; Visual and Performing Arts. Usually, credential candidates take their exams during their last year in college or shortly after they graduate.

In the credential program, prospective teachers learn about teaching methods, but not about the content subject matters. Content is presumed to be known. Usually pre-service teachers take separate classes on how to teach mathematics, reading, history, arts, and sciences. Therefore, if a candidate didn't learn the necessary content matter in his/her undergraduate program, very little is done to fulfill this gap. The underlying assumption is that if a candidate scored satisfactorily on the CSET, the candidate has enough content knowledge to teach in K-8. However, what constitutes enough content knowledge, and in particular enough depth of content knowledge, is a matter of ample debate. For example, in mathematics Liping Ma observed that the US teachers lagged behind their Chinese counterparts, despite the fact that on average the US teachers had more formal math education, often including advanced math courses. What the US teachers lacked was in-depth knowledge of elementary mathematical concepts, such as area or the different meanings of division. This more advanced knowledge without enough depth into the foundational aspects of the subject was responsible for most of the difficulties that US teachers had in the classroom (Liping Ma, 1999).

Adding to the problem of lack of in-depth knowledge, elementary school teachers have a hard time integrating different subject matters. The elementary school classroom is integrative in nature, and good lesson plans evolve around a theme where different subjects are used to shed light on. But if integration among the different subjects does not exist, students will see the differing approaches to a problem as disconnected and unrelated. If teachers compartmentalize their lessons without using a context that allows for integration of knowledge, they run the risk of fostering among their students the concept that knowledge itself is naturally divided into compartments. As an example of the consequences of this view of knowledge, we repeat Jacobs's remark about the fourth grader Mike, who said that math "is something we do in the mornings" (Jacobs, 1989).

When we think about elementary school teacher preparation, in addition to covering a plethora of subject areas (hopefully in-depth), we should also think about how to integrate those subject areas, training future teachers to see knowledge as interdisciplinary and not as discipline specific. In fact, teacher preparation programs that foster interdisciplinary thinking are shown to have positive effects on teachers graduating from such programs (Jacobs, 1989). And as with the subjects themselves, the integration of different disciplines should be transparent and effortless from the part of the future teacher. We may argue that integrative interdisciplinary skills are like a window: the best windows allow you to see the outside, without distracting you, integrating the internal space with the external. In the same way, the ability to integrate knowledge from different disciplines should enable you to see the problem from different angles, and not distract you with the attempt of integration itself. Therefore, we advocate teaching future teachers how to build windows of knowledge by not only teaching them about the literature of interdisciplinarity—its terminology and methodology—but also by providing opportunities for future teachers to do integration themselves.

3. Our Background

As part of the California State University (CSU) system, San Francisco State University created its Liberal Studies (LS) undergraduate major in 1972 as a multi-disciplinary general liberal arts degree that organized itself along the four teacher preparation areas discussed above. The only noticeable difference was that foreign languages were grouped with English and Communication, instead of Humanities and the Creative Arts. According to Goldsmith (2009), Liberal Studies at San Francisco State has been an unusual program within the CSU system, as it has "always attracted an almost equal number of students who choose it for the pre-professional training or for an interest in the multidisciplinary curriculum and even some who choose it for both reasons." On most other CSU campuses, well over 90% of the students majoring in Liberal Studies intend to become teachers (Goldsmith, 2009).

Throughout its history at San Francisco State, Liberal Studies has always been a popular major. At its peak in 1999, its enrollment exceeded 1,100 students (Goldsmith, 2009). Despite its popularity, the Liberal Studies program had no faculty of its own, and it was not housed in a department or college. Instead, it was located in the Division of Undergraduate Studies. Faculty from all over the university advised Liberal Studies majors, and a program coordinator coordinated it. A Liberal Studies Council created under the umbrella of the Academic Senate and composed of elected faculty from all colleges on campus (except the College of Business) oversaw curricular decisions.

Numerous educational and university developments arose that required curricular and program changes. San Francisco State did not escape from nation-wide trends in assessment, but without its own faculty the program was unable to do any. The explosion of interest in interdisciplinary research and education trickled down to elementary and secondary education, where integrated approaches to subject matter ranging from important figures as Abraham Lincoln and Martin Luther King, Jr. to California history became increasingly deployed. At San Francisco State, a longstanding interdisciplinary social science program was discontinued, jeopardizing the two interdisciplinary social science courses required for the Liberal Studies major. A change in university-wide upper-level writing requirements to include a writing course in every major, however, necessitated the hiring of faculty, and in 2007, much to the University's amazement, Provost John Gemello authorized the hire of five tenure-track faculty to teach new courses, one of which would satisfy the writing requirement. The faculty were hired with the expectation that they would revamp the existing curriculum by making a multidisciplinary degree more interdisciplinary, as well as more coherent in the advising process. There was hope that the new faculty would oversee each of the four areas. Moreover, the faculty were expected to do program assessment, something that had not occurred in its over thirty-year existence as a degree program.

Five faculty were hired: Matthew Luskey for Area I (English and Communication), Acacio de Barros for Area II (Math and Sciences), Tendai Chitewere and Logan Hennessy for Area III (Social Sciences), and Tanya Augsburg for Area IV (Creative Arts and Humanities). Two faculty were hired in Area III, as those faculty would also be teaching the jeopardized interdisciplinary social sciences courses since there were no longer any tenured or tenure-track faculty teaching them.

Matt Luskey is a composition specialist with his doctorate in English. He brought to the program knowledge and passion about writing. The two social scientists hired have complementary areas of expertise. Logan Hennessy is an environmental geographer who received his doctorate in environmental studies. He teaches SS301, an interdisciplinary social sciences course that introduces students to Economics, Political Science, and Geography. Tendai Chitewere is a cultural anthropologist with a water engineering background who studies consumption in green communities. She teaches SS300, the other required interdisciplinary social sciences courses, which introduce students to cultural anthropology, psychology, and sociology. The authors of this paper are also interdisciplinarians. Tanya Augsburg's undergraduate degree is in dramatic literature, and her doctorate is in the interdisciplinary humanities. She taught writing composition and public speaking at an engineering university. She spent ten years building an interdisciplinary studies program at a large public university that did not have any tenure lines. She also brought to the program experience working with student portfolios. Acacio de Barros' doctorate and undergraduate degrees were in Physics, but his interests have always been interdisciplinary. In his dissertation he integrated concepts in the philosophy and foundations of mathematics and logic to gravitation. After working in the US on the foundations of quantum mechanics, de Barros went back to Brazil, where he accepted a faculty position and later became the Physics credential program director. This contact with future Physics teachers led to his physics education research.

4. Working Together for Changing the Major

In addition to students interested in being schoolteachers, the LS program at San Francisco State has a significant amount of majors seeking an interdisciplinary Liberal Arts education. For that reason, some of the core curriculum courses in the LS program are essentially interdisciplinary, within a major area of knowledge such as the aforementioned SS300 and SS301 courses. However, even though those courses are interdisciplinary, because LS students do not have the strong tradition of a discipline (they are not Physics majors, or Anthropology majors, or English majors), they struggle to see the integration of different disciplines in those courses, simply because they struggle to see the disciplines themselves. For that reason, the LS faculty decided to include two new courses that teach about disciplines, interdisciplinarity, and integration of knowledge.

The first course created is usually taken when students reach their Junior year, and it is called Perspectives in Liberal Studies (LS300, for short). We often refer to LS300 as our gateway course. One of the main goals of this course is to introduce students to the concept of disciplines, and to show how different disciplines approach questions and consider evidence in different ways. But we also discuss how some questions do not have a natural disciplinary domain, and how it is necessary to draw from multiple disciplinary perspectives to have a better understanding of those questions (Augsburg, 2006). The second course is taken at the last semester before graduation, and is called Liberal Studies Senior Seminar (LS690). The main goal of this course is to provide students with the opportunity to apply the knowledge they acquired during their education towards culminating integrative projects, and to look back at their education and assess it critically. In the next section we will focus on the process of creating LS300, as well as the reasoning that led to our main choices. We will not discuss LS 690, as neither of us will be teaching it when it is taught for the first time Fall 2009.

4.1. *The Process of Creating LS 300*

During the summer of 2007 Robert Cherny, the Dean who hired the new LS faculty, introduced all of us to each other over email, encouraging us to share our CVs before we even met. He furthermore told us that we needed to think about changes to the major what would be useful toward assessment. Our first meeting was a social event, rather than strictly work related. We learned about each other in a relaxed atmosphere, with the sharing of food, wine, and conversation.

Once the Fall 2007 semester began the faculty met regularly and started working on the initial goals of creating program learning objectives and a gateway class. It is important to note that we were given time to work together—while we did not have a complete teaching release, we had a reduced teaching load that permitted us to meet weekly to work on program and curriculum development. While discussing program goals and learning objectives, the disciplinary traditions of each faculty became evident. For example, while one faculty member pushed for goals that could be measured, another defended goals that could be more open to interpretation and not necessarily subject to clear performance-based outcomes.

The faculty were making good headway on program objectives and were planning to complete them before creating the class; however, the Academic Senate calendar dictated otherwise, and the faculty succeeded in having their course proposal approved by the designated university committees, the library and the Academic Senate so that we could launch it Fall 2008. Thus we ended up giving up on the goals (temporarily), in order to create two courses: LS300 and LS690. The idea behind having two courses was that one would be the program entrance point, and the last the exit point, thus allowing for assessment. Moving quickly whilst abandoning the work on program goals was necessary in order to launch LS300 in the Fall 2008 semester. The consequence was that our discussion of program goals and learning objectives was truncated and deferred.

The essential goals of the gateway course are the following. First, it should introduce students to the different modes of inquiry among the disciplines and to foster interdisciplinarity. Second, it should also be a writing intensive course, where students learn writing within the discipline (a tricky definition for Liberal Studies). Finally, it would be desirable for LS300 to facilitate program advising by encouraging students to reflect on their education and goals.

Early in our course development the faculty made two key decisions consensually:

1. To introduce students to the concepts of disciplinarity and interdisciplinarity through a common theme, climate change.
2. To use an anchor text: Al Gore's *An Inconvenient Truth.*

Why did we choose climate change as a topic? First, climate change is an extremely complex problem, with ramifications that extend well beyond climate science. For that reason, no single disciplinary approach can address but the simplest questions in climate change. For example, to understand the greenhouse effect and the role CO_2 plays required the collaboration and methods of Physicists, Chemists, and Geologists, just to name a few disciplines used, and today's politics and social effects of climate change are undeniably beyond the scope of any scientific approach. Therefore, because of its complexity, the subject certainly qualifies as an interdisciplinary, if not transdisciplinary, problem.

Another reason for choosing climate change is its increasing importance, and corresponding presence, in today's world. As a consequence, it is a topic that is increasingly being taught in schools. In fact, in California there was a recent attempt by the legislative body to mandate the teaching of climate change (in combination with environmental education) throughout the K-12 curriculum. The State Senate approved the bill in January 30th, 2008, but on July 26th Governor Arnold Schwartzenegger vetoed it. Nevertheless, the possibility of a future state mandate to teach climate change in schools is not farfetched.

Furthermore, as the possible consequences of climate change mount, citizens will need to make tough but informed political decisions that may have a deleterious impact for future generations. Given the high level of political engagement of San Francisco State students, we thought that climate change would be a topic of general interest among our students, and not only something useful to future teachers. Anecdotally we found that this was indeed the case.

Why did we select *An Inconvenient Truth* as the anchor text? We realize that it is neither academic nor scholarly. This text, accompanied by the award-winning film, serves as an accessible and popular introduction to climate change. It has been already adopted for use in schools worldwide. Because it evolved from years of Al Gore trying to increase popular awareness of the urgency of dealing with the climate consequences of our current behavior, it is a very accessible way to introduce climate change as a scientific problem, as a social problem, as a political issue, a moral problem, and a theme for the creative arts. This accessibility allows the use of the text as a starting point to address how particular modes of inquiry. *An Inconvenient Truth* opens up the complexity of climate change and shows that it needs to be approached from several different angles (or disciplinary perspectives).

Because Gore writes in the book about the interconnections between his political career and his personal life, and because the book had an important effect in the collective consciousness of climate change, we can examine this text from multiple perspectives. For example, *An Inconvenient Truth* can be seen as an autobiographical work. Gore uses his personal story to illustrate issues related to climate change. And he makes it clear that throughout his life, climate change was one of the main threads tying all his work together. But you can also see it as a political work, where his intent is to shape policies and to establish interests and relations. Finally, you can see the text from the point of view of a very effective piece of rhetoric, and with its graphic design, photographs, and graphs you can also appreciate it as a fine example of visual communication.

In addition to Gore's book, we used other ancillary texts that went in depth into some of the processes discussed on *An Inconvenient Truth*. For the scientific aspects of it, de Barros suggested Spencer Weart's *The Discovery of Global Warming* (Weart, 2003). Instead of discussing the science of climate change in detail, something well beyond the scope of our course, Weart's book allows us to examine the history of the science behind global warming. By looking into this history, we could focus on providing our students with a very good look into the process that is science, instead of getting lost into the endless (and very specialized) details of scientific theories. In other words, we could concentrate on how scientists approach problems, weigh evidence, and answer questions.

Furthermore, we found it intriguing that Weart highlights the interdisciplinary nature of the discovery of climate change. For the humanities component of the course, Augsburg found Intuit personal narratives that were useful in considering the impact of climate change on difficult cultures in various geographical regions (see International Institute for Sustainable Development, 2000). The faculty later realized that these Intuit personal narratives were better suited for introducing the social sciences than the humanities. The narratives were extremely useful in introducing climate change's impact on various geographical locations and local cultures. Other texts such as those on environmental criticism, and climate change's impact on mining, were shared among the faculty, but not all faculty elected to use them.

The process of choosing the course-reading list was interesting in itself. One of the major challenges we had was to limit the number of texts from each area. For example, at some point one of the faculty suggested to limit to two the number of texts per major areas, i.e., Natural Sciences, Communications and Literature, Creative Arts and Humanities, and Social Sciences. Even though everybody agreed, when it came to their contributions to the list, exceptions were asked and more than two texts were presented. This is not surprising, as each one of us believe that our (disciplinary) approach is important, and therefore deserves additional consideration.

Another challenge was in developing an overall structure for the course that had some invariant characteristics across semesters and sections. The reason for seeking this invariance was grounded on the necessity to create courses that could be used for future program assessments. Because all LS students are required to take LS300, and this is the first interdisciplinary course they take, it would make sense to use it as a baseline for any program assessments.

Despite weekly meetings about developing LS300, consensus could not be reached on the details of the course. The main reason was that each faculty had strong feelings about what was important to teach and how much time should be spent on each component. During those meetings, we often had "spirited discussions" between how to think about the selected complex problem, and how to approach it in the classroom.

Another issue was how much time in LS300 do we devote to the examination of climate change? For example, one of the problems was to balance enough discussions of a discipline with an analysis of the contributions of this discipline to the overall question. Ultimately the faculty agreed that an important amount of discussions should be about the ways different disciplines view climate change. In so doing faculty could present various disciplinary approaches to a complex problem, allowing students to see through the lenses of multiple disciplinary perspectives.

During a faculty retreat late Spring 2008, the faculty met to discuss the remaining details of developing LS300. The initial goal was to prepare our courses together, but we eventually moved from this goal to how to make sure that our assignments and requirements could be compared and used in the future for assessment. The retreat concluded with agreed upon core assignments and an overall structure with which everybody felt comfortable. Our shared assignments, such as an integrative essay and team presentations, were designed to "prompt" students to integrate different modes of inquiry. Only after the Fall 2008 semester, while we were conducting research for American Association of Colleges and Universities (AAC&U) VALUES Initiative in early January 2009, did we discover that each of us had interpreted each of the shared assignments differently.

Once the course was created, it was challenging for all of us to teach it. First of all, none of the faculty were experts on climate change. Not only were we not experts on climate change, we were not experts on our colleagues' areas. For example, de Barros found himself quite comfortable when the topic was on the methods and questions of science, and fairly comfortable with the social sciences, but when the subject moved to eco-criticism, he felt very uneasy. So, we all found ourselves teaching outside one's comfort zone and areas of expertise. Yet all faculty recognized the importance of teaching the topic, hence our commitment to trying.

This commitment at times clouded our overall teaching objectives: the course was not about climate change per se; climate change was used as an example for students to examine how various modes of inquiry address an interdisciplinary complex problem, which was climate change. We realized quickly that our initial design of the course foregrounded the content of climate change, thus overpowering the course objectives in introducing students to the study of interdisciplinarity. In Augsburg's section of the course, students were initially confused: they thought the class was a class about climate change. It took much fancy footwork on the instructors' part to help students understand what we were trying to accomplish: we were teaching students the common elements of every discipline, familiarizing them with concepts, theories, methods, assumptions, and approaches to problems in order for students to understand differences among disciplinary perspectives.

Interestingly, there was a perceived initial challenge we faced teaching the class that turned out to be not the case: what we fondly refer to as "the odd couple" syndrome. What, you may ask, is the odd couple syndrome? Our definition can be found in the answer to the following: Did you know the joke about the physicist and performance artist sharing an office? Well, Acacio and Tanya didn't know it either, but we have been sharing an office since the fall of 2007. To situate a physicist and a humanities scholar/performance artist in the same office initially struck some people as odd. It was as if we were the latest incarnation of the popular Neil Simon play that later became a film and successful 70s American television show. Each of us was asked repeatedly how we were getting along. Each of us found the question odd, since, from the very start, we were able to find commonalities, such as an interest and commitment to education, and we also discovered that we have similar work styles. From both of our experiences in academia we know that similar work styles do not always occur within disciplines. We are both thriving from our workspace situation, as we have learned to appreciate the differences in our epistemologies and learning "styles." Acacio, the physicist, is more analytical and evidence-based, while Tanya, the humanities scholar and occasional performance artist, is more intuitive and interpretive. Instead of conflict, we find complementarity and synergy. Much had to do with the fact that the Liberal Studies faculty invested time and effort from the start to get to know each other and communicate, and we believe that the initial investment has yielded profitable dividends, as evidenced by all of our joint projects completed in the first two years of working together: our curriculum development; shared assignments, which included the design and implementation of a program electronic portfolio; a joint conference panel at a national conference; a national grant, and this conference presentation in which Tanya Augsburg and Acacio de Barros co-wrote this paper.

5. Future Challenges

In 2009 the state of California faced a severe budget crisis. In order to close an approximate $28 billion deficit, approximately half a billion was slashed from the California State University budget. This amount represented approximately 20% of the total CSU budget, a nontrivial amount. To cope with these cuts, 40 thousand less students would be admitted during the 2009-10 and 2010-11 academic years with no Spring 2010 admissions. Tuitions were raised by more than 30% in 2009, and university employees (including faculty) were furloughed twice per month during the 2009-10 academic year. How all this will impact teacher preparation is unknown, but surely it will be detrimental. Given the extent of the cuts at our campus, our ability to work "full steam ahead" has been compromised. After the furloughs were announced in July 2009 we lost one faculty member, Matt Luskey, who accepted a position elsewhere. This has been a blow to the program for a myriad of reasons, including that his departure will hamper the rest of the faculty moving forward on a future joint project concerning *An Inconvenient Truth*.

As mentioned above, the faculty jointly presented "An Introduction to A Convenient Text: Utilizing *An Inconvenient Truth* in an Interdisciplinary Gateway Course at San Francisco State University," a panel at the annual conference of the Association for Integrative Studies in Springfield, Missouri, in October 2008. From our work on this panel, we have discussed the possibility of a book project that would enable us to share our ideas about teaching *An Inconvenient Truth* in both multidisciplinary and interdisciplinary manners. With the scarcity of time and overabundance of work, we have limited time to start research projects of this magnitude when existing projects await completion and publication. The pressure to publish is very real for probationary faculty, which is why Augsburg made the painful decision to return to her areas of expertise rather than to continue focusing on climate change for the near future. In so doing, she hopes to model more effectively interdisciplinary knowledge and practice.

6. Conclusion

In this paper, we argued for the introduction of interdisciplinarity into the undergraduate teacher preparation curriculum in California. We have also recounted how we transformed a longstanding multidisciplinary undergraduate degree into a more interdisciplinary one by introducing two new core classes into the major. We have briefly mentioned the addition of a teacher preparation track into San Francisco State's Liberal Studies curriculum. And we have discussed the challenges of faculty with different academic disciplinary backgrounds working together in the design and implementation of a new course intended to introduce students to disciplinarity and interdisciplinarity. In so doing, we have provided a partial case study for undergraduate interdisciplinary studies curriculum development and program building.

Since the teacher preparation track has been implemented Spring 2009 semester, students can take courses designed specifically for future teachers, such as Oral interpretation of Children's Literature. de Barros's Physical Science for Elementary School Teachers, a course he has designed and taught, has been very successful. As long as these classes can continue to be offered, we have confidence that teacher preparation at San Francisco State has been improved. We have anecdotal evidence that this confidence seems to be shared by our students, who seem to now think that our major is a consistent whole, and not a collection of unrelated courses thanks to the addition of the two new Liberal Studies courses. In our estimation students are now better prepared to become teachers as the result of the Liberal Studies faculty interdisciplinary collaborations that began in 2007, and are ongoing. We plan to do further research to quantify and validate our initial preliminary findings.

References

1. Tanya Augsburg, *Becoming interdisciplinary: An introduction to interdisciplinary studies*, 2nd Edition, Kendall/Hunt Publishing Co, Dubuque, IA, 2006.

2. Tanya Augsburg and Stuart Henry (eds.), *The Politics of Interdisciplinary Studies: Essays on Transformations in American Undergraduate Programs*, McFarland & Co., Inc, Jefferson, N.C. and London, 2009.

3. Dorph, R., Goldstein, D., Lee, S., Lepori, K., Schneider, S., Venkatesan, S. (2007). The status of science education in the Bay Area: Research brief. Lawrence Hall of Science, University of California, Berkeley; California.

4. Helen Goldsmith, "Interdisciplinarity and Teacher Preparation at San Francisco State University," in Tanya Augsburg and Stuart Henry (eds.), *The Politics of Interdisciplinary Studies: Essays on Transformations in American Undergraduate Programs*, McFarland & Co., Inc, Jefferson, N.C. and London, 2009, 133-143.

5. Heidi Heyes Jacobs, *Interdisciplinary Curriculum: Design and Implementation*, Association for Supervision and Curriculum Development, Alexandria, VA, 1989

6. Hari P. Koirala and Jacqueline K. Bowman, “Preparing Middle Level Preservice Teachers to Integrate Mathematics and Sciences: Problems and possibilities,” School Science and Mathematics 103(3), 145-154 (2003).

7. International Institute for International Development. Intuit Observations On Climate Change. http://www.iisd.org/casl/projects/inuitobs.htm (accessed September 17, 2009).

8. Lewenstein, Morris. California State University/San Francisco, The School of Behavioral and Social Sciences and the Preparation of Elementary School Teachers. A Working Paper. 15 September, 1972.

9. Liping Ma, *Knowing and Teaching Elementary Mathematics: Teacher's Understanding of Fundamental Mathematics in China and the United States*, Lawrence Elrbaum Associates, New Jersey, 1999.

10. Alan Repko, *Interdisciplinary Research: Process and theory, Sage Publications*, Thousand Oaks, CA, 2008.

11. Spencer R. Weart, *The Discovery of Global Warming*, Harvard University Press, Cambridge, Massachusetts, 2003.

12. Faye Ong and Janet Lundin, Science Framework for California Public Schools: Kindergarten Through Grade Twelve, California: Bureau of Publications, sales unit, California Department of Education, Sacramento, CA, 2002.

ON THE MORPHOLOGY OF UNCERTAINTY IN HUMAN PERCEPTION AND COGNITION

MADAN M. GUPTA
Intelligent Systems Research Laboratory, College of Engineering, University of Saskatchewan, Saskatoon, Saskatchewan S7N 5A9, Canada
E-mail: Madan.Gupta@usask.ca

ASHU M. G. SOLO
Maverick Technologies America Inc., Suite 808, 1220 North Market Street, Wilmington, Delaware 19801, U.S.A.
E-mail: amgsolo@mavericktechnologies.us

Human cognitive and perception processes have a great tolerance for imprecision or uncertainty. For this reason, the notions of perception and cognition have great importance in solving many decision making problems in engineering, medicine, science, and social science as there are innumerable uncertainties in real-world phenomena. These uncertainties can be broadly classified as either uncertainties arising from the random behavior of physical processes or uncertainties arising from human perception and cognition processes. Statistical theory can be used to model the former, but lacks the sophistication to process the latter. The theory of fuzzy logic has proven to be very effective in processing the latter. The methodology of computing with words and the computational theory of perceptions are branches of fuzzy logic that deal with the manipulation of words that act as labels for perceptions expressed in natural language propositions. New computing methods based on fuzzy logic can lead to greater adaptability, tractability, robustness, a lower cost solution, and better rapport with reality in the development of intelligent systems.

1. Introduction

For a long time, engineers and scientists have learned from nature and tried to mimic some of the capabilities observed in humans and animals in electrical and mechanical machines. The Wright brothers started their work on the first airplane by studying the flight of birds. Most scientists of the time thought that it was the flapping of wings that was the principle component of flying. However, the Wright brothers realized that wings were required to increase the buoyancy in air.

In biomedical engineering, the principles of natural science and engineering are applied to the benefit of the health sciences. The opposite approach, *reverse biological engineering,* is used to apply biological principles to the solution of engineering and scientific problems. In particular, engineers and scientists use this reverse engineering approach on humans and animals in developing intelligent systems.

The principle attributes of a human being can be classified in three categories (3 Hs): hands, head, and heart. The hands category refers to the physical attributes of humans. These physical attributes have been somewhat mimicked and somewhat improved on to surpass the restrictive physical limitations of humans through such mighty machines as the tractor, assembly line, and aircraft. The head category refers to the perception and cognition abilities of the brain. The restrictive reasoning limitations of humans have been supplemented through the ongoing development of microprocessors. However, the challenge of creating an intelligent system is still in its incipient stages. Finally, the heart category refers to emotions. Emotions have not yet been mimicked in machines. It is debatable whether or not they can be replicated and whether or not replicating them is even beneficial if they could be replicated.

One of the most exciting engineering endeavors involves the effort to mimic human intelligence. Intelligence implies the ability to comprehend, reason, memorize, learn, adapt, and create. It is often said that everybody makes mistakes, but an intelligent person learns from his mistakes and avoids repeating them. This fact of life emphasizes the importance of comprehension, reasoning, learning, and the ability to improve one's performance autonomously in the definition of intelligence.

There are essentially two computational systems: carbon-based organic brains, which have existed in humans and animals since their inception, and silicon-based computers, which have rapidly evolved over the latter half of the twentieth century and beyond. Technological advances in recent decades have made it possible to develop computers that are extremely fast and efficient for numerical computations. However, these computers lack the abilities of humans and animals in processing cognitive information acquired by natural sensors. For example, the human brain routinely performs tasks like recognizing a face in an unfamiliar crowd in 100-200 ms whereas a computer can take days to perform a task of lesser complexity. While the information perceived through natural sensors in humans is not numerical, the brain can process such cognitive information efficiently and cause the human to act on it accordingly. Modern day computers fail miserably in processing such cognitive information.

This leads engineers to wonder if some of the functions and attributes of the human sensory system, cognitive processor, and motor neurons can be emulated in an intelligent system. For such an emulation process, it is necessary to understand the biological and physiological functions of the brain. Hardware can be developed to model aspects of neurons, the principle element of the brain. Similarly, new theories and methodologies can be developed to model the human thinking process.

Many advances have been made in mimicking human cognitive abilities. These advances were mostly inspired by certain biological aspects of the human brain. One of the intriguing aspects of human perception and cognition is its tolerance for imprecision and uncertainty [1-7], which characterize most real-world phenomena.

2. Certainty and Precision

The excess of precision and certainty in engineering and scientific research and development is often providing unrealizable solutions. Certainty and precision have much too often become an absolute standard in design, decision making, and control problems. One of the fundamental aims in science and engineering has been to move from perceptions to measurements in observations, analysis, and decision making.

Through the methodology of precise measurements, engineers and scientists have had many remarkable accomplishments. These include putting people on the moon and returning them safely to Earth, sending spacecraft to the far reaches of the solar system, sending rovers to explore the surface of Mars, exploring the oceans depths, designing computers that can perform billions of computations per second, developing the nuclear bomb, mapping the human genome, and constructing a scanning tunneling microscope that can move individual atoms. However, the path of precision, as manifested in the theories of determinism and stochasticism, has often caused engineers to be ineffectual and powerless as well as lose scientific creativity.

3. Uncertainty and Imprecision in Perception and Cognition

The attribute of certainty or precision does not exist in human perception and cognition. Alongside many startling achievements using the methodology of precise measurements, there have been many abysmal failures that include modeling the behavior of economic, political, social, physical, and biological systems. Engineers have been unable to develop technology that can decipher sloppy handwriting, recognize oral speech as well as a human can, translate between languages as well as a human interpreter can, drive a car in heavy traffic, walk with the agility of a human or animal, replace the combat infantry soldier, determine the veracity of a statement by a human subject with an acceptable degree of accuracy, replace judges and juries, summarize a complicated document, and explain poetry or song lyrics. Underlying these failures is the inability to manipulate imprecise perceptions instead of precise measurements in computing methodologies. Albert Einstein wrote, "So far as the laws of mathematics refer to reality, they are not certain. And so far as they are certain, they do not refer to reality [8]."

There are various types of uncertainty. However, they can be classified under two broad categories: *type one uncertainty* and *type two uncertainty.*

3.1. *Type One Uncertainty*

Type one uncertainty deals with information that arises from the random behavior of physical systems. The pervasiveness of this type of uncertainty can be witnessed in random vibrations of a machine, random fluctuations of electrons in a magnetic field, diffusion of gases in a thermal field, random electrical activities of cardiac muscles, uncertain fluctuations in the weather pattern, and turbulent blood flow through a damaged cardiac valve. Type one uncertainty has been studied for centuries. Complex statistical mathematics has evolved for the characterization and analysis of such random phenomena.

3.2. *Type Two Uncertainty*

Type two uncertainty deals with information or phenomena that arise from human perception and cognitive processes or from cognitive information in general. This subject has received relatively little attention. Perception and cognition through biological sensors (eyes, ears, nose, etc.), perception of pain, and other similar biological events throughout our nervous system and neural networks deserve special attention. The perception and cognition phenomena associated with these processes are characterized by many great uncertainties and cannot be described by conventional statistical theory. A person can linguistically express perceptions experienced through the senses, but these perceptions cannot be described using conventional statistical theory.

Type two uncertainty and the associated cognitive information involve the activities of neural networks. It may seem strange that such familiar notions have recently become the focus of intense research. However, it is the relative unfamiliarity of these notions and their technological applications in intelligent systems that have led engineers and scientists to conduct research in the field of type two uncertainty and its associated cognitive information.

4. Perception and Cognition

The development of the human cognitive process and environmental perception starts taking shape with the development of imaginative ability in a baby's brain. A baby in a cradle can recognize a human face long before it is conscious of any visual physical attributes of humans or the environment. Perception and cognition are not precise or certain. Imprecision and uncertainty play an important role in thinking and reasoning throughout a person's life.

For example, when a person is driving a car, his natural sensors (in this case, the eyes and ears) continuously acquire information and relay it through sensory neurons to the cognitive processor (brain), which perceives environmental conditions and stimuli. The cognitive processor synthesizes the control signals and transmits them through the motor neurons to the person's hands and feet, which activate steering, accelerating, and braking mechanisms in the car.

The human brain can instantly perceive and react to a dangerous environmental stimulus, but it might take several seconds to perform a mathematical calculation. Also, the brain has an incredible ability to manipulate perceptions of distance, size, weight, force, speed, temperature, color, shape, appearance, and innumerable other environmental attributes.

The human brain is an information processing system and neurocontroller that takes full advantage of parallel distributed hardware, handling thousands of actuators (muscle fibers) in parallel, functioning effectively in the presence of noise and nonlinearity, and capable of optimizing over a long-range planning horizon. It possesses robust attributes of distributed sensors and control mechanisms.

The perception process acquires information from the environment through the natural sensory mechanisms of sight, hearing, touch, smell, and taste. This information is integrated and interpreted through the cognitive processor called the brain. Then the cognitive process, by means of a complex network of neurons distributed in the central nervous system, goes through learning, recollecting, and reasoning, resulting in appropriate muscular control.

Conventional mathematical tools, both deterministic and stochastic, are based on absolute measures of information. However, the perception and cognition activity of the brain is based on relative grades of information acquired by the human sensory system. Human sensors acquire information in the form of relative grades rather than absolute numbers. Fresh information acquired from the environment using natural sensors and information (experience, knowledge base) is stored in memory. The action of the cognitive process also appears in the form of relative grades. For example, a person driving an automobile perceives the environment in a relatively graded sense and acts accordingly. Thus, cognitive processes are based on graded information.

Conventional mathematical methods can describe objects such as circles, ellipses and parabolas, but they are unable to describe mountains, lakes, and cloud formations. One can estimate the volume of snow, the heights of mountains, or the frequencies of vibrating musical strings, but conventional mathematics cannot be used to linguistically describe the feelings associated with human perception. The study of such formless uncertainties has led engineers and scientists to contemplate developing morphology for this amorphous uncertainty. In the past, engineers and scientists have disdained this challenge and have chosen to avoid this gap by devising theories unrelated to perception and cognitive processes.

Silicon-based computers can outperform carbon-based computers like the human brain in numerical computations, but perform poorly in tasks such as image processing and voice recognition. It is widely believed that this is partly because of the massive parallelism and tolerance for imprecision and uncertainty embedded in biological neural processes. Therefore, it can be very beneficial to have a greater understanding of the brain in developing intelligent systems.

5. Fuzzy Logic

The theory of fuzzy logic is based on the notion of relative graded membership, as inspired by the processes of human perception and cognition. Lotfi A. Zadeh published his first famous paper on fuzzy sets [9] in 1965. Since then, fuzzy logic [10-13] has proven to be a very promising tool for dealing with type two uncertainty. Stochastic theory is only effective in dealing with type one uncertainty.

Fuzzy logic can deal with information arising from computational perception and cognition that is uncertain, imprecise, vague, partially true, or without sharp boundaries. Fuzzy logic allows for the inclusion of vague human assessments in computing problems. Also, it provides an effective means for conflict resolution of multiple criteria and better assessment of options. New computing methods based on fuzzy logic can be used in the development of intelligent systems for decision making, identification, recognition, optimization, and control.

Measurements are crisp numbers, but perceptions are fuzzy numbers or fuzzy granules, which are groups of objects in which there can be partial membership and the transition of a membership function is gradual, not abrupt.

A granule is a group of objects put together by similarity, proximity, functionality, or indistinguishability.

Fuzzy logic is extremely useful for many people involved in research and development including engineers (electrical, mechanical, civil, chemical, aerospace, agricultural, biomedical, computer, environmental, geological, industrial, mechatronics), mathematicians, computer software developers and researchers, natural scientists (biology, chemistry, earth science, physics), medical researchers, social scientists (economics, management, political science, psychology), public policy analysts, business analysts, jurists, etc. Indeed, the applications of fuzzy logic, once thought to be an obscure mathematical curiosity, can be found in many engineering and scientific works. Fuzzy logic has been used in numerous applications such as facial pattern recognition, washing machines, vacuum cleaners, antiskid braking systems, transmission systems, control of subway systems and unmanned helicopters, knowledge-based systems for multiobjective optimization of power systems, weather forecasting systems, models for new product pricing or project risk assessment, medical diagnosis and treatment plans, and stock trading. This branch of mathematics has instilled new life into scientific disciplines that have been dormant for a long time.

6. Computing with Words and Computational Theory of Perceptions

Computing has traditionally been focused on the manipulation of numbers and symbols. On the other hand, the new methodology called computing with words in its initial stages of development is focused on the manipulation of words and propositions taken from natural language. The computational theory of perceptions is based on the methodology of computing with words [14-15]. Computing with words and the computational theory of perceptions are subsets of fuzzy logic. Words act as the labels for perceptions, and perceptions are expressed in natural language propositions.

The methodology of computing with words is used to translate natural language propositions into a generalized constraint language. In this generalized constraint language, a proposition is expressed as a generalized constraint X is$r$ R where X is the constrained variable, R is the constraining relation, and is$r$ is a variable copula in which r is a variable that defines the way X is constrained by R. The basic types of constraints are possibilistic, probabilistic, veristic, random set, Pawlak set, fuzzy graph, and usuality.

There are four principle reasons for using computing with words in an engineering or scientific problem:

1. The values of variables and/or parameters are not known with sufficient precision or certainty to justify using conventional computing methods.
2. The problem can't be solved using conventional computing methods.
3. A concept is too difficult to be defined using crisp numerical criteria.
4. There is a tolerance for imprecision or uncertainty that can be exploited to achieve greater adaptability, tractability, robustness, a lower cost solution, or better rapport with reality in the development of intelligent systems.

The computing with words methodology is much more effective and versatile for knowledge representation, perception, and cognition than classical logic systems such as predicate logic, propositional logic, modal logic, and computational intelligence techniques for knowledge representation and natural language processing. The computational theory of perceptions and computing with words will be extremely important branches of fuzzy logic in the future.

7. Examples of Uncertainty Management in Perception and Cognition for Linguistic Evaluations

The following examples of uncertainty management in computational perception and cognition for linguistic evaluations illustrate applications of fuzzy sets, linguistic variables, fuzzy rule bases, fuzzy math, the computational theory of perceptions, and computing with words.

A database of students' grades on a test is kept by a professor:

Database = Students [*Name, Grade*]

The students are assigned a linguistic grade for their test performance. These linguistic grades are *fail, poor, satisfactory, good, excellent,* or *outstanding.* A crisp set is created for the *fail* linguistic qualifier and a fuzzy set is created for each of these remaining linguistic qualifiers, as shown in Figure 1. Each of these crisp and fuzzy sets corresponds to a range of numeric grades.

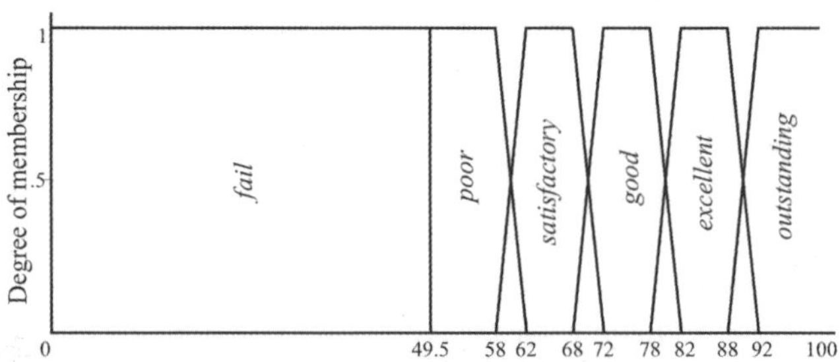

Figure 1. Fuzzy Sets for Test Grades

The *fail* crisp set is represented by a Z-shaped membership function that assigns a membership value of 1 to all grades below 49.5% and a membership value of 0 to all grades above 49.5%. The *outstanding* fuzzy set is represented by an S-shaped membership function that assigns a membership value of 0 to grades below 88%, a membership value increasing from 0 at 88% to 1 at 92%, and a membership value of 1 to grades above 92%.

The *poor, satisfactory, good,* and *excellent* fuzzy sets are represented by trapezoidal membership functions. The *poor* fuzzy set assigns a membership of 1 from 49.5%-58%, membership transitioning from 1 at 58% to 0 at 62%, and 0 membership below 49.5% and above 62%. Membership in the *satisfactory* fuzzy set transitions from 0 at 58% to 1 at 62%, is 1 from 62% through 68%, and transitions from 1 at 68% to 0 at 72%. Membership in the *good* fuzzy set transitions from 0 at 68% to 1 at 72%, is 1 from 72% through 78%, and transitions from 1 at 78% to 0 at 82%. The *excellent* fuzzy set assigns a membership of 1 from 82%-88%, membership transitioning from 0 at 78% to 1 at 82%, and membership transitioning from 1 at 88% to 0 at 92%.

The absolute values of the slopes is the same for the trailing edge of the *poor* fuzzy set, the leading edge of the *outstanding* fuzzy set, and the leading and trailing edges of the *satisfactory, good,* and *excellent* fuzzy sets. Thus, membership is not biased toward one fuzzy set over another and if a student's test grade has membership in two fuzzy sets, the summation of its membership in both of these sets will be 1. For example, for a grade of 70%, there is a membership of 0.5 in the *satisfactory* fuzzy set and 0.5 in the *good* fuzzy set. For a grade of 79%, there is a membership of 0.75 in the *good* fuzzy set and a membership of 0.25 in the *excellent* fuzzy set.

Linguistic qualifiers by themselves can be restrictive in describing fuzzy variables, so linguistic hedges can be used to supplement linguistic qualifiers through numeric transformation. For example, a *very* hedge can square the initial degree of membership μ in a fuzzy set:

$$\mu_{very\ good}(\text{Tom}) = [\mu_{good}(\text{Tom})]^2$$

That is, if a student named Tom has a membership of 0.6 in the *good* fuzzy set, he will have a membership of 0.36 in the *very good* fuzzy set.

Following are some other possible linguistic hedges:

- somewhat
- slightly
- more or less
- quite
- extremely

- above
- below
- fairly
- not

A fuzzy rule base can be developed to decide what to do with individual students based on their grades on a test. A sample fuzzy rule base follows:
If a student receives a grade of *fail* on a test, then phone her parents and give her additional homework.
If a student receives a grade of *poor* on a test, then give her additional homework.
If a student receives a grade of *satisfactory* or *good* or *excellent* on a test, then do nothing.
If a student receives a grade of *outstanding* on a test, then give her a reward.

If a student receives a grade that has membership in two different fuzzy sets, this will potentially assign two different crisp consequents for dealing with the student. For example, if a student receives a grade of 59%, then this grade has membership in two fuzzy sets:

1. *poor* with a degree of membership of 0.75 or
2. *satisfactory* with a degree of membership of 0.25

The maximum (MAX) operator is used to resolve conflicting fuzzy antecedents or fuzzy consequents to fuzzy rules. The MAX operator is used to determine the membership for a fuzzy antecedent resulting from an OR operation between two or more fuzzy antecedents. Similarly, the MAX operator is used to determine the membership for a fuzzy consequent resulting from an OR operation between two or more fuzzy consequents. The MAX operator in fuzzy set theory used for an OR operation is a generalization of the MAX operator in crisp set theory used for an OR operation. In the following equation, μ represents the degree of membership; x represents the domain value; $k1, \ldots, kn$ represent fuzzy sets in which there is membership for a domain value of x; and k represents the fuzzy set in which there is greatest membership for a domain value of x:

$$\mu_k(x) = \text{MAX}\left\{\mu_{k1}(x), \ldots, \mu_{kn}(x)\right\}$$

One fuzzy antecedent or the other can be used for a grade of 59%. Using the MAX operator, the antecedent used in the fuzzy rules above is *poor* with a degree of membership of 0.75. This results in the student getting additional homework.

Linguistic grades possibly modified by linguistic hedges can be defuzzified using the center of gravity defuzzification formula to obtain a single crisp representative value. In the following equation for center of gravity defuzzification, μ is the degree of membership, x is the domain value, A is the membership function possibly modified by linguistic hedges, and x_0 is the single crisp representative value:

$$x_o = \frac{\int_x \mu_{\tilde{A}}(x)x\,dx}{\int_x \mu_{\tilde{A}}(x)\,dx}$$

An instructor can assign linguistic grades to students for a series of tests. Each of these linguistic grades can be defuzzified using the center of gravity defuzzification formula, so a crisp overall numerical average can be calculated. Then this crisp overall grade can be refuzzified for an overall linguistic grade.

A numerical average is not always useful in determining overall performance. For example, there could be two critical exams that test two very different criteria, and performance in each of these exams must be considered both individually and together in making a decision on an individual's advancement. In such a case, a fuzzy rule base can be developed to evaluate an individual's performance on different exams and make a decision on the individual's advancement.

A sample fuzzy rule base is shown below that evaluates the performance of a soldier on a special forces academic exam and a special forces athletic exam, and then determines the degree of desirability of that soldier for special forces training. The same fuzzy sets shown in Figure 1 and described above are used for the antecedents of fuzzy rules in this fuzzy rule base. The fuzzy sets shown in Figure 2 are used for the consequents of fuzzy rules in this fuzzy rule base to describe the degree of desirability of soldiers for special forces training.

The sample fuzzy rule base follows:

If a soldier receives a grade of *fail* or *poor* on the special forces academic exam or a grade of *fail* or *poor* on the special forces athletic exam, then he is *undesirable* for special forces training.

If a soldier receives a grade of *satisfactory* on the special forces academic exam or a grade of *satisfactory* on the special forces athletic exam, then he has *low* desirability for special forces training.

If a soldier receives a grade of *good* or *excellent* on the special forces academic exam and a grade of *good* or *excellent* on the special forces athletic exam, then he has *medium* desirability for special forces training.

If a soldier receives a grade of *good* or *excellent* on the special forces academic exam and a grade of *outstanding* on the special forces athletic exam, then he has *high* desirability for special forces training.
If a soldier receives a grade of *outstanding* on the special forces academic exam and a grade of *good* or *excellent* on the special forces athletic exam, then he has *high* desirability for special forces training.
If a soldier receives a grade of *outstanding* on the special forces academic exam and a grade of *outstanding* on the special forces athletic exam, then accept him for special forces training.

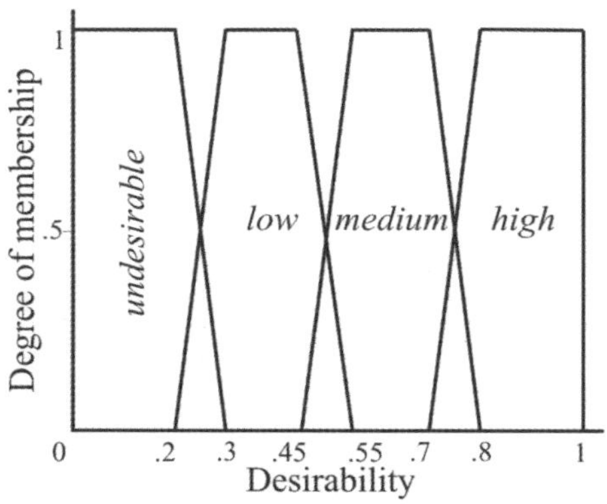

Figure 2. Fuzzy Sets for Desirability

Consider a student who receives the following numerical grades:

1. 96% on the special forces academic exam and
2. 81% on the special forces athletic exam

This maps to the following fuzzy grades:

1. *outstanding* with a membership of 1 on the special forces academic exam and
2. *excellent* with a membership of 0.75 on the special forces athletic exam

A soldier with a grade of *outstanding* on the special forces academic exam and a grade of *excellent* on the special forces athletic exam has *high* desirability for special forces training.

The minimum (MIN) operator is used to determine the membership for the fuzzy consequent resulting from an AND operation between two or more fuzzy antecedents. The MIN operator in fuzzy set theory used for an AND operation is a generalization of the MIN operator in crisp set theory used for an AND operation. In the following equation, μ represents the degree of membership; *k1*, . . ., *kn* represent fuzzy sets in which there is membership; *x1*, . . ., *xn* represent the domain values in those fuzzy sets; *k* represents the fuzzy set in which there is least membership for the domain value in that set; and *x* represents the domain value in the fuzzy set in which there is least membership:

$$\mu_k(x) = \text{MIN}\left\{\mu_{k1}(x1), \ldots, \mu_{kn}(xn)\right\}$$

One antecedent has a degree of membership of 1 and the other antecedent has a degree of membership of 0.75. Using the MIN operator, the consequent has a degree of membership of 0.75. That is, the soldier's desirability for special forces training is high with a degree of membership of 0.75.

The overall desirability for special forces training of soldiers in a platoon can be determined to evaluate the performance of a platoon leader. To calculate this, the desirability for special forces training of each individual soldier in the platoon is defuzzified. Then the crisp overall numerical average can be calculated and refuzzified for a single linguistic grade describing the overall desirability for special forces training of soldiers in a platoon.

The computational theory of perceptions and computing with words can be used to form propositions and queries on linguistic grades. A professor keeps the grades of N students in a database. The name and grade of student *i* where $i = 1 \ldots, \text{N}$ is indicated by $Name_i$ and $Grade_i$, respectively.

Consider the following natural language query: What proportion of students received *excellent* grades on the test? The answer to this query is given by the sigma count:

$$\sum \text{Count} \frac{(excellent.students)}{students} = \frac{1}{\text{N}} \sum_i \mu_{excellent}(Grade_i)$$

Consider another natural language query: What proportion of students received *good* or better grades on the test? The answer to this query is given by the following:

$$\sum \text{Count} \frac{(good.students + excellent.students + outstanding.students)}{students}$$
$$= \frac{1}{\text{N}} \sum_i [\mu_{good}(Grade_i) + \mu_{excellent}(Grade_i) + \mu_{outstanding}(Grade_i)]$$

Consider the natural language proposition: *Most* students did *good* on the test. The membership function of *most* is defined as in Figure 3. This natural language proposition can be translated to a generalized constraint language proposition:

$$\frac{1}{\text{N}} \sum_i \mu_{good}(Grade_i) \text{ is } most.$$

Finally, a fuzzy rule base can be developed to decide whether to make adjustments to students' test grades. A sample fuzzy rule base follows:

If *most* students receive a grade of *fail* on the test, then give them a makeup test and take teaching lessons.
If *most* students receive a grade of *fail* or *poor* or *satisfactory* on the test, then boost the grades so *most* students receive a grade of *good* or *excellent* or *outstanding* on the test.
If *most* students receive a grade of *good* or *excellent* or *outstanding* on the test, then hand back the tests with these grades.

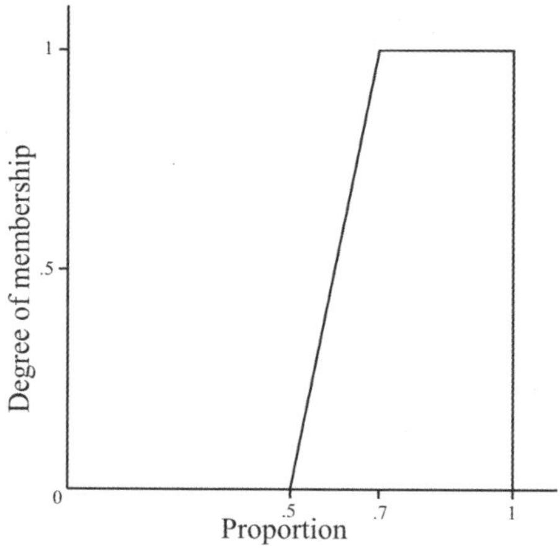

Figure 3. Membership Function of *Most*

8. Conclusion

Uncertainty is an inherent phenomenon in the universe and in peoples' lives. To some, it may become a cause of anxiety, but to engineers and scientists it becomes a frontier full of challenges. Engineers and scientists attempt to comprehend the language of this uncertainty through mathematical tools, but these mathematical tools are still incomplete. In the past, studies of cognitive uncertainty and cognitive information were hindered by the lack of suitable tools for modeling such information. However, fuzzy logic, neural networks, and other methods have made it possible to expand studies in this field.

References

1. M. M. Gupta, "Cognition, Perception and Uncertainty," in *Fuzzy Logic in Knowledge-Based Systems, Decision and Control,* M. M. Gupta and T. Yamakawa, eds., North-Holland, New York, 1988, pp. 3-6.
2. M. M. Gupta, "On Cognitive Computing: Perspectives," in *Fuzzy Computing: Theory, Hardware, and Applications,* M. M. Gupta and T. Yamakawa, eds., North-Holland, New York, 1988, pp. 7-10.

3. M. M. Gupta, "Uncertainty and Information: The Emerging Paradigms," *International Journal of Neuro and Mass-Parallel Computing and Information Systems,* vol. 2, 1991, pp. 65-70.
4. M. M. Gupta, "Intelligence, Uncertainty and Information," in *Analysis and Management of Uncertainty: Theory and Applications,* B. M. Ayyub, M. M. Gupta, and L. N. Kanal, eds., North-Holland, New York, 1992, pp. 3-12.
5. A. M. G. Solo and M. M. Gupta, "Perspectives on Computational Perception and Cognition under Uncertainty," *Proceedings of IEEE International Conference on Industrial Technology (ICIT) 2000,* Taleigaon, Goa, India, Jan. 19-22, 2000, vol. 1 of 2, pp. 221-224.
6. G. J. Klir, "Where Do We Stand on Measures of Uncertainty, Ambiguity, Fuzziness and the Like," *Fuzzy Sets and Systems,* Special Issue on Measure of Uncertainty, vol. 24, no. 2, November 1987, pp. 141-160.
7. G. J. Klir, "The Many Faces of Uncertainty," in *Uncertainty Modelling and Analysis: Theory and Applications,* B. M. Ayyub and M. M. Gupta, eds., North-Holland, New York, 1994, pp. 3-19.
8. A. Einstein, "Geometry and Experience," in *The Principle of Relativity: A Collection of Original Papers on the Special and General Theory of Relativity,* Dover, New York, 1952.
9. L. A. Zadeh, "Fuzzy Sets," *Information and Control,* vol. 8, 1965, pp. 338-353.
10. L. A. Zadeh, "Outline of a new approach to the analysis of complex system and decision processes," *IEEE Transactions on Systems, Man, and Cybernetics,* vol. 3, pp. 28-44, 1973.
11. M. Ayyub and M. M. Gupta, eds., *Uncertainty Analysis in Engineering and Sciences: Fuzzy Logic, Statistics and Neural Networks Approach,* Kluwer Academic, Boston, 1997.
12. M. M. Gupta, L. Jin, and N. Homma, "Fuzzy Sets and Systems: An Overview," in *Static and Dynamic Neural Networks: From Fundamentals to Advanced Theory,* John Wiley & Sons, Hoboken, N.J., 2003, pp. 636-644.
13. A. M. G. Solo, M. M. Gupta, N. Homma, and Z.-G. Hou, "Obama, McCain, and Warren Needed Fuzzy Logic to Define 'Rich' by Income," *Proceedings of the 2009 International e-Learning, e-Business, Enterprise Information Systems, and e-Government (EEE'09),* Las Vegas, July 13-16, 2009, pp. 265-270.
14. L. A. Zadeh, "Outline of a Computational Theory of Perceptions Based on Computing with Words," in *Soft Computing & Intelligent Systems: Theory and Applications,* N. K. Sinha and M. M. Gupta, eds., Academic, New York, 2000, pages 3-22.

THE CONSILIENT EPISTEMOLOGY: STRUCTURING EVOLUTION OF OUR LOGICAL THINKING

CHANDRASEKHAR ROYCHOUDHURI

University of Connecticut and Femto Macro Continuum, CT, USA
E-mail: croychoudhri@earthlink.net

It is argued that the current principles of modern culture are not congruent with the demands for long-term sustainable evolution. Even our modern democracy is run by the "Golden Rule" favoring "Gold" owners. To assure our peaceful and sustainable evolution, we need to replace this "Golden Rule" by the true democratic rule, or "Knowledge & Debate Rule" through the development of an over-arching *consilient epistemology,* now enabled by the global internet system. Inseparable connection between the consilient epistemology and our successful evolution is justified from an understanding that, for many millennia, the rapid evolution of human minds is being driven dominantly by the cultural-selection (intra-cultural and inter-cultural conflicts and pressures) rather than by the Darwinian natural-selection.

1. Why Do We Need Such An Over-arching Consilient Epistemology?

Let us first justify the necessity of pondering over such an over-arching methodology of thinking when the human specie appears to have conquered nature and no other species can threaten our competitive superiority for over at least 100 thousand years. It is good to start with a recognizable global problem - Global Warming. Interestingly, our scientific knowledge behind the complex and interlocked working processes behind the living and evolving biosphere is so limited that we do not even know how to completely reverse the Global Warming; and restoring it is beyond our current engineering comprehension. In other words we still have not learnt how to carry out sustainable terra-forming, as evidenced by our recent experiment called "Biopsher-2" [1]. The foundational principle behind the current "working" Capitalism is continuous growth. But the biosphere is a finite closed system designed to function through 100% recycling of everything.

Over the last ten thousand years, as we succeeded in eliminating any serious threats from other species and as our political and social cultures, under the strong-hand guidance of the tribal rulers, succeeded in becoming a reasonable safety-net for survival and belonging to, we slowly started to deviate from the necessity of living in harmony with nature. Now we are irresponsibly overriding our genomically programmed knowledge for such harmony. Let us quickly list a few of the issues from diverse fields that might convince the readers that human thinking has to change profoundly to undo the damages done by our non-harmonic behavior.

1.1 Cultures, Concepts and Theories Contradict our Sustainability

- **Limits of current economic theories:** "Successful" capitalist system is driven by continuous growth and exploitation & control of nature. We certainly have not mastered the terra-forming technologies. In the past, Capitalism meant control and management of finance capital to create profit for the owner of the capital by producing socially congruent goods and services. Thus, in spite of so called "exploitation" of human labor and natural resources (biospheric capital), there was a tacit understanding that Capitalism serves the overall greater good and collective well being of the human society. Now the key *purpose* of Capitalism has become maximizing the profit for the capital owners, not necessarily for the production of socially congruent goods and services. Supporting greater social good and promoting collective well being are no longer underscored in major business schools even though they are essential for our sustainable evolution.
- **Limits of current political ethics:** Socio-politico-econo system is still under the control of less than 1% of the tribal leaders just as it was 100 thousand years ago. Private wealth, whether accumulated by force or by "Golden" rules enforced by the "Gold Owners", is still controlling the governments around the world, just as it had in ancient times. Physical and mental well being of well over 50% of humans are challenged daily as they are forced to live on a subsistence level. In a Knowledge Age, even business leaders will agree that this is a very unproductive use of many billions of potentially creative minds.
- **Limits of current epistemology of Physics:** In spite of staggering advancements in our technologies utilizing knowledge created by physics, its century old epistemology will eventually slow down the progress of technology unless we focus our attention to model the invisible interaction processes, instead of modeling experimental data only. It is by emulating the processes behind natural phenomena that we create new technologies to assure our sustainable evolution.

- **Limits of present religions in guiding societies:** When we accept any human organized body of knowledge as the final and inviolable ultimate truth, whether meant for complex social engineering or nature's technological engineering, we consciously court the freezing of our mental and material evolution. Such cultural behavior is not congruent with our sustainable evolution.
- **Limit of the life of our Sun:** We may delay the "Global Warming" by some changes in our behavioral economics, but the insurmountable "Solar Warming" is coming to dry up the earth in less than a billion years! We must learn to proactively nurture our biosphere now to buy time, while preparing for terra-forming other planets of our Solar system and then in other stars to assure continued human evolution in some other planets [2].
- **Limit of purposeless evolution of humans:** We are evolving inside the vast cosmo-sphere and biosphere but keeping our minds totally wrapped under our limited, narrow-vision cultural spheres. But, have we ever defined a collective purpose for the evolution of the human specie? Should we? Can we? Or, will we grow more divisive and become Knowledge Age Neanderthals? We cannot keep on evolving forever and become successful space travelers without a *collective endeavor guided by some collective purpose*.

In Section-2 below we demonstrate that the *epistemology behind modern physics* is not geared towards proactively supporting sustainable human evolution. The result is that we have been diverted in inventing elegant theories rather than discovering real interaction processes that are behind the cosmic evolution. In Section-3 we summarize our *core assumptions for this paper* that are analyzed further. Section-4 rationalizes that the rate of human *evolution due to cultural-selection far out-weighs that due to Darwinian natural-selection*. Section-5 defines the value and significance of our proposal, consilient epistemology, towards directing our evolution with a purpose.

2. Is the Epistemology of Physics Leading us in an Evolution Congruent Direction?

In this section we will establish that even the epistemology behind immensely successful modern physics is seriously deficient in providing leadership to other branches of knowledge that use various logics of physics as their foundation. Some readers with expertise in physics might challenge the very legitimacy of the question that heads this section. This will be especially true for those who believe that "science and arts are for the sake of science and arts" only.

Scientific thinking is supposed to be perpetually enquiring. Could, treating the current formulation of quantum mechanics as complete and inviolable just because it is working, threaten the evolution of our minds, our sciences, our technologies, and hence our very existence? While emulation of a success model brings more successes in the short run, falling into the success rut can stall our evolution. It is worth pondering. Recall that a good number of dedicated physicists at the end of the nineteenth and early twentieth centuries put their best efforts to find out the limits of classical physics, which generated most of the critical foundational experiments and concepts that gave birth to quantum mechanics (QM) [3]. Within a few decades, a flood gate of new knowledge of the micro world became available to us through the application of QM. We have become so mesmerized with this staggering degree of successes that the Copenhagen Interpretation essentially declared that QM represents a complete (final) theory of the micro universe. The enquiring minds of multiple generations of physicists have been forced to create non-causal concepts to explain "single photon (particle) interference" like non-locality of interaction processes, teleportation of particles, generation of multi universes, etc.[4].

Such "mystical interpretations" by physicists are already developing serious doubt on our long-term leadership capabilities. Finally, some well known theoretical physicists have begun to speak out through their books [5-8] that there are serious problems with the current paradigm, philosophy and direction of physics! Like Ptolemy's time, when they were enamored with the spherical symmetry centering the Earth, we are now worshipping esthetically pleasing mathematical elegance. The author believes that this prolong cultural shift has taken place because of the absence of any serious and consistent measurement epistemology – how much information is available to an experimenter & how to structure that limited information into a causal theory, so we can iteratively keep on advancing towards creating better theories that can keep on extracting the missing information.

2.1 Measurement/observation Epistemology

2.1.1. All interactions are "local"

A self consistent measurement epistemology leads us to the following realizations: All interactions are local; or more accurately, they all are *superposition effects* taking place in a small region defined within the instrument. Measured *transformations* are produced by *exchange of energy* between the space-finite interactants guided by one of the four allowed forces. As per the Copenhagen Interpretation, quantum mechanical superposition effects between elementary particles, photons, atoms and molecules, are nonlocal. A significant portion of all activities in the biological world, including human thinking, is driven by molecular interactions guided by rules well formulated by quantum mechanics. Could we have been using our brains (thinking) independently of each other if all molecular interactions were nonlocal? It can be nonlocal only in the limited sense that atoms and molecules are not geometric point objects. They have a finite physical extension and they influence each other beyond their physical size, but they are limited to spatial ranges dictated by force laws they must obey. Figure 1 provides the logical flow diagram justifying *locality* of interactions as physical superposition of different interactants within their sphere of influence.

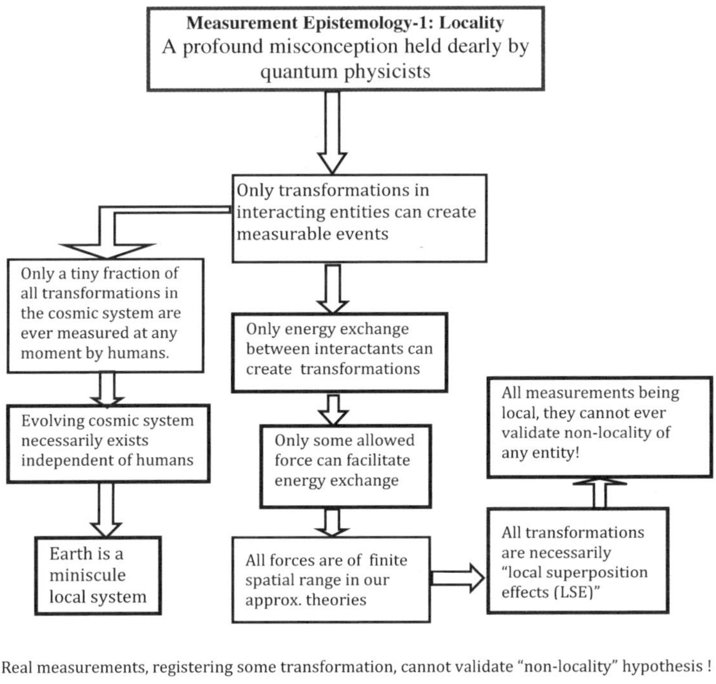

Fig. 1. The logical flow chart of local interactions.

2.1.2 Incomplete Information Challenge (IIC)

Let us now look into the incomplete information challenge (IIC). This is a well known and eternal challenge to human minds and the genius thinkers have been filling up those information gaps by wisely creating new hypotheses that make their new theories logically self consistent. But these hypotheses may not be the ultimate and the best, which our future research may elucidate. All experiments give incomplete information about the interactants under investigation. Thus, all "working" theories are necessarily incomplete and must be subjected to iterative improvements and/or re-structuring. The logical flow diagram of Fig. 2 underscores the relevant points. We should remember that we had started with incomplete information about the interaction under study right at the very first stage of information *generation.* It is due to intrinsic limitations of our complex instruments and their capability to relay faithfully all the information related to the transformation we are studying. A simple example would be band-limited frequency responses of all electrical circuits connected to photon detectors. The only way to incrementally overcome such fundamental incomplete information challenge (IIC) is to indirectly gather information about the same interactants but from very different fields of studies, which we are defining as a consilient epistemology [9, 10].

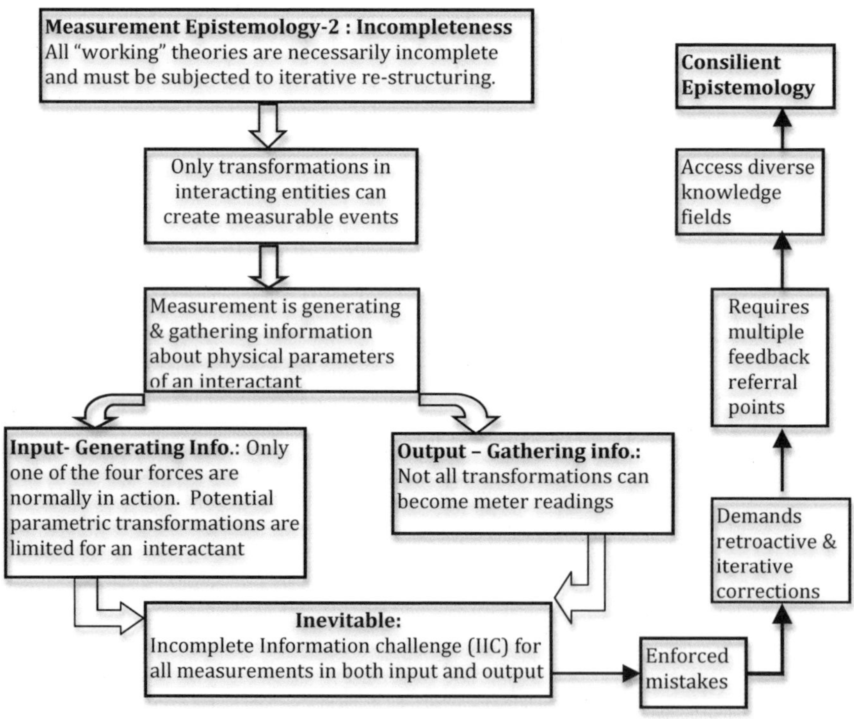

Fig. 2. The logical flow chart of Measurement Epistemology.

2.1.3 The NIW principle; Non-Interference of Waves

Propagating waves do not exchange energy with each other unless mediated by some material media to display the superposition effect. This is a universal principle followed by all propagating waves. Let us assume that you have dropped two small stones separated from each other by about one meter on a quiet water surface in a pond. You are watching the evolving circular wave packets propagate out. A segment of the circular arcs of each wave group will eventually cross through each other and then continue propagating out as if they have never experienced the presence of the other. This is the principle of Non-Interference of Waves (NIW) [see Fig. 3]. The phenomenon is true for all propagating wave forms, whether they are material based (string waves, sound waves, etc.) or "vacuum" based electromagnetic waves. Coming back to the two circular water waves – a careful observation will reveal that *during the temporal duration and within the physical domain* [note the locality], where the two wave groups were just crossing through each other, the water surface clearly displays enhanced and suppressed surface undulations resembling what we call "interference pattern" due to "local" and real physical superposition of the two wave groups. These patterns should be called *superposition effects*, rather than *interference patterns*, to underscore the simultaneous presence (superposition) of the two periodic surface undulations of different phases in the same "local" space and time volume. Only the presence of material media can display such superposition effects due to propagating waves. Since, light travels through the cosmic medium, or the cosmic tension field [3], which we have not yet learnt to make visible, we need to insert separate material substance, capable of simultaneously interacting with all the superposed wave groups to make the superposition effects observable. If the detector can respond to only one frequency of the two frequencies of the two superposed waves, there will be no superposition effect.

We can hear each other even in a very crowded place because the sound waves generated by my friend's mouth arrived unperturbed into my ears in spite of being crossed by innumerable other sound waves. We image many distant stars on our retina as separate spots unperturbed by each other even though their light beams have crossed through each other while forming independent images on our retina and also crossed by light beams from billions of other stars during their journey through the vast cosmic space. The same NIW principle enables us to clearly recognize a desired face unperturbed by innumerable other crossing light beams generated by surrounding sceneries. We have been neglecting this NIW principle for centuries perhaps because our current culture of mathematical formulation does not pay attention to the physical processes that are the key behind generating measurable transformation. This is a profoundly important conceptual mistake of classical physics. Quantum physicists accepted this wrong notion of "interference of light" from classical physics which, in turn, promotes belief in a wide variety of unnecessary mystical interpretations of QM mentioned earlier [3]. It is interesting to note that the NIW principle is already built into the proper mathematical formulation of the classical Huygens-Fresnel's diffraction principle for the propagation of waves. One can derive the "final" complex amplitude at any forward plane by summing the complex amplitudes (secondary wavelets) from any arbitrary previous plane. This clearly indicates that the secondary wavelets never "interfere" to reorganize a locally new energy distribution. QM almost found the NIW principle when they discovered that "photons" behave like non-interacting Bosons. Unfortunately, to accommodate the mistake of classical physics, that light beams interfere by themselves, QM declared "photon interferes only with itself", which then forced generations of physicists to concoct various non-causal concepts to explain "single photon interference"!

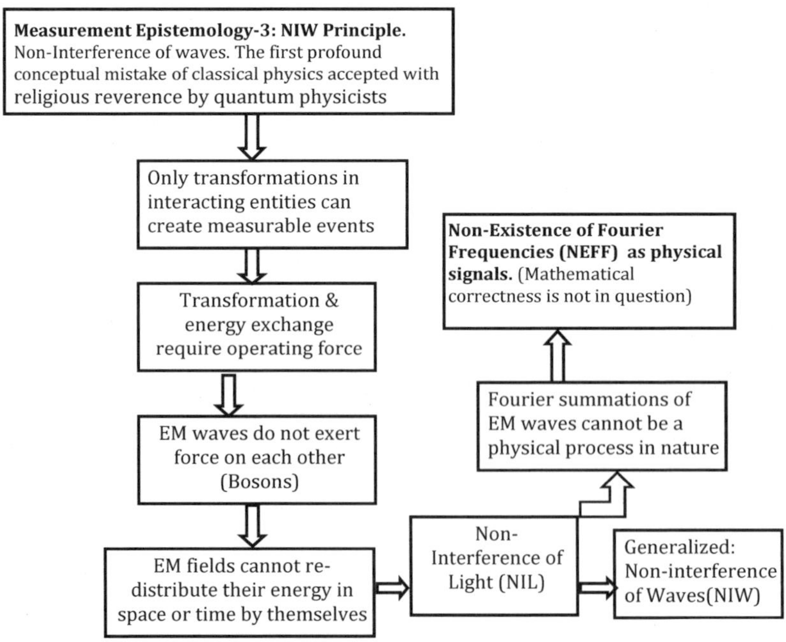

Fig. 3. The logical flow diagram of the principle of Non-Interference of Waves and Non-Existence of Fourier Frequencies.

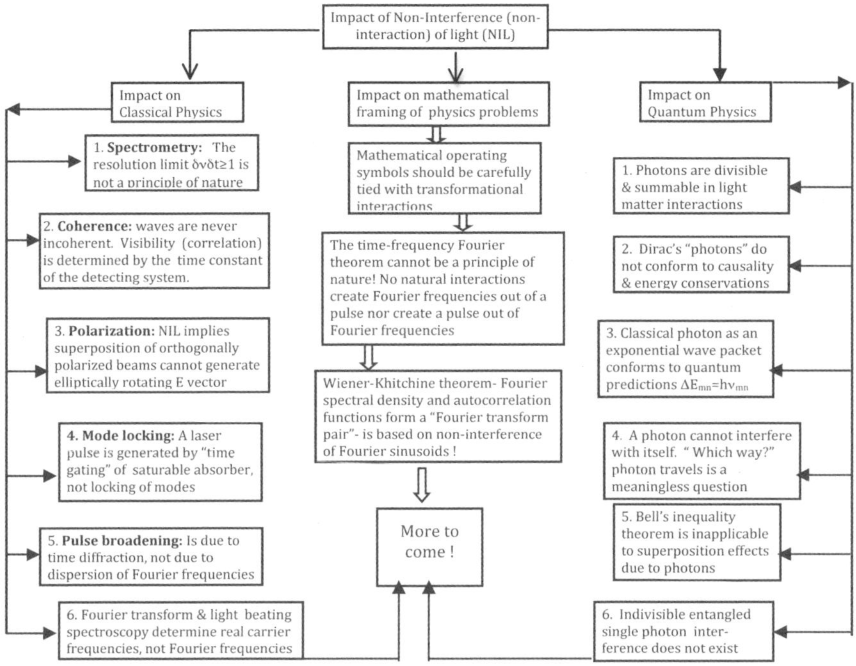

Fig. 4. List of consequences of NIL principle, non-interference of light (waves, in general) in both classical and quantum optics.

Readers familiar with the time-frequency Fourier theorem will recognize that the NIW principle demands that we must not use this widely used theorem as a *principle* of nature. We are now mathematically deriving "superluminal" velocity of light pulses while propagating "monochromatic" Fourier frequencies as real waves with real frequencies [11,12]. This should have been obvious to us from the law of conservation of energy - no waves can exist in all space and for all times. A list of important impacts in classical and quantum optics is given in Fig. 4 as a consequence of neglecting the NIL principle. See the author's earlier papers [3, 13].

2.2 Fusion of Reductionism and Emergentism

Descartes reductionism from 1600' is still the driving force behind modern Physics. Most successful mathematical models comprise of wave and field theories. We have known that so-called "impenetrate-able" masses are not immutable objects in nature, $m=E/c^2$. Under suitable conditions, some of the elementary particles can be completely transparent to other "material" particles.

Particles are convertible to energy as is demonstrated by our energy producing nuclear reactors and bombs. We know that framing the question determines the answer we generate and since the answers are always incomplete, we never know whether the current "correct" answers will withstand the test of time as our knowledge advances further. So, while emulating a success model does bring more successes, it could eventually lead us into a mathematically elegant but misguided rut. A theory can become very successful only if it has grasped some real cosmic logic. But, because, all successful theories, by definition, are incomplete, it is worth periodically challenging the foundational hypotheses of the most successful theories. Discovery of any breakdown point will lead us to illumination about the cosmic reality.

Energy is at the root of everything that is manifest in our observations and measurements. It is more fundamental than electromagnetic waves and particles. So it is worth employing emergentism[8] to model the emergence of protons, neutrons and electrons etc. out of the field that permeates the cosmic volume, the cosmic tension field (CTF)[3, 13], instead of only seeking "mass producing" elementary particles [15]. Most likely, we will need to iterate back and forth multiple times between reductionism and emergentism to find another major break-through in physics (Quantum Physics) as we have experienced during the second half of 1920's after systematic challenges to various aspects of the then most successful physics, now known as Classical Physics.

General Lessons:

- Same physical problem can be modeled "correctly" by different mathematical logics. No logic or theory is unique or "final". This is truer when we model a small segment of a vast complex logical system built out of a few simple rules.
- Correctly structured mathematics is never wrong. But correctness does not assure that it represents *the* reality of nature, or it has captured *the* cosmic logic behind the phenomenon.
- Mathematics must be structured to map the interaction processes in nature, not just the measurable transformation. It is a tool like the magnifying glass.
- Quantum mechanics, being enormously successful, has more realities built into it than the Copenhagen Interpretation has allowed us to extract out of it. Enquiring minds should be nurtured to ask more probing questions: What are the processes going on in between the preparation & detection of the transformation produced by the quantum entities?
- Equations cannot speak for themselves; humans do. Our thinking is dominated by evolution-congruent subjectivity rather than scientific objectivity. So, no interpretation should be taken as the final truth.

3. Foundational Assumptions and Challenges

Let us now summarize the key logical assumptions behind the paper. The human species has reached the stage of evolution to understand that we are a very fragile species. Our biosphere with moderate temperature, even if we slow down the current Global Warming, will succumb to extreme heat as the solar corona keeps on expanding with its age, which will happen well within a billion years [16]. In our psyche, most of us are genetically programmed with a strong desire to *live forever through our progenies*. So we must learn to think how to proactively construct a purposeful evolutionary path for ourselves on earth and then on other planets, supported by our science and engineering [4].

3.1 Causal Cosmic System

Based on our broad successes in understanding diverse simple and complex phenomena of this universe using pure human logics and mathematical theories, we have reasons to assume that the cosmic system is a causally evolving system. In other words, inter-related and inter-connected *cosmic logics* are inseparably behind all interaction *processes* that are creating incessant changes supporting the cosmic evolution. Our objective is to discover and understand this complex system, while visualizing the invisible interaction processes using *diversity of human logics and approaches*. By iterative approaches to better model nature, our human logics will incrementally tend towards the actual cosmic logics. The correct path to this refinement is guided by our successes in creating new technologies by emulating nature's rules and processes, which are essential for our sustainable evolution. We do science for our purposeful and successful evolution, not just for the sake of doing science alone.

3.2 Cosmic Complexity Challenge (C^3)

It constitutes two fold challenges. Everything influences everything in this universe somehow or other and the more complex and/or large a system, the more complex emergent behavior it displays, which cannot be described by equations guided by our current reductionist epistemology [8,15]. Two of the four forces, weak and strong nuclear forces, are of rather short range, confined within the nuclear dimension. The other two, Gravitational & Electromagnetism, are of very long range. After all, the galaxies are influencing each other! Can the stars, and the atoms in them, be completely free from influencing each other just because they belong to different galaxies? Regarding emergentism [8,17] even though MRI images of the brain indicate that human thinking is based on molecular interactions and transitions, our current quantum mechanics is not good enough to write down an equation for the emergence of our intelligent and creative consciousness (thinking). Thus, we must acknowledge that (i) the logical elegance of a theory, and (ii) its capability of correctly predicting a wide variety of experimental outcomes does not assure us that the theory has reached the level of being the ultimate "God's Equations", despite the fact that it may have correctly captured some segments of cosmic logics. Then, how can one keep on correcting and advancing a "working" theory to higher and higher levels of capturing cosmic logics?

3.3 Incomplete Information Challenge (IIC)

This point has already been discusses in Section-2.1 in the context of discussing measurement epistemology. None of our theories can be complete because they are necessarily built upon conjectures which are based on incomplete information. Limits of these imprecise conjectures must be identified by pushing all "successful" but provisional theories to their limits and thereby open up platforms for the development of new theories.

3.4 Subjective Interpretation Challenge (SIC)

Successful biological evolution has given us hardwired genomic programs to make quick and subjective interpretations of our sensorial inputs. The objective realities, the laws of nature, are to be extracted out by using our evolutionary gift of free-will, provided we are conscious about our hardwired subjective interpretation propensity, which still pervades our modern epistemologies. Section-4 will elaborate this point.

3.5 Pre-Adaptive-State Challenge (PASC)

Our current scientific culture accepts successful theories as final ignoring the challenges, C^3 & IIC, mentioned above. We have also noted in Section-1 the limits of the foundational assumptions behind almost all of our well organized bodies of knowledge due to their incongruence with the sustainability needs of the biosphere. Can this incongruence be attributed to any common epistemological root? Could it be that our scientific thinking is driven by the Darwinian pre-adaptive-state model? In Darwinian DNA evolution, the current "successful" state, achieved through previous natural selection, becomes the pre-adaptive-state (PAS) for the next level of natural selection. The bio-genomic evolution is forced to make advancements and improvements only pro-actively, not retro-actively, because the core molecular foundation has already been permanently set by the structure of the DNA helix. So DNA can evolve in consecutive steps building upon the available "current" pre-adaptive-state. Since it does not know the distant-future state for natural-selection, it adopts the optimum changes to acquire the traits suitable to overcome the current adversity. This adoption may appear not to be the best choice in the distant future, but it cannot make any changes in the DNA *retro-actively*. It can only improve upon what it already has. The DNA evolution is stuck by the pre-adaptive-state-challenge (PASC) at every step of its evolution. Is that why we develop next generation theories by building upon the foundation of the previously "successful" theory because even our free-will suffers from this genetic challenge, PASC, like our DNA? Human free-will has developed the capacity to connect information from distant-past and distant future (like the death of the Sun) using our theoretical predictions. So, we should feel comfortable to retro-actively correct our past "successful" theories to conform to newer understanding rather than taking it as the correct foundation for the next higher level theory.

3.6 Retroactive Iterative Corrections (RIC)

This must be applied to all working theories to overcome inherent limitations imposed on human logics by the cosmic challenges, IIC & C^3, and by the genomic challenge PASC. But we need an unchanging referral "guiding polar star" for reliable feedback for iterative improvements on our conjectures. We are proposing that the diverse interaction *processes* in nature should be used as our "guiding polar star". Nature being causal, iterative corrections on our conjecture and mathematical logics will help us make increasingly better visualizable models of the interaction *processes* and steadily reduce the gap between the actual cosmic logics and our human logics (theories). The gap will always exist due to the challenges IIC, C^3 and PASC. The conscious exercise of human free-will will help us liberate ourselves from these challenges and completely re-build theories from a new foundation, whenever necessary, rather than just improvising the right kinds of stones to fit into the current edifice or the "pyramid of science".

3.7 Human Evolution is now Dominantly Driven by Cultural-selection Rather than the Darwinian Natural-selection

The role of Darwinian natural-selection on the rate of human evolution due to environmental pressure and competition has become insignificant compared to the pressures produced by our concepts (knowledge) driven human cultures. The dominance of cultural-selection, of course, started when humans (i) outsmarted competition from all other animals and (ii) technologies behind agriculture, animal husbandry and skills of house building assured our physical survival on an even keel. There was no more serious environmental pressure for any need of physical evolution. Tribal leaders realized that sustainability of their leadership now depends upon controlling the minds of the tribal members by delivering right kind of pleasures, pains and fears supported by justifiable *concepts or social theories*. Socio-political or cultural conflicts began to accelerate, and ever since, our mental evolution is being driven by pressures from intra- and inter-cultural pressures. This is discussed further in Section-4.

3.8 Evolution is a Collective Endeavor

The core survival mechanism and the biological evolution are managed collectively by the four musketeer molecules GACT (Guanine, Adenine, Cytosine and Thiamine) through their hydrogen bonding [G-C and A-T] to create DNA's, which are protected inside a double helix chain. So evolution is a collective endeavor. This is further discussed in Section-4.

3.9 Consilient Epistemology (CE)

The power of Retroactive Iterative Corrections (RIC) will be significantly enhanced once we realize that all the challenges mentioned above can be broached if we proactively accept that the different bodies of knowledge (biological and social sciences) and their knowledge gathering processes can be of immense help to hard sciences, and vice versa. For social and biological sciences only emergent complex behaviors are observable, even though, in the final analysis, only complex atomic and molecular interactions are at the root of all emergent properties, including human minds. Obviously, our immensely successful reductionist theories are as yet unable to precisely model these emergent properties. Just as the living biosphere thrives on diversity by proactively leveraging potential symbiotic & synergistic inter relationships between everything and every specie, so will human epistemology thrive by proactively looking for symbiotic & synergistic inter relationships between all of our knowledge systems as one science, irrespective of how we have been characterizing them so far, as hard or soft sciences. *Just as intra- and inter-species interactions have been advancing the evolution of the living biosphere, so will the diverse concepts behind diverse fields of knowledge, when treated as "living concepts" assure our collective evolution driven dominantly by our cultural-selection.* Of course, to respect various concepts behind human cultures as "living specie", they must, at their very foundation, be congruent to our sustainable evolution through creation of higher and higher levels of technologies and mind-broadening cultures. Such an approach will be the key to developing a mutually beneficial Consilient Epistemology (CE) [9]. The potential ability to shape the direction of our evolution consciously and purposefully is the key higher level ability possessed by the humans compared to all other species.

4. Darwinian Natural-selection vs. Human Cultural-selection

In Section-3.7 we have briefly introduced the concept that the evolutionary pressures for changes in human behavior are now coming dominantly from pressures due to inter-cultural and intra-cultural conflicts as a result of diverse types of human generated concepts (collectively, a culture) trying to shape our collective behavior in favor of the group that promulgates the concept. Accordingly, the human mind, hence the consequent genomic programming, has been changing perceptibly under our cultural pressures and conflicts. We are calling the process cultural-selection to rhyme with Darwinian natural-selection. Various human concepts that constitute a culture are products of our minds guided by our neural network constructed by our genome. Thus our culture could be considered as a repository of collective genomic soup cooked by us and drunk by us for our continued growth and evolution. So, it is our prime responsibility to assure that all the concepts that are driving, or added to our culture, are definitely congruent with our sustainable evolution.

4.1 Evolution and Entrepreneurship are an Inseparable Collective Enterprise

Let us briefly review the modern understanding of DNA [18] that is at the root of both the Darwinian natural-selection and modern cultural-selection. At the fundamental level, starting from about 3.5 billion years past, the evolution has been a marvelously creative and collective enterprise, not an individualistic one. As mentioned earlier, the DNA double-helix has been invented by "four musketeers" GACT (Guanine, Adenine, Cytosine and Thiamine). The evolution, as we know now, is driven by the DNA's. We are then forced to conclude that at the very foundation, evolution has originated as a *collective effort* by the GACT's. While the weakly bound GACT's have protectively enclosed themselves within the central region of the double-helix, the outside molecules in the neighborhood are easily chemically accessible, through the mostly open helical structure, so they can get hold of them and process them to send out new information, and/or to organize new desired molecules, effectively creating a wide variety of "wealth" for their purposeful evolution.

Demonstrated diverse wealth creation propensity of human DNA can be safely identified into two groups: (i) material and technical wealth and (ii) cultural and spiritual wealth. The first one covers all material wealth producing activities including complementary science and engineering activities. The second one corresponds to processing and analyzing information to understand the meaning and the purpose of the evolving universe, which can provide the long-term guidance to the first wealth creating groups so that the *means* for our material evolution does not become its *end*. Human minds are continuously evolving to higher planes of understanding the complexities behind the evolutionary processes by becoming more and more efficient in processing information with steadily enhancing efficiency, expending less and less energy per unit mass per unit time [20]. Does it imply, at least scientifically, that there could be a vectorial direction of biological evolution – towards higher levels of information processing capability with increasingly higher energy efficiency to become capable of understanding the ultimate meaning and the purpose behind the cosmic evolution?

4.2 Darwinian Natural-selection vs. Cultural-selection

That the slow Darwinian natural-selection due to pressures of habitat has been taken over and is being accelerated by our cultural-selection**,** is apparent from the following dichotomy. From the standpoint of evolution of the human specie, our militarism has been obsolete since we can only eliminate ourselves by waging wars. Organized fight or militarism is a smart genomic program for all species that are still evolving by natural-selection due to various habitat pressures, foods and related competitions. Humans have won the competition against all other species, probably, a few hundred thousand years ago. But we still have not given up this trait because we still are victims of our hardwired genomic program to fight and kill when we perceive a mortal threat. It is thus essential for us to proactively nurture our enquiring free-will by continuously refining our cultures so as to promote selectively those thinking traits that are congruent with our collective and sustainable evolution. Nurturing obsolete genomic war programs only diverts our energy away from progressive evolution.

It is important to recognize that the GACT's in our DNA have invented a physical and a mental process, the combination of which gives us a powerful tool to accelerate our evolution towards a better quality of material and spiritual life without the need to wait for the slow genetic mutations. The first one is the selective sexual exchange of DNA components containing preferred progressive traits. The second one is the proactive creation of a culture which is a collective repository of our genome-created concepts contributed by all members of the society, rather than remaining subservient to the culture promulgated by the tribal leader class.

In reality, almost all species have a culture for their collective well being. Obviously, because of highly developed diversity of means of communications, human culture is the most matured one of all species. In fact, over the last few millennia, the evolution of the human mind (information gathering, analyzing and organizing) has been dramatically accelerated through the cultural pressures, while the physical evolution has remained almost stagnant. The scientific component of our culture has advanced to the level that we can now model, reasonably well, our biological past to some 3.5 billion years, and our cosmological past to almost 14 billion years. As we model our future, we see clear termination of the biological evolution on earth for most of the specie in about one or two billion years. Physical models for the evolution of stars, supported by registered data for a good number of supernova explosions, imply that our Sun would die in another 4 to 5 billion years. Well before that time its excessive heat will dry up the earth causing the termination of our biosphere as we know it.

Since culture is a product of our genomic thinking (concepts), it also dictates the allowable modes of thinking, or our epistemologies, to develop further concepts, which must conform to the well being of the ongoing culture! To over ride the stagnancy of our cultural evolution, our DNA has also invented a substantial amount of free-brain to assure positive evolution through our traits of enquiry and debate. And the ruling classes have always recognized the power and threat of this free-brain and developed the counter force to co-opt them by permeating the society with well structured educational and "public" media systems to make the members conform. But, as we have underscored in the beginning, due to shortsightedness and focus on immediate gains, the current epistemologies behind most of cultural forces (dominant concepts) are incongruent with our sustainable evolution. Fortunately, the collective wisdom of the GACT's has succeeded in creating the unusual trait in a limited set of our members who can never be completely brain washed by the rulers of the society. History bears this out. Rulers have never succeeded in brain washing all the people, all the time, all over the world. Such few thinkers and philosophers time and time again helped the masses to identify whether the ongoing cultural (ruling) concepts are congruent with their sustainable collective evolution.

Our cultural identities are now so strong that an appreciable percentage of members belonging to a specific culture, unhesitatingly give their lives to save their culture even though their physical well being is not threatened. However, our current thinking process (epistemology) is still dominated by hardwired genomic programs acquired over 3.5 billion years of biological evolution for physical survival. Perhaps, anticipating this coming dichotomy, the human genome has helped evolve a significant amount of analytical information processing neural network, the free-will component of our mind. We are supposed to use this analytical free-will to generate, evaluate and nurture only those concepts in our culture that are congruent with our sustainable collective evolution. Since ancient times, our tribal leaders recognized the existence of this free-will, along with the traits of faith and belief, in every human being. So they have proactively and successfully utilized them by saturating the minds of the tribal members with dominant tribal concepts to organize, manage and rule their societies. We should recognize that the complex traits, faith and belief, are genomically hardwired in the genes of all species, from bacteria to humans. Without the faith in nature, even a bacterium would not have the courage and will to struggle for life under the pressures of persistent threats to its life. Without the belief in self, the bacterium would not display the confidence that it can win the struggle to get to the food and eat it too and live for another day! Human traits of faith and belief are magnified by virtue of powerful free-will. Of course, great messiahs recognized the existence of these traits in human beings and adeptly utilized them to help create many independent religious concepts and implanted great ongoing cultures of *social engineering*. So, we cannot just "preach" to throw out all religions. We must learn to filter and protect all those useful concepts out of all religions that are congruent and sustainable with our biospheric evolution.

Fortunately, no culture has ever succeeded in brain washing all the people all the time all over the world. This is now evident from the intense "social networking" activities on the low-cost global internet system, albeit, built for profit by the big businesses to access the global market. The internet has now become the global dissemination medium for all possible concepts, besides those desired by the ruling tribes. This is, of course, healthy because now the masses are beginning to appreciate and utilize their personal free-will, which is a critical component for our mental evolution through cultural-selection. They are now beginning to challenge their own faith and belief traits to refine their objective thinking. Perhaps within this century we will see the realization of the desire: Blessed are enterprising common folks, for they shall inherit the Earth. We should note here that, it is not the great heroes of different fields; it is rather the collective wisdom of common folks, which are at the root of sustaining our rich and evolving human cultures.

4.3 Nurturing Our Concepts to Assure Sustainable Cultural Evolution

We would like to propose that we must now learn to accept and appreciate that diverse human concepts in the global human concept-sphere are equivalent to diverse species in our bio-sphere. We know that it is the mutual interactions between the diverse species, from all bacteria to all plants and animals, which have been assuring our sustainable evolution, even though we do not yet know the exact detailed steps and processes by which the living biosphere is carrying this out. Similarly, we do not yet know all the detailed processes by which different competing concepts from different cultures are facilitating the evolution of new and better concepts for our future well being. What are the real road-blocks against the highly desirable future transformations by which common people will eventually invent a methodology of collective-leadership and become the "inheritor of the earth"?

5. The Consilient Epistemology

The more we analyze our understanding about the biological and the cosmological evolution, the more we find logical orders and inseparable connectivity in every direction. Naturally, it is safe to assume that working principles behind all evolutions are logically congruent to each other. So, we should not face serious conceptual difficulties in discovering a consistent logical connection or congruence or consilience between different fields of human knowledge that maps different aspects of evolution. This should also be obvious from the fact that we humans are products of natural biological evolution. After all, our thinking is engineered by our neural network based on incessant molecular interactions driven by precise laws of physics and chemistry. We can model and write down equations for simple molecular reactions, however, we still cannot carry out such modeling for our emergent behaviors, like our thinking. It is safe to assume that we cannot do any better than thinking like reverse-engineers until we figure out all the cosmic logics. Let us then take the liberty of classifying human thinkers as Simplicity Engineers (SE) and Complexity Engineers (CE).

Simplicity Engineers of hard sciences deal with the understanding and modeling of the most foundational cosmic rules that governs the *interaction processes* behind the material universe of atoms and molecules, built out of protons, electrons and neutrons. Their approach, so far, has been to use the epistemology of reductionism [17]. The fruits of their efforts constitute: First, the rapidly advancing technologies that are based on emulation of the rules of nature, which are essential for the survival of more than six billion humans today within the thin biospheric shell. Second, the modern science that provides us with the preliminary understanding of the rules behind the cosmic evolution, has alerted us that our Sun has a finite life and our desire to continue living "forever" through our progenies will require us to invent new technologies (i) for terra-forming other planets and (ii) for deep space travel that are much more sophisticated than what we have now.

Complexity Engineers identify and model emergent rules of behaviors of complex and large (animate and inanimate) systems built out of very large numbers of molecules and/or sub-systems. The examples are: sustainability of the living biosphere, global climate, individual and collective human psychologies, politics of social governance, laws of economics, rules of industries, management of human social interactions through religious morality, etc. The epistemology of complexity engineers can be characterized as emergentism. Emergentism identifies self-consistent behaviors of complex systems and models them to understand and predict further behaviors of these systems. Even though the emergent rules of behavior of complex systems, on a fundamental level, are built out of simple assemblage of atoms and molecules, we are, as yet, unable to construct any theoretical continuity equations for a direct interface between simple and complex systems [17]. However, we can safely assume that the same set of cosmic logics (laws) is behind both the simple- and the complex-systems.

The reductionism and the emergentism, both these schools of epistemologies are finding the limits in their progress. We now need to develop logical steps to discover conceptual continuity and logical congruence between the two epistemologies. In the living biological world, genetic and DNA engineers are beginning to help us bridge the gap by discovering the steps through which simple molecular association and/or dissociation between GACT gives rise to the most complex "coding properties" of DNA-complexes guiding the entire biological evolution. In the inanimate world, physicists with deep knowledge of heat-flow, thermodynamics, fluid dynamics, etc., are beginning to model very complex long-term global climate and the future implications of global warming. While such scientific approaches are essential for our further progress, we should accelerate the process by imposing the following three over-arching guiding concepts:

1. Purposeful evolution: We raise our children by helping them create a purpose for their individual lives. This helps them to keep on asking the right questions to find the right solutions, which keep them always moving towards their life-long goals in pursuit of happiness in the human society.

Similarly, human society needs to create a purpose for our collective evolution. Then we can draw guidance to persistently ask pertinent questions related to (i) the purpose behind the evolving universe and (ii) the role the human can and should play in it. Then the followers of reductionism and emergentism will be inspired to become consilient, rather than disregard each other's epistemology. Diversity of approaches to understand the same complex system is the best scientific approach.

2. Sustainable evolution: If we are truly inspired by the dream of perpetual human evolution, then we must figure out how to emulate the behaviors of both the simple and complex natural systems to create all the necessary technologies and ideas (concepts) that can assure our sustainable evolution. Simplicity and Complexity Engineers must collaborate.
3. Visualization (understanding) of invisible interaction processes: Surprisingly, all modern scientific methodologies have essentially remained geared towards modeling the observables both in the complexity and the simplicity world. We have not been paying focused attention to appreciate the "invisible" interaction processes that are giving rise to the final outcomes or results. In physics, our result-driven mathematical model fools us to accept the superposition principle as a mathematical principle of nature rather than as a tool to peer deep into the interaction process between multiple superposed beams and a detector that can simultaneously respond to all the superposed beams. As a result, we have missed the NIW-principle. The waves can generate the *superposition effect* if there is an interacting intervening material medium. Attention and respect for the real physical processes behind all phenomena will not only stabilize the society, it will also bring the Simplicity and Complexity Engineers closer to each other.

6. Conclusions

Global warming is a timely warning to us by our biosphere. The eventual demise of the biosphere due to the aging Sun is inevitable. We must accept the reality that we are mortals and that our body-mind is subservient to the material laws of the universe. However, we have the gift of the biological evolution, analytical and critical-thinking free-will, to become outstanding reverse-engineers. Our desire to live indefinitely has to be through our progenies. Then we must hand over to them a sustainable biosphere for the time being and a technologically and spiritually driven culture of *consilient epistemology* for them to thrive into the distant future. Then they will have the opportunity to become reverse engineers of the highest capacity and will succeed in uncovering the complex rules behind the evolving universe and then develop the technologies to travel to distant planets and then terra-form them as their new habitat.

Continued and *SAFE* human evolution lies with collective organizational efforts to develop socio-cultural interaction processes and nurturing systems that assure Self Actualization For Everybody (*SAFE*)! So, we would like to propose that all cultures build their own versions of *SAFE* organizations and actively work towards *PEACE* (Purposeful Evolutionary Achievements with Collective Efforts)! Of course, we will have to learn to leverage our hardwired genomic programs of fear, faith and belief from being *pawns of cultural control* to *self empowering* tools for purposeful evolution.

References

1. Biosphere 2: http://en.wikipedia.org/wiki/Biosphere_2. See also: http://www.biospherics.org/
2. S. Brand, *Whole Earth Discipline: An Ecopragmatist Manifesto* (Viking Adult, 2009). See also Dennis Overbye ,"Setting Sail Into Space, Propelled by Sunshine", NYTimes, Tuesday, Nov.11, 2009.
3. C. Roychoudhuri, "Inevitable incompleteness of all theories: an epistemology to continuously refine human logics towards cosmic logics", Ch.6 in *The nature of light: What is a photon?* Eds., C. Roychoudhuri, A. Kracklauer & K. Creath, CRC Press (2008).
4. C. Roychoudhuri, "Locality of superposition principle is dictated by detection processes", Phys. Essays **19** (3) (2006).
5. L. Smolin, *Trouble with Physics* (Houghton Mifflin, 2006).
6. R. Penrose, *Road to Reality*, Alfred Knopf (2005).
7. P. Woit, *Not Even Wrong* (Basic Books, 2007).
8. R. Laughlin, *A Different Universe: Reinventing Physics from the Bottom Down* (Basic Books, 2006).
9. E. O. Wilson, *Consilience, the Unity of Knowledge* (Alfred Knopf, 1998).
10. H. Hermann, *From Biology to Sociopolitics: Conceptual Continuity in Complex Systems* (Yale University press, 1998).
11. R. Boyd, "Slow and fast Light", Ch.6 in *Progress in Optics*, Ed. Emil Wolf (Elsevier Science, 2002).
12. C. Roychoudhuri, N. Tirfessa, C. Kelley & R. Crudo; "If EM fields do not operate on each other, why do we need many modes and large gain bandwidth to generate short pulses?" SPIE Proc. Vol.**6468**, paper #53 (2007).

13. C. Roychoudhuri, "The NIW (Non-Interference of Waves) principle removes wave-particle-duality and "spooky" conjectures", invited paper at the conference, "Quantum Theory: Reconsideration of Foundation-4" (Vaxjo, Sweden, 2009).
14. C. Roychoudhuri and C. V. Seaver, "Are dark fringe locations devoid of energy of superposed fields?" Proc. SPIE Vol. **6285**-01 (2006).
15. F. Wilczek, *Lightness of being: Mass, ether and unification of forces*, (Basic Books, 2008).
16. See web videos: http://video.google.com/videosearch?q=Death+of+the+sun&oe=utf-8&rls=org.mozilla:en-US:official&client=firefox-a&um=1&ie=UTF-8&ei=BtcBS6fwIMy2lAefo_iTCw&sa=X&oi=video_result_group&ct=title&resnum=4&ved=0CBgQqwQwAw#
17. S. A. Kaufman, *Reinventing the Sacred* (Basic Books, 2008)20. Vision book
18. Richard Dawkins, *The Evidence of Evolution, The Greatest Show on Earth* (London: Transworld Publishers, 2009).
19. T. S. Kuhn, *The Structure of Scientific Revolutions*, 3rd.Ed. (University of Chicago Press, 1996).
20. E. J. Chaisson, *Cosmic Evolution* (Harvard University Press, 2001).

SCIENTIFIC ORTHODOXIES: MOVING CHALLENGE TOWARD REVOLUTION

GERALD H. POLLACK

Dept. of Bioengineering, Box 355061,
University of Washington, Seattle WA 98195, USA
E-mail: ghp@u.washington.edu

1. Introduction

Challenges to well-established scientific orthodoxies are thought to be rare. Particularly in fields beyond their own, scientists presume that any evidence seemingly in conflict with an established school of thought has been carefully vetted by relevant experts, who have appropriately modified their views and have thereby drawn ever closer to ground truth. Head-on challenges to established orthodoxies are presumed to be nil.

In fact, the situation is opposite. I present a representative sampling of the many foundational challenges that currently pervade today's science. Each one arises out of some senior investigator's lifetime effort, detailed not only in multiple papers but also in one or more books. Although only a minority of scientists embraces each challenge, that minority often includes people of some distinction who are moved by the evidence.

However, the majority of the orthodox community sets itself apart from these challenges. It collectively dismisses each one as unworthy, and considers the scientists advancing them to be naïve, stubborn, and/or misguided. And perhaps not surprisingly, the challenge areas themselves have become out of bounds to investigation — as though the challenger had somehow tainted his/her sphere of investigation with a deadly plague.

Stumbling upon these diverse challenges has come about largely because of the unconventional nature of my own research. Colleagues seem perpetually eager to alert me to others who have followed along similarly unconventional lines. Some of those challenges have struck me as unworthy, others potentially interesting, still others compelling enough to make me sit up and think. A few have come from scientists of great distinction: Linus Pauling had unconventional views on the role of Vitamin C, and both he and Albert Szent Györgyi had unconventional views on cancer. Multiple Nobel Prizes notwithstanding, skeptical mainstream scientists dismissed these proposals as forays into the improbable, in much the way that today's scientists dismiss the challenges described below.

But such challenges do occasionally prevail. A stunning example is the concept of "jumping genes" pioneered by Barbara McClintock. Shunned by her colleagues and left in isolation for decades, McClintock was ultimately vindicated in the early 1980s with the award of a Nobel Prize. Her work on transposons was finally recognized as one of the momentous discoveries of modern biology. Even in the popular literature, McClintock now stands as a symbol of dissent against an intolerant scientific establishment.

Another example is Peter Mitchell. Mitchell won the Nobel Prize in 1978 for chemi-osmosis, which describes how cells transduce energy in the form of ATP. Like McClintock, Mitchell was for many years dismissed by mainstream scientists and ridiculed as an eccentric iconoclast. Yet, he could press on in his home laboratory because of his wealth. Eventually, his view prevailed, and Mitchell became something of a scientific folk hero.

Whether any such scenarios could play out in a similar way today is an issue I consider at the end. Although iconoclasts of the past could prevail in the right circumstances, times have changed: the culture has become more conservative. It has become less tolerant of alternative views. This dismissive attitude is reinforced by the scientific system itself, which poses unintended obstacles, whose nature should become clear from the examples presented below — as should the route through which these obstacles can be overcome. Simple measures, albeit perhaps less than obvious *a priori*, could open the door to the long-awaited conceptual leaps.

2. A Sampling of Challenges to Orthodox Views

The three representative examples presented below all lie within the field of cell biology, the field I know best. All deal with foundational concepts.

2.1 Nerve Conduction Arises from a Phase Transition in the Peripheral Cytoskeleton

Trained in the laboratory of Alan Hodgkin, who won the Nobel Prize jointly with Sir Andrew Huxley for their work on nerve conduction, Ichiji Tasaki left that laboratory in despair. He saw evidence that the Hodgkin-Huxley interpretation could not be correct, and spent his career at the NIH replacing their mechanism with an alternative mechanism supported by more evidence. Actively working on a daily basis until his recent passing at age 96, he continued to produce two or three papers per year in addition to the several books he had written. In deference to his superior scientific attributes, Japanese scientists for many years referred to him as "the emperor." Tasaki was evidently a major player.

Tasaki's thesis is that the nerve impulse has nothing to do with currents driven by concentration gradients across the cell membrane. It has to do with the peripheral cystoskeleton, a gel-like layer of tissue that envelops the nerve cell (and other cells), just beneath the cell membrane. Tasaki's experiments have shown that the cytoskeleton undergoes a reversible transition from a condensed state to an expanded state. In so doing, it abruptly changes the cell's potential. This change propagates along the cytoskeleton, creating the propagating signal known as the action potential — the basis for cell signaling.

Tasaki did not advertise himself. Those willing to invest the time to read his books/papers will see what seems a compelling presentation of supporting evidence. Whether he is right is not clear, but what is clear is that his arguments are at least as cogent as those of the prevailing orthodoxy. Yet, with the overwhelming dominance of the membrane-channel orthodoxy, Tasaki's approach has drawn progressively less attention, and is now all but forgotten — except for a few scattered and largely unsupported groups springing up to revisit his hypothesis with more modern approaches. If Tasaki is correct, much of neurobiology will be revolutionized.

2.2 The Cell Membrane Is Less Important Than Generally Thought

According to prevailing views, much of the cell's functional machinery is controlled by elements lying within the cell membrane. Among the most significant of these are membrane pumps and channels. Pumps drive ions into or out of the cell against their concentration gradients in order to maintain the required ion balance. Channels allow ions to pass selectively down their respective concentration gradients into or out of the cell. With more and more of the cell's machinery revealed to lie in this vital envelope, the membrane has become critical to almost every aspect of cell function — so much so that any compromise of its integrity is anticipated to spell instant disaster for the cell.

Not so, says Gilbert Ling. Ling's five books spanning almost a half-century offer numerous reasons why these views may be fundamentally flawed. One of the most compelling of them is the evidence that membrane pumps require more power than the cell could conceivably muster. At present, more than 100 different pumps have been identified, with new ones continually added to the list. The first one to be identified was the sodium pump, which consumes about 1/3 of the cell's energy. For many scientists this figure has seemed surprisingly large, just for pumping an ion. But according to Ling, that's just the beginning. If one adds to this figure the energy required to power the 100+ additional pumps that have since been discovered, one is inevitably led to the conclusion that for ion pumping alone, the cell needs more energy than is actually available.

Ling deals with this anomaly by suggesting the "pump" proteins do not really pump. Their role is different. He argues that pumping is unnecessary — that the observed partitioning of ions and solutes between inside and outside of the cell occurs naturally, as a consequence of the physical attributes of the cytoplasm. His "association-induction" hypothesis builds on this idea, and purports to explain diverse fundamentals of cell function ranging from cellular potentials to enzyme action.

Reading any one of his books (e.g., *A Revolution in the Physiology of the Living Cell*, Krieger, Malabar, FL, 1992; *Life at the Cell and Below-Cell Level,* Pacific Press, NY, 2001) reveals the wealth of evidence that Ling brings to these issues, as well as the logic of his arguments. His books are not effortless reading; nevertheless, the strength of his arguments emerges lucidly for those willing to make the investment.

Ling has not had an easy time of it. Chosen to be among the first group of three Chinese scholars to study in the US after the Second World War (one of whom, Yang, won the Nobel Prize), Ling quickly gained recognition by perfecting the glass microelectrode, a seminal technique widely used in electrophysiology and predecessor to the patch clamp technique. (The latter has led to several Nobel Prizes.) More impacting have been his contributions in the conceptual arena. Because of them, he was chosen to be included in a select cohort along with Linus Pauling, destined to receive lifetime NIH support. Unfortunately for Ling (and several others), the program was quietly withdrawn when it became clear to the House Un-American Activities Committee that Pauling allegedly had communist leanings. Rather than singling out Pauling, the entire program was scuttled.

Ling's views had been actively debated until about 25 years ago, when the membrane-pump orthodoxy finally forced him out of business. He now resides within the Fonar Corporation, set up by his former collaborator Raymond Damadian. There he dwells with a skeleton staff of one, still generating papers and books, but lacking any serious influence on the field's direction. Except for a band of devotees, older scientists consider Ling's paradigm dead, while younger ones know nothing of it. Yet, it is a potentially transformative paradigm that, if right, could revolutionize cell biology at its core.

2.3 Biological Motion does not Arise from the Action of "motor proteins," but from Contraction of Protein Filaments

Current views on the molecular mechanism of muscle contraction derive from the work of Sir Andrew Huxley and Hugh Huxley (no relation), who published back-to-back papers in *Nature* more than half a century ago. They found that contraction arose out of interdigitating protein filaments aligned parallel to the muscle axis: sliding increased filament overlap, thereby accounting for muscle shortening. It was hypothesized that the increase of filament overlap was driven by the action of numerous molecular "motors" projecting comb-like from one filament and driving the other filament to slide past. This half-century old idea has held firm. Indeed, it has served as the prototype for understanding virtually all forms of biological motility.

Impressed by the elegance of the theory, I entered this field early in my career with considerable eagerness. I wanted to learn more about how it worked. But the myth of adequacy was soon shattered when a colleague, Tatsuo Iwazumi, showed me with a surprisingly simple argument that the proposed mechanism could not possibly work because it was intrinsically unstable; each time it was activated to contract, the molecular apparatus would fly apart. To this date, no serious response to Iwazumi's challenge has appeared, and a forthcoming book of his outlines attempts to fix matters by introducing rigorous physics into considerations of how muscle contracts.

A second shock came a few years later when I found an obscure paper showing that the filaments themselves might actually contract. How could this be? The prevailing mechanism rested on the sliding of constant-length filaments. That obscure paper just had to be wrong. Combing the literature, I was astonished to find reports of filament shortening in a substantial number of additional papers. The main exceptions, it seemed, were studies carried out by the two Huxleys, which seemed to show the opposite — although a careful reading seemed to leave open the possibility that they too may have found the same. For me, this was a turning point. My 1983 muscle-contraction review paper lists more than 30 studies on the issue of filament shortening published during the 30-year span between the emergence of the sliding filament theory and the preparation of the review. Those reports included studies from many laboratories using a variety of different techniques on a variety of different specimens. Almost all showed thick filament shortening — exactly the opposite of the prevailing view.

This story is laid out in my 1990 book, *Muscles and Molecules: Uncovering the Principles of Biological Motion*. Anyone can judge the evidence for him / herself. For the non-expert, more accessible may be the newer, 2001 book, *Cells, Gels and the Engines of Life*, which devotes a chapter to these issues within a larger context. If the arguments are valid, then the molecular basis of biological motion laid out in the textbook may be fundamentally flawed. Biological motion could arise from a different set of principles.

I have outlined three existing challenges to fundamental biological orthodoxies. A small minority of scientists considers each to be potentially valid, but the establishment accords them scant attention.

The three challenges are representative of a much larger number. I am aware of at least a half dozen more in closely allied biological fields. Still others are likely to exist in biological areas with which I'm less familiar, let alone in areas completely beyond my ken such as some of the various physical sciences. Additional ones are presented in the book edited by Mentre and Thomas, entitled *Scientists Talking to the Wind*. Thus, serious challenges to orthodox thinking easily number in the scores, probably in the hundreds. Nobody knows for sure, but the number is substantially larger than commonly thought.

The reason these potentially revolutionary challenges are not so broadly known is that they are seldom publicized. What orthodoxy would be motivated to publicize a serious challenge to its own authority? As a result, the messages do not easily get out, and people remain unaware. Contributing further to this lack of awareness is scientists' narrowness: So specialized have scientists become that seldom is there time left to invest in fields beyond one's own. Hence, it is little surprise that most scientists are unaware of challenges outside their immediate field, sometimes even those within it.

For any potentially meritorious challenge to gain strength, there must be the right *gestalt*, and the period several decades ago when the Mitchell/McClintock ideas began to gain momentum, intellectual contributions were considered the apogee of scientific achievement. Big brains counted. Prevailing orthodoxies certainly had power, but not enough to prevent revolutionary challenges with compelling experimental support to prevail. The cream eventually rose to the top.

But today things have changed. Dominating grant-review panels by virtue of their perceived expertise, orthodoxies now have the power to choke their challengers by cutting off their supply of funds. Whereas dissenters of the past could press on with their challenge because their work did not depend on the grant system — it could proceed either by dint of a wealthy benefactor, by support from family money, or by the fact that it simply did not require large sums of money — progress today depends critically on grants. Without money, little gets done. But the very people under challenge can cut the grant lifeline, for it is the orthodoxy that reviews the proposal of the challenger. In today's system, the defendant judges the plaintiff.

Out of such an oddly tilted system of judgment comes the inevitable default sentence: scientific death by exhaustion of funding. In ways never before imagined, orthodoxies have gained power to quash the opposition. It is no wonder that in recent times the bold conceptual advances that can turn fields upside down have become so rare.

3. Solution: Unorthodox Measures for Unorthodox Ideas

This counterproductive *gestalt* evidently needs remediation, and an element of the solution must involve reducing the power of prevailing orthodoxies to suppress the opposition. Dissenters must be granted the wherewithal to press on with their challenge. That much seems clear.

But merely enabling qualified dissenters would be an incomplete, even self-defeating, solution. As the empowered challenger's case is strengthened by fresh evidence, the establishment's resistance will persist — perhaps even intensify as the threat to their standing looms ever more serious. A few thoughtful scientists might buy in to the challenge, but those who reap benefits from their membership in the prevailing orthodoxy will be less likely to capitulate. When it is money and power that count most in today's scientific society, those who realize maximum benefits will not easily surrender their hard-earned positions. It's like imagining the dominant political leaders conceding that the other party may have a better plan, and offering them the opportunity to run the show. Not a very likely scenario.

Thus, no matter how compelling the enabled challenger's case may become, real gains of momentum seem unlikely. So long as the field's opinion makers retain the power to choose between anointing the challenge and maintaining their own standing, the choice will be easy. So long as the establishment can cling to their views with impunity, conceptual breakthroughs will rarely advance into the mainstream.

What is needed is a way of assuring that deserving challenges can gain momentum. Serious challenges are not in short supply; what is in short supply is their influence. The pathway to influence is blocked by the granting agencies themselves, who have inadvertently conferred power on the establishment to choke the opposition at every stage. If these agencies are to help revitalize science, measures may be needed to revise the power balance; establishments cannot be allowed to block the deserving challengers' gain of momentum and influence. The necessary measures to achieve this need not be complicated, although they would require willingness to try something new.

What I believe is needed is a scientific forum, within which the most significant challenge paradigms can compete openly with their orthodox counterparts. With competing schools of thought debated side-by-side in a civilized manner, the strengths and weaknesses of each side will soon become obvious and the truth will quickly emerge. Genuine advances will replace random walks.

To implement such a forum, *Step 1* would identify the most promising of opposing schools of thought. This could be done through solicitations. Responses to these solicitations would be evaluated by a body of reviewers selected not only for their grounding in mainstream science but also for their receptivity to unconventional thinking. The reviewers' fields would need to be close enough to the applicant's field for the arguments to be appreciated, but far enough to avert potential conflict-of-interest issues. Challenger and orthodoxy would be invited to present their arguments side-by-side, either orally, in writing, or both, and transcripts of the debate might be made available to the public on the web. The debate could quickly single out those challenges with exceptional promise.

Step 2 would provide funding. The funds would fuel the pursuit of those challenges with exceptional promise — money enough to fund *multiple* grants on each theme selected. With multiple grants, a cohort of talented investigators from diverse laboratories would be pursuing each promising challenge paradigm, creating momentum enough to demand the field's attention. In this way the challenge could be bootstrapped into real competition with the prevailing view. The two views would be debated side-by-side on equal footing. A winner should emerge in short order, and should the winner be the challenge paradigm, then this process will have brought a revolution.

A potentially effective way of accomplishing this goal is to set up a separate granting institute, chartered expressly to invest in creative ideas that challenge conventional views. The institute could be an independent arm of an existing agency, but I believe that a fully independent institute, unencumbered by past philosophies and procedures, would function best. The new institute would be charged with identifying promising *status-quo*-challenging ideas in all areas of science, and investing heavily in each one with multiple grants. A proposal of this nature has been delivered to President Obama for his consideration, as the "Institute for Venture Science."

I am convinced that a program of this sort could revitalize science. Scientific meetings could transform from routine to riveting. The challenging of staid paradigms could gain fresh legitimacy. An increasingly moribund scientific culture could be very quickly reinvigorated.

This program should yield significant outcomes. If a promising challenge paradigm proves flawed, as most will, it will die a natural death; money may seem to have been misspent, although having survived the challenge, the orthodox view will emerge newly strengthened. On the other hand, should a challenge paradigm be validated, the mill could quickly begin to churn. Scientists would soon be chasing questions never before dreamed of, with the very real prospect of breakthroughs never before imagined. In much the same way as quantum mechanics a century ago turned physics on its ear, validation of a challenge paradigm could turn today's science on its ear.

All of this could happen with some immediacy, for the challenges in question are not merely vague dreams of future creativity, but existing constructs just waiting in the wings for proper hearing. They are available here and now — ripe fruit waiting to be plucked.

4. Prospects

The main impediment to revolution is not the dearth of creative ideas; it is the reluctance of the entrenched scientific establishment to consider those ideas seriously. It is this sociological issue that needs to be addressed if revolutions are to come. Thus far, no granting institute anywhere in the world, to my knowledge, has grappled with this issue.

As outlined above, an "Institute for Venture Science" could be designed to deal with the natural resistance of the orthodoxy. In the same way as venture capitalists hedge their bets with multiple investments, science administrators could hedge their bets with similarly broad investments —in the form of "high-risk" scientific ventures. Most such ventures will fail, but the ones that succeed will bring the long-sought scientific revolutions. The key is to create critical masses of challenge investigators, compelling the orthodoxy to pay attention.

Creating such an institute would be a bold and unprecedented endeavor — inevitably subject to the kicking and screaming of those now in commanding scientific positions. Orthodoxies resist anything that smells of challenge. However, if science is to be revitalized, then something meaningful must be done, and I believe that the Institute for Venture Science, or equivalent, is the necessary vehicle.

DETERMINATION OF HOST-GUEST BINDING SITES FOR β-CYCLODEXTRIN URETHANE COPOLYMERS

LEE D. WILSON*, MOHAMED H. MOHAMED

Department of Chemistry, 110 Science Place (Rm. 156), University of Saskatchewan, Saskatoon, SK. Canada S7N 5C9

and

JOHN V. HEADLEY

Aquatic Ecosystems Protection Research Division, Water Science and Technology Directorate, 11 Innovation Boulevard, Saskatoon, SK. Canada S7N 3H5

BEST STUDENT POSTER PRESENTATION

In this paper, we propose a dye based spectrophometric technique as quantitative analytical method for estimating the available inclusion sites for β-cyclodextrin (β-CD) based urethane copolymers. The method relies on the absorbance changes (decolourization) of phenolphthalein (phth) in the visible region (λ=552 nm) at pH 10.5 in aqueous solution. Various types of urethane copolymers composed of β-CD and five different diisocyanate linkers (*i.e.*, 1,6-hexamethylene diisocyanate (HDI), 4,4'-dicyclohexyl diisocyanate (CDI), 4,4'-diphenylmethane diisocyanate (MDI), 1,4-phenylene diisocyanate (PDI), and 1,5-naphthalene diisocyanate (NDI)) at variable β-CD: linker reactant ratios; 1:1, 1:2, and 1:3, respectively. The dye decolourization studies provided estimates of the 1:1 polymer/dye binding constant (K_1), inclusion site accessibility, and the Gibbs free energy of sorption. It was concluded that the values of K_1 for the supramolecular polymers with readily accessible β-CD inclusion sites had comparable values of K_1 as compared with native β-CD. The accessibility of the β-CD inclusion sites for the copolymer materials ranged between 1-100% and the variability was attributed to steric effects in the annular hydroxyl region of the β-CD macrocycle. The Gibbs free energy of complex formation (ΔG^o) and the site occupancy (θ) of the inclusion sites for the copolymer materials was estimated independently using the Sips isotherm model. The ΔG^o values ranged between -27 to -30 kJ mol^{-1} and are in agreement with the Gibbs free energy for the 1:1 β-CD/phth complexes (-27 kJ mol^{-1}). The phth decolourization technique provides a simple, low cost, and versatile method for the estimation of the available inclusion sites in β-CD based urethane copolymer materials. This method is proposed to have extensive analytical applications in materials research and for the rational design of novel functional sorbent materials containing β-CD.

* Corresponding Author: lee.wilson@usask.ca

1. Introduction

Cyclodextrins (CDs) are cyclic carbohydrate compounds consisting of six, seven, or eight α-D-glucopyranose units connected by α-(1→4) linkages commonly referred to as α-, β-, and γ-CDs, respectively.[1] β-CD possesses a characteristic toroidal shape with a well-defined lipophilic cavity and a hydrophilic exterior that is suitable for the inclusion binding of appropriate sized guest compounds (*cf.* Scheme 1). CDs are of interest, in part, because of their ability to form stable inclusion complexes in aqueous solution.[2-7]

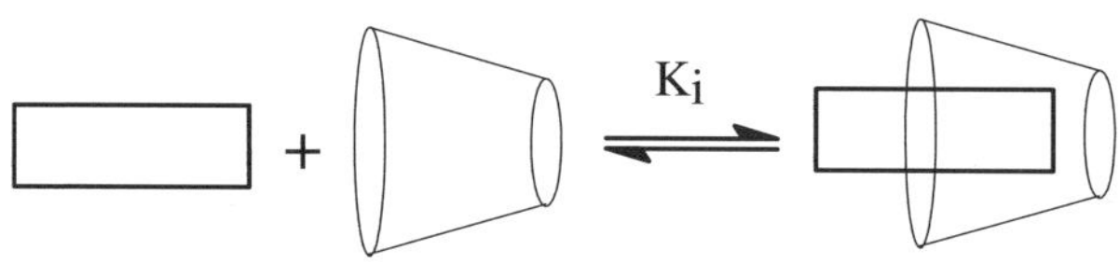

Scheme 1: The formation of an host-guest complex is shown for a β-CD (toroid) and a guest molecule (rectangle) according to an equilibrium process where K_i is the 1:1 equilibrium binding constant and the solvent has been omitted for clarity.

β-CD based copolymer materials are commonly prepared by crosslinking β-CD with a variety of appropriate linker molecules (*e.g.*, epichlorohydrin, glutaraldehye, succinyl chloride, diisocyanates, diacid chlorides, dicarboxylic acids, cyanuric chloride).[7-11] Such copolymer materials have been utilized for the sequestration of organic compounds from the gas and condensed phases. The relative accessibility of the β-CD inclusion sites is an important aspect in the design of β-CD based copolymer materials. The corresponding surface area and pore structure properties are important parameters known to affect the sorption properties of porous materials; particularly in the case of amorphous and non-templated copolymer materials.[12] The determination of the available inclusion binding sites is anticipated to provide an understanding of the sorption mechanism. In particular, the relative role of macrocyclic and linker domains in sorption remains an important question. Conventional nitrogen porosimetry provides estimates of the total surface accessible domains (micro- to macropores) because typical backfill gases (*e.g.*, N_2 or Ar) do not discriminate between the different adsorption sites provided there are no steric effect. In contrast, techniques such as XRD are less discriminatory and estimate the total surface accessible and inaccessible pore domains. Dye based sorption methods may provide complementary information about the surface area and pore structure properties of polymeric materials; however, the estimates refer only to the surface accessible domains. In particular, UV-Vis spectrophotometry[19-27] measures the absorbance changes of a suitable organic dyes such as phenolphthalein (phth), methyl orange (MO) and *p*-nitrophenol (PNP) in the presence of β-CD[20-29] exhibit specific molecular recognition properties according to the nature of the CD/dye inclusion complex. The 1:1 β-CD/phth inclusion complex has a relatively large binding constant ($K_1 \sim 10^4\ M^{-1}$)[20-25] as evidenced by its specific host-guest inclusion geometry.[3-5]

Recent studies have examined the interaction between phenolphthalein (phth) and β-CD copolymer materials; however, there were no detailed quantitative studies reported at the time of this publication[34-38]. Studies that utilize phth as a probe for the analysis of polymeric materials containing β-CD groups have been reported.[34-38] Topchieva *et al.*[34] estimated the number of binding sites for β-CD based nanotubes; however, the assumption of a non-zero molar absorptivity (ε) for the bound form of phth (*i.e.* β-CD/phth) in the equilibrium binding model contradicts the parameter estimates ($\varepsilon \approx 0$) reported by other researchers, *vide infra*. Velaz *et al.*[35] concluded that the changes in the absorbance of phth provided evidence of the limited accessibility of β-CD inclusion sites in their polymer materials. Fontanova *et al.* indicated that the accessibility and binding sites of the β-CD polymers may be estimated using the decolourization of phth[37] while Uyar *et al.* recently reported the degree of decolourization for composite polystryrene fibers.[38]

In this paper, we report a detailed quantitative study that provides estimates of the surface accessible β-CD inclusion sites for a series of urethane based copolymer materials using a dye based method. The materials investigated are urethane copolymers comprised of β-CD and several types of diisocyanate cross linker molecules at 1:1, 1:2, and 1:3 reactant (β-CD:linker) mole ratios, respectively. The diisocyanates are as follows: 1,6-hexamethylene diisocyanate (HDI), 4,4'-dicyclohexyl diisocyanate (CDI), 4,4'-diphenylmethane diisocyanate (MDI), 1,4-phenylene diisocyanate (PDI), and 1,5-naphthalene diisocyanate (NDI). We conclude that the 1:1 binding constants for the β-CD inclusion sites in copolymers that possess good surface accessibility are similar in magnitude to the values for native β-CD. The 1:1 binding constants decrease as the relative inclusion accessibility of β-CD decreases. Finally, the relative importance of inclusion binding sites vs. non-inclusion linker domains in the sorption mechanism are commented upon and the application of this dye based method are discussed.

2. Experimental Section

2.1. *Materials*

Phenolphthalein, phth (BDH Chemicals Ltd), sodium hydrogen carbonate (BDH Chemicals Ltd) and sodium hydroxide (Alfa Aesar) was used to prepare aqueous phth dye solutions. The materials and procedure for the synthesis of urethane based β-CD materials was described in a previous work with an additional Soxhlet extraction step after methanol with ethyl ether.[10] Polymers of β-cyclodextrin, β-CD (VWR) and D(+)-glucose monohydrate (EMD) were synthesized via cross linking reactions with diisocyanate cross linker molecules. These linkers include the following; HDI (Fluka), CDI (Aldrich), MDI (Aldrich), PDI (Aldrich) and NDI (TCI).

The spectroscopic characterization of the copolymer materials was described previously.[8,10] The content (w/w, %) of carbon (C), hydrogen (H), and nitrogen (N) was measured by Perkin Elmer 2400 CHN Elemental Analyzer with a detection limit ± 0.3%. The results were uncorrected according to the estimated water/solvent content. The presence of water/solvent mixtures in the polymers was confirmed by ^{1}H NMR (500 MHz Brüker), and ^{13}C NMR for water soluble and insoluble materials, respectively. The residual solvent content (*e.g.*, water and DMA) was estimated using a thermogravimetric analyzer, TGA (Q50 TA Instruments).

2.2. *Solution Preparation*

All solutions were prepared by volume in a 0.1 M sodium hydrogen carbonate buffer adjusted to pH 10.5 with 6 M sodium hydroxide. The concentration of phth (C_{phth}) was maintained at ~ 3.6 x 10^{-5} M in all experiments. A stock solution of phth in ethanol was made and aliquots were utilized to prepare aqueous solutions of phth in buffer. The ethanol/water (0.04%; v/v)[6] solution was used to increase the solubility of phth. All aqueous solutions were freshly prepared and run within 24 h to ensure that absorbance changes due to any instability of phth did not contribute to experimental artefacts. All absorption measurements were carried out at λ=552 nm and no change in the shape of the visible absorption band with increasing $C_{\beta\text{-CD}}$ was observed at this wavelength.

2.3. *Polymer Sorption*

7 mL of aqueous solution containing phth (~3.6 x 10^{-5} M) were added to vials containing variable mass amounts of sorbent (*e.g.*, β-CD, glucose, glucose polymer and CD polymers). The mixtures were shaken for 24 h, centrifuged

(Precision Micro-Semi Micro Centricone, Precision Scientific Co.) at 1550 rpm, and the absorbance of the supernatant was measured using a double beam spectrophotometer (Varian CARY 100) at room temperature (295 ± 0.5 K) to monitor the absorbance changes at λ_{max} of 552 nm.

2.4. *Data Analysis*

A non-linear least squares (NLLS) fitting procedure was used to determine the 1:1 equilibrium binding constants (K_1) between β-CD and phth.[6-7] The method utilizes the Beer-Lambert law and the assumption that the molar absorptivity of the β-CD/phth complex is zero.[2-4] The formation of the 1:1 complex between β-CD and phth at equilibrium, and the mass-balance relations are given below

$$\beta\text{-CD} + \text{phth} \overset{K_1}{\rightleftharpoons} \beta\text{-CD/phth} \quad (1)$$

$$[\text{phth}]_o = [\text{phth}] + [\text{CD-phth}] = [\text{phth}](1+K_1[\text{CD}]) \quad (2)$$

$$[\text{CD}]_o = [\text{CD}] + [\text{CD-phth}] \quad (3)$$

The terms $[CD]_0$, [CD], and [CD-phth] refer to the total, unbound, and 1:1 complexed forms of β-CD, respectively. The 1:1 equilibrium binding constant (K_1) for β-CD and phth is expressed as follows

$$[CD - phth] = [phth]_o \left(1 + \frac{1}{K_1[CD]}\right)^{-1} \quad (4)$$

The values for [CD-phth] and [phth] were obtained using the Beer-Lambert law for phth and Eq. (2-3). The estimates of K_1 using Eq. 4 employ the Beer-Lambert law for the unbound fraction of phth and the assumption that the molar absorptivity of the β-CD/phth complex is zero.[2-4] The criterion utilized for the best fit in the NLLS procedure involved the minimisation of the sums of the squares of the residuals (SSR) according to the relation,

$$SSR = \sum_i [(A_{calc})_i - (A_{expt})_i]^2,$$

where A_{calc} and A_{expt} are the calculated and experimental absorbance values, respectively.

The inclusion accessibility or site occupancy (θ) for phth onto the polymer framework and ΔG^o values for sorption (complex formation) were determined independently from the fit parameters, Q_m and K_{eq}, using the Sips equilibrium isotherm model[39-40] (*cf.* Eq. 5). The Sips isotherm accounts for the surface heterogeneity (*i.e.*, macrocyclic inclusion sites and linker domains) of the adsorbents and the interactions between phth and the copolymer adsorption sites.

This model provides a better fit over the Langmuir model and reaffirms that other adsorption sites (*i.e.*, linker domains) may contribute as well as the β-CD inclusion sites. The linker domains are not anticipated to be significant; however, they are taken into account using the Sips isotherm model.

$$Q_e = Q_m \frac{K_{eq} C_e^{\,n_s}}{1 + K_{eq} C_e^{\,n_s}} \qquad (5)$$

Q_e is the amount of phth adsorbed by the polymer (mol phth/g copolymer), Q_e is the maximum amount of phth adsorbed by the polymer, C_e is the equilibrium amount (M) of phth in aqueous solution, K_{eq} is the 1:1 equilibrium binding constant (M^{-1}), and n_s is the Sips constant (where $n_s>1$). In the case of homogeneous adsorption, it is worth to note that the Sips model converges with the Langmuir model when n_s is equal to unity. The inclusion accessibility or site occupancy (θ) of the dye and the corresponding value of ΔG^o for complex formation is defined as follows

$$\theta = \frac{Q_e}{Q_m} \qquad (6)$$

$$\Delta G^o = -RT \ln K_{eq} \qquad (7)$$

R is the gas constant ($Jmol^{-1}K^{-1}$), T is temperature in K, and K_{eq} is defined as in Eq. 5. The Sips parameters are forthwith interpreted in terms of the formation of 1:1 β-CD/phth inclusion complexes (*cf.* Eq. 1); where $K_{eq} \approx K_1$.

3. Results and Discussion

3.1. *Characterization of the CD polymers*

^{13}C solid state NMR and IR have been reported previously where the copolymer materials were fully characterized.[8] Elemental analysis provided the estimates of the linker composition since the N composition increased as the linker ratio increased, as shown in Table 1. The presence of residual water and/or solvent residues within the copolymer materials after extensive drying under vacuum were not accounted for because quantitative estimates of the relative amounts of solvents in the products were difficult to establish. However, the contribution of total solvent ranged between ~0.3 - 2% as indicated by TGA. The presence of water/solvent residues was confirmed with ^1H NMR where DMA and water spectral signatures were observed (results not shown). The presence of trace residual solvent residues was attributed to occlusion within the polymer framework during polymerization. The elemental analyses results in Table 1 for the theoretical composition of β-CD and glucose were corrected by varying the hydrate water content. Whilst the percentages of C and N increase as expected, H does not decrease as predicted and corrections due to presence of hydrate water provide better agreement between experimental and the calculated composition analyses.

3.2. *Sorption of Phenolphthalein*

A low ethanol composition (0.04%(v/v)) was chosen to eliminate possible interferences due to competitive binding by ethanol and to limit undue solvent effects.[6,41-43] The solution pH at 10.5 was chosen in order to optimize ε_{phth}, to provide greater sensitivity for absorption measurements, and to minimize the potential for deprotonation of the hydroxyl groups of β-CD ($pK_a \cong 12$). The equilibrium structures of the dianion of phth in aqueous solution are shown in Scheme 2.

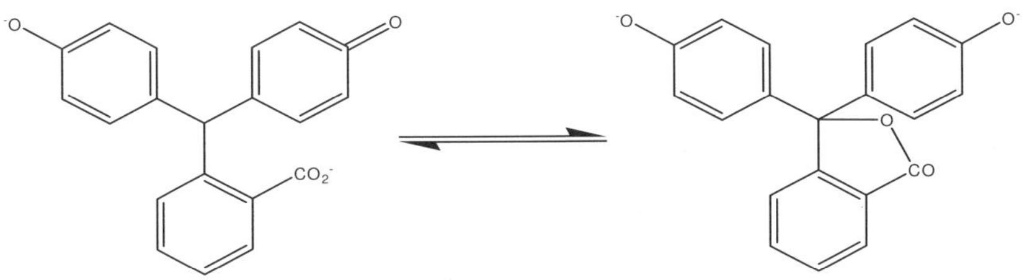

Scheme 2: The molecular structure of the two equilibrium forms of the red colored phenolphthalein dianion in aqueous solution at pH 10.5. The left hand structure represents the quinoid form and the right hand structure is the benzenoid dianion form.

Bertau *et al.*[44] previously studied the interaction of low molecular weight saccharides with phth and it was concluded that decolourization of phth was induced. Decolourization was attributed to non-specific enclathration and H-bonding interactions between the saccharides and phth. Such interference effects indicate the potential limitation of phth as a molecular probe for estimation of the inclusion binding sites. To evaluate these potential interferences, a glucose based urethane copolymer material was studied (*i.e.* glucose:CDI (1:3)) and the copolymer/dye interactions were studied. Figure 1 illustrates the change in absorbance (Abs) of phth against the mole concentration of glucose for a glucose:CDI (1:3) polymeric material. There is a minor but gradual decolourization of phth (~ 0.1 absorbance units) over a 10 mM concentration range. The weak interaction between phth and glucose:CDI (1:3) as compared with β-CD is supported by the lower magnitude of the 1:1 binding constant (K = 19.6 M^{-1}; *cf.* Table 2) and consistent with the observations noted by Bertau *et al.*[44]

Table 1. Elemental Analysis (C, H, N) Results for β-CD, Glucose, and the Corresponding Copolymers at 1:1, 1:2, and 1:3 β-CD:linker Reactant Ratios.

Material	Theoretical[1]			Experimental			Solvent/Water Mixture %
	%C	%H	%N	%C	%H	%N	
β-CD	38.4	6.90	0.00	38.2	6.81	0.00	11.6
Glucose	36.4	7.12	0.00	35.9	7.24	0.00	6.65
Glucose:CDI (1:3)	31.9	4.09	4.37	59.8	8.18	8.03	0.841
β-CD:HDI (1:1)	46.1	6.34	6.89	41.8	6.99	2.68	0.874
β-CD:HDI (1:2)	47.4	6.44	3.81	43.0	6.68	3.73	0.700
β-CD:HDI (1:3)	48.4	6.52	5.13	44.0	6.89	5.00	0.633
β-CD:CDI (1:1)	49.0	6.64	2.00	45.3	7.25	2.31	0.889
β-CD:CDI (1:2)	52.1	6.92	3.38	46.6	7.25	3.03	0.607
β-CD:CDI (1:3)	54.4	7.13	4.37	49.0	7.51	3.87	0.302
β-CD:MDI (1:1)	49.4	5.82	2.02	43.4	6.21	3.70	0.569
β-CD:MDI (1:2)	52.9	5.55	3.43	51.1	5.73	4.21	0.846
β-CD:MDI (1:3)	55.4	5.34	4.46	52.6	5.90	4.50	0.645
β-CD:PDI (1:1)	46.4	5.76	2.16	41.4	6.20	2.37	1.09
β-CD:PDI (1:2)	47.9	5.40	3.85	45.5	5.75	2.90	0.875
β-CD:PDI (1:3)	49.1	5.12	5.20	44.6	5.43	4.90	1.16
β-CD:NDI (1:1)	48.2	5.69	2.08	40.5	6.01	0.76	0.990
β-CD:NDI (1:2)	51.0	5.31	3.6	46.5	5.60	3.68	1.60
β-CD:NDI (1:3)	53.1	5.02	4.76	54.4	5.64	7.47	1.63

[1]Based on the synthetic feed ratios used in synthesis of polymer material. β-CD and glucose theoretical values were corrected for the amount of hydrate water whereas results for polymeric materials are uncorrected.

Figure 2 outlines the amount of phth removed (%) from solution by monomeric glucose and the glucose:CDI (1:3) copolymer. The attenuation of the absorbance of phth for each material is very similar and may be attributed to the presence of glucose. The mole content of pure glucose exceeds that of the glucose:CDI (1:3) copolymer and shows correspondingly greater decolourization. Thus, the results in Figs. 1-2 support that the interaction between glucose and phth attenuates the dye absorbance through a non-inclusion interaction; whereas, the linker (*i.e.* CDI) contributes negligibly.

Figure 3 illustrates a typical plot of absorbance *vs.* $C_{\beta\text{-CD}}$ for a fixed concentration of phth. The line through the data points represents the calculated absorbance values obtained from the NLLS fitting procedure (*cf.* Eq. 4). The sharp decrease in absorbance as the $C_{\beta\text{-CD}}$ increases represents the formation of the transparent β-CD/phth complex since its molar absorptivity is regarded as zero.[2-4]

The calculated lines represent the best-fit NLLS curve according to the 1:1 equilibrium binding model. The average 1:1 equilibrium binding constant (K_1) for the β-CD/phth complex binding is estimated to be $K_1=2.66\pm0.3 \times 10^4\ M^{-1}$ with a standard error of 7.0% from four independent experimental trials. The latter result is in good agreement with independent estimates.[1-3,23,24]

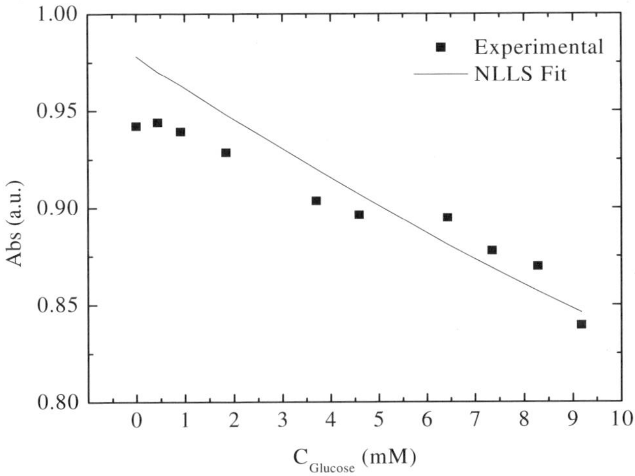

Figure 1: Phenolphthalein sorption with variable amount of insoluble Glucose:CDI (1:3) polymer at 295 K and pH 10.5 in 0.1 M $NaHCO_3$ buffer solution. The solid line refers to the NLLS best-fit according to Eq. 4.

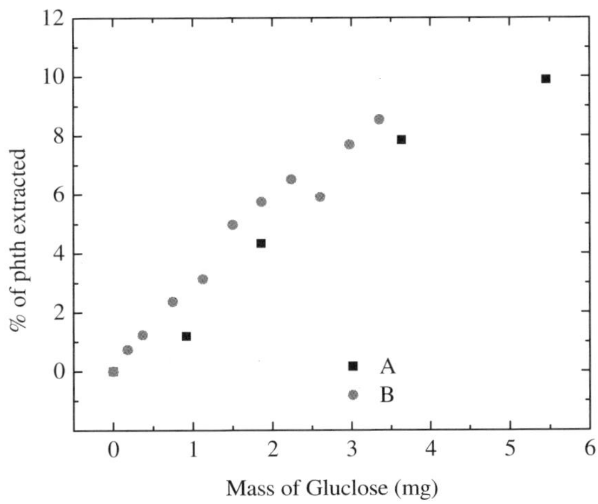

Figure 2: Phenolphthalein removal (decolourization) from solution using with variable amount of material at 295 K and pH 10.5 in 0.1 M $NaHCO_3$ buffer solution at fixed concentration of phth (3×10^{-5} M); A) D-(+)-Glucose, and B) Glucose:CDI((1:3) copolymer.

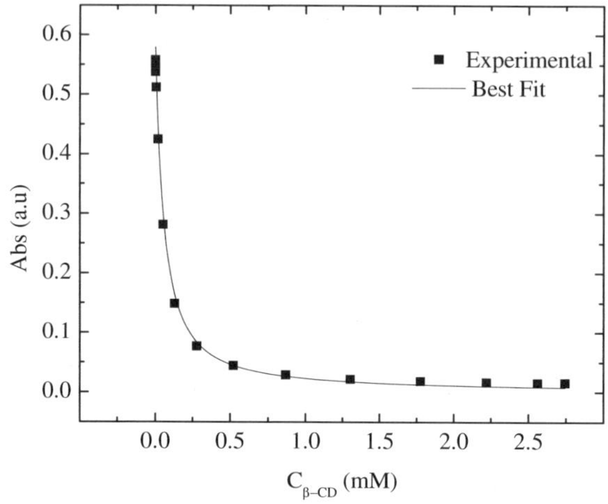

Figure 3: Absorbance (Abs) of phenolphthalein with variable concentration of β-CD ($C_{\beta\text{-CD}}$) in its native form in aqueous 0.1 M $NaHCO_3$ buffer solution at pH 10.5 and 295 K. The solid line refers to the NLLS best-fit according to Eq. 4.

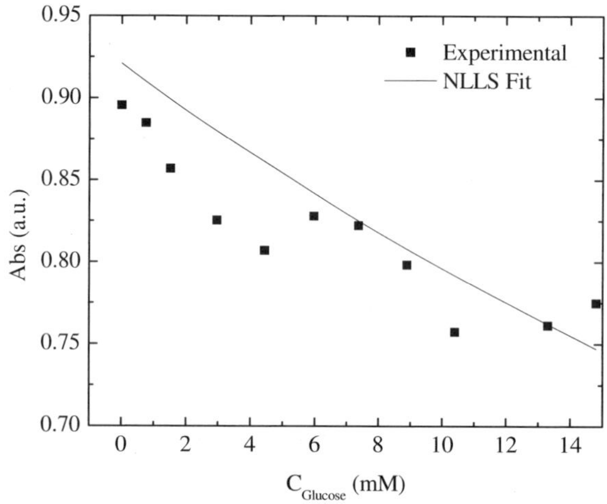

Figure 4: Phenolphthalein sorption with variable amount of glucose at 295 K and pH 10.5. The solid line refers to the NLLS best-fit according to Eq. 4.

Figure 4 represents a plot of Abs *vs.* $C_{Glucose}$ at a fixed concentration of phth. In contrast to the results obtained for β-CD, Figure 4 shows a more gradual decrease in absorbance for glucose in accordance with its lower binding constant, as compared with β-CD. The maximum absorbance change for glucose was ~15% and the 1:1 glucose/phth binding constant was 15.7 M^{-1}. This relatively weak non-inclusion binding is attributed to H-bonding and van der Waals interactions between glucose and phth.[42]

3.3. *Sorption of phth with CD polymers*

The decolourization of phth increases as the amount of CD polymer increased for a constant value of C_{phth}. The decolourization results provide support that phth forms 1:1 inclusion complexes between phth and β-CD within the copolymer framework. The results from Figure 2 indicate that H-bonding interactions between glucose and phth contribute to the observed decolourization effects;[40] however, the effect is small in comparison with the observed inclusion effects for native β-CD (*cf.* Figure 3).

Figure 5a-e illustrates the removal of unbound phth vs. the mass of β-CD urethane copolymer. In Fig. 5a, the β-CD:HDI copolymer (1:3) shows substantial removal of phth, as shown by the sharp increase in bound phth; whereas; the 1:2 and 1:3 materials are correspondingly less effective for the removal of phth from aqueous solution. The attenuated removal observed for the 1:3 copolymer is attributed to the decreased accessibility of the β-CD inclusion sites. At lower CD:linker ratios (*i.e.* 1:1) the greater decolourization of phth implies 1:1 complexes are formed more readily with the β-CD inclusion sites in the copolymer framework. An increasing cross link density may reduce the sorption properties of the copolymer materials for two possible reasons; *i*) steric crowding of the annular hydroxyl region of β-CD, and *ii*) limited access to the polymer framework and the β-CD inclusion sites. According to the space filling models presented by Taguchi,[4] the inclusion of the benzenoid form of phth (*cf.* Scheme 2) and H-bonding between the phenolate moieties of phth and the hydroxyl groups of β-CD are important and concluded result in the transparent 1:1 β-CD/phth dianion complex (*cf.* Scheme 2, ref. 4). The increasing presence of diisocyanate substituents in the annular hydroxyl region of β-CD inhibits the H-bonding interactions and inclusion binding of phth. Therefore, the decolourization of phth with highly cross linked copolymer materials is attenuated, as observed in Figure 5a-e. In addition, copolymers with greater cross link density may restrict the access of phth to the inclusion domains within the copolymer framework containing β-CD.

In Figure 5b, the β-CD:CDI (1:1) copolymer displays the greatest phth removal; whereas the 1:2 and 1:3 materials show attenuated effects. A comparison of CDI and HDI linkers indicate that the former is a bulkier linker molecule and the anticipated steric effects for CDI are more pronounced (*cf.* Fig. 5a). Figs. 5c-e illustrates the removal efficiency for three types of polymeric materials containing aromatic linker molecules (*e.g.*, MDI, PDI, and NDI) at variable reactant mole ratios. In each case, similar complexation of phth is observed for the 1:1 and 1:2 materials, whereas; complex formation is more pronounced in the case of the 1:3 copolymers. Steric effects between phth and the annular hydroxyl groups are anticipated to be greatest for the 1:3 copolymers, and this is consistent with the observed attenuation of the inclusion binding of phth. The copolymers with different profiles of bound phth are shown in Figs. 5a-e and illustrate the differences in the relative accessibility of the β-CD inclusion sites.

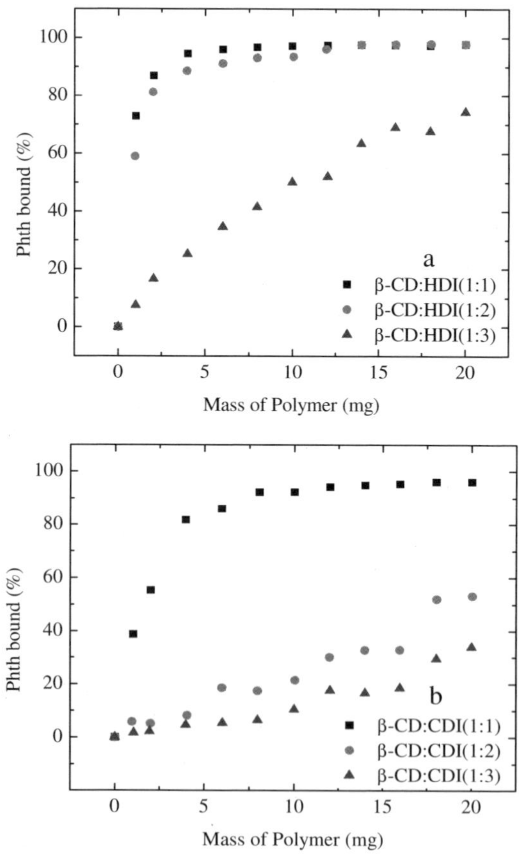

Figure 5 a-e: Percentage of phenolphthalein bound from aqueous solution with β-CD based copolymers vs. mass of polymer (mg) in 0.1 M $NaHCO_3$ buffer at pH 10.5 and 295K: a) β-CD:HDI, b) β-CD:CDI, c) β-CD:MDI, d) β-CD:PDI, and e) β-CD:NDI at 1:1, 1:2, and 1:3 reactant mole ratios, respectively.

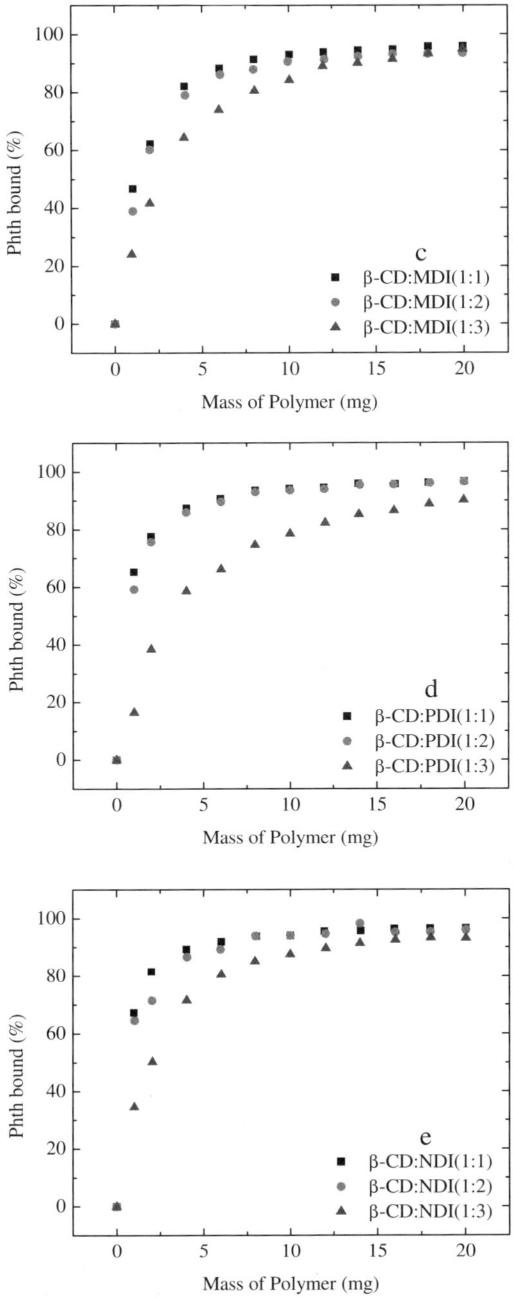

Figure 5 (Continued)

Figure 6a-e illustrates the change in absorbance (Abs) of phth vs $C_{\beta\text{-CD}}$ in β-CD copolymers at a fixed concentration of phth (C_{phth}). The solid lines through the experimental data represent the calculated Abs values according to the NLLS fitting routine according to Eq. 4. The decrease in Abs as $C_{\beta\text{-CD}}$ increases becomes more gradual as the linker ratio increases for all of the polymers studied. In the case of CDI copolymers (Fig. 6b), inclusion binding of phth is reduced even at intermediate (1:2) linker ratios. The greatest attenuation of phth binding is observed at the greatest (1:3) linker ratios for both the aliphatic (*i.e.* HDI and CDI) and aromatic (NDI, PDI, and MDI) linkers. The decolourization effect appears to correlate with the molecular size of the linker. The degree of decolourization of phth with the 1:1 copolymer materials are listed in descending order for each linker as follows: HDI>NDI>PDI>MDI≈CDI. Thus, the relative size of the linker affects the accessibility of the β-CD inclusion sites within the polymer framework. Figs. 6a-b illustrates the behaviour for polymers with aliphatic linkers, whereas; Figs. 6c-e represents results for aromatic linkers. It is important to note that the aliphatic and aromatic linkers are anticipated to exhibit structural differences because of their relative numbers of conformational isomers. Large, aromatic, rigid linkers, such as NDI and MDI, may form more open lattice structures in the copolymer framework. Aliphatic linkers such as HDI and CDI, may form collapsed and reduced lattice frameworks because these linkers have more conformational isomers resulting in a more compact pore network structure. This hypothesis is supported by the fewer accessible inclusion sites in copolymers with aliphatic vs. aromatic linkers (*cf.* Figs. 5-6).

Scheme 3 outlines the effect of increasing the substitution of the annular hydroxyl region of β-CD for a diisocyanate linker over the range of reactant mole ratios (1:1, 1:2, and 1:3 β-CD:linker) investigated. At the 1:1 ratio, the primary hydroxyl groups at the narrow end of the β-CD annulus are expected to react preferentially due to their greater reactivity, as compared with the secondary hydroxyl groups. In the case of bulky diisocyanates, the estimated degree of substitution of substituents in the annular hydroxyl region is estimated to be three per annular face (*cf.* Scheme 3) for bulky groups.[45-48] Once three substituents are added, steric crowding occurs at the narrow end of the β-CD annulus, and subsequent groups may react at the wider secondary annular region to overcome additional steric crowding at the primary hydroxyl region. Thus, the number of substituents at each face of the β-CD annulus is estimated ~ 3 substituents. Scheme 3 illustrates the distribution of substituents from left to right (a-c) in accordance with the CD:linker ratios from 1:1 to 1:3, respectively. The surface accessibility data for the inclusion sites (*cf.* Table 2) support this hypothesis and good accessibility is observed at the 1:1 linker ratio. According to Glazyrin *et al.*,[46] a study of various modified forms of β-CD indicated that a four- to fourty-fold decrease in the 1:1 binding constant was observed for acetylated and hydroxypropyl substituted β-CD with an average degree of substitution of ~ 3.8 and 3.6, respectively. They concluded that the binding of β-CD and some of its water-soluble derivatives may form complexes with phth depending on the structure, site of substitution, and degree of substitution at the C_2-, C_3-, and C_6- hydroxyl positions, respectively.

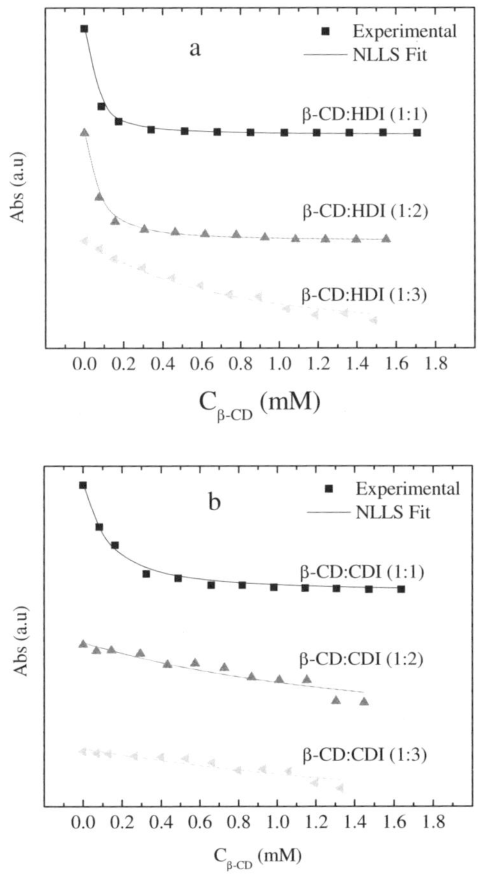

Figure 6 a-e: Absorbance changes for phenolphthalein as a function on increased mole content of β-CD ($C_{\beta\text{-CD}}$) for various insoluble copolymers at variable synthetic feed ratios at pH 10.5 in 0.1 M $NaHCO_3$ and 295K: a) β-CD:HDI, b) β-CD:CDI, c) β-CD:MDI, d) β-CD:PDI, and e) β-CD:NDI. The solid line refers to the NLLS best-fit according to Eq. 4 where K_1 = 2.66 $\times 10^4$ M^{-1} and the fraction bound of β-CD (CD-phth) is an adjustable parameter between 0 to 100 mol%.

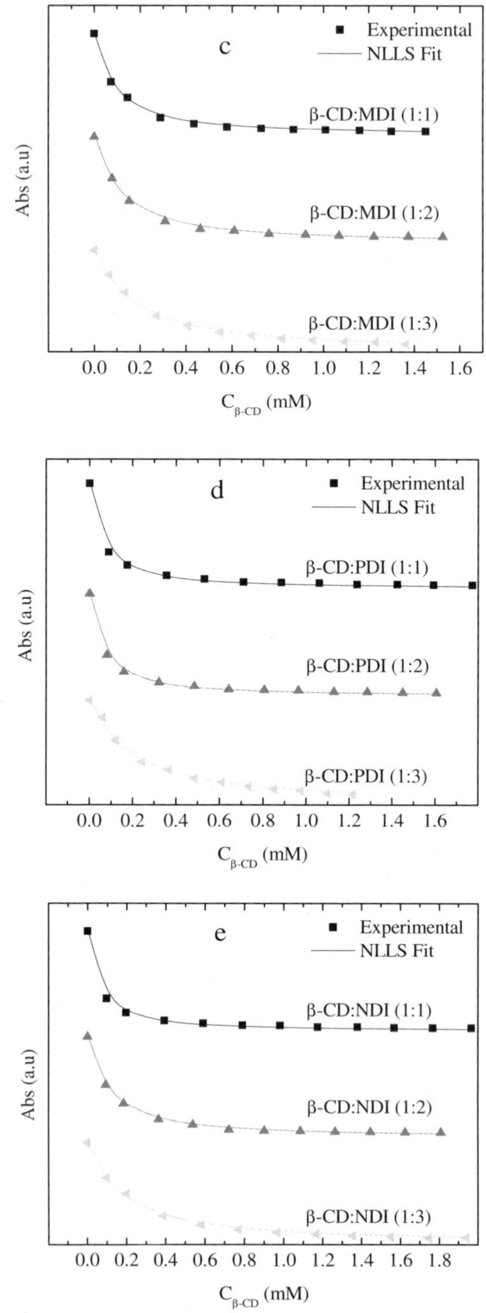

Figure 6 (Continued)

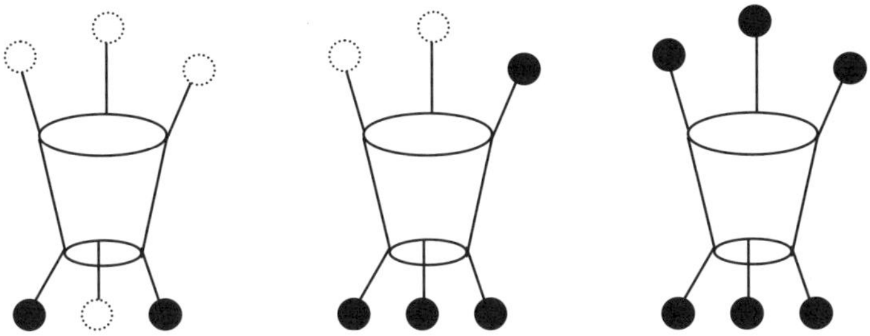

Scheme 3: Illustration of the sites of substitution of diisocyanate cross linker units to the primary (narrow end) and secondary (wide end) hydroxyl groups in the annular regions of β-CD: a) β-CD:linker (1:1), and b) β-CD:linker (1:2), and c) β-CD:linker (1:3) reactant ratios. The solid spheres represent covalently attached sites and open spheres represent available (unreacted) sites.

3.4. *Accessibility of β-CD Inclusion Binding Sites in the Polymers*

The estimation of accessible inclusion binding sites of β-CD requires certain assumptions: *i*) polymers possess variable accessibility due to differences in the cross link density, *ii*) phth binds exclusively to β-CD as a 1:1 inclusion complex, and *iii*) the 1:1 binding constant is assumed similar to that of native β-CD in the absence of steric effects. The third assumption is supported from experiments that compare the binding between glucose and glucose based polymers. It was concluded that the urethane linkage does not markedly affect the magnitude of the 1:1 binding constant for the glucose-CDI copolymer. The presence of available hydroxyl groups strongly affects the decolourization of phth as observed in aqueous solutions containing glucose and are in agreement with the conclusions of Bertau *et al.*[42]

The percentage of the accessible β-CD was calculated by varying the percentage of bound β-CD (varying [CD-phth] in Eq. 2-4) for copolymers containing β-CD. The best-fit between the calculated Abs and experimental values (*cf.* Figure 6 a-e) provided estimates of the accessible inclusion sites. Equation 4 was applied using a fixed value for K_1 for the copolymer materials, where the value for 1:1 β-CD/phth complex is assumed to be similar to that for the copolymer/phth complex. Table 2 shows the estimates of the surface accessible β-CD inclusion sites for each copolymer at different values of cross link density. The relative surface accessibility of the inclusion sites decreases as the cross link density increases. The relative ordering of the steric effects at high cross link density correlates inversely with the size of the linker as follows: HDI> CDI> PDI > MDI ≈ NDI. Steric crowding and the reduction of the pore volume within the framework are anticipated to retard 1:1 inclusion complex formation between the β-CD sites and phth. For polymers with low cross link density, the β-CD:HDI (1:1) copolymer has reasonable surface accessibility whereas the β-CD:CDI (1:3) copolymer exhibits the lowest surface accessibility. The linker domains within the framework do not contribute to complex formation with phth to any appreciable extent (*cf.* Figure 2). The decolourization effect of the glucose-CDI polymer is less than 0.1% (*cf.* Table 2).

Table 2. Phenolphthalein Based Estimates (%) of the Surface Accessible β-CD Inclusion Sites[1] in β-Cyclodextrin Urethane Copolymer Materials.

Material	β-CD:linker Mole Ratio	β-CD (% mole content)	Accessible Inclusion Sites (% β-CD)
β-CD	-	100	100
Glucose	-	0	5.91×10^{-2}
Glucose:CDI	1:3	0	6.39×10^{-2}
β-CD:HDI	1:1	87.1	106
	1:2	77.1	66.7
	1:3	69.2	4.78
β-CD:CDI	1:1	81.2	33.2
	1:2	68.4	2.03
	1:3	59.1	1.04
β-CD:MDI	1:1	81.4	38.0
	1:2	69.4	31.6
	1:3	60.2	18.6
β-CD:PDI	1:1	87.6	68.3
	1:2	78.0	57.4
	1:3	70.3	14.1
β-CD:NDI	1:1	84.4	77.6
	1:2	73.0	39.2
	1:3	68.4	24.2

[1]Based on the assumption that the binding constant of the β-CD copolymer material/phth is similar to that of β-CD/phth where the magnitude of $K_1 = 2.66(0.3) \times 10^4\ M^{-1}$.

Table 3: Fractional Coverage, θ^a, at 1mg of the polymer and Gibbs free energy change[b], ΔG^o of complex formation with phth in aqueous solution at pH 10.5 and 295 K.

Polymer	Ratio β-CD:Linker	
	1:1 (θ, ΔG° kJ/mol)	**1:2 (θ, ΔG° kJ/mol)**
HDI	0.773, -29.7	0.741, -28.7
CDI	0.619, -27.0	NR[c]
MDI	0.838, -28.3	0.818, -28.3
PDI	0.633, -27.9	0.501, -27.3
NDI	0.598, -28.4	0.500, -26.7

θ^a – refer to Eq. 6, ΔG^{ob} –refer to Eq. 7, and NR[c] – not reported because the surface accessibility was relatively low resulting in poor fits according to the Sips isotherm model (*cf.* Eq. 5)

Table 3 lists the Gibbs free energy change of complex formation (ΔG^o) for copolymer/phth complexes. The relative similarity of the 1:1 binding constants for the native β-CD/phth and the copolymer/phth complexes suggest that the observed decolourization effects observed are attributed to the formation of inclusion complexes. The better fit obtained using the Sips model instead of the Langmuir ($n_s = 1$) model suggests that the linkers do affect the decolourization of phth; however, the formation of 1:1 inclusion complexes between phth and β-CD are the predominant mechanism. In Table 3, the site occupancy decreases as the cross linking density increases, and correlates with the decreased accessibility of the inclusion sites. The values of ΔG^o calculated depend on the relative site occupancy of the sorption sites and they are directly proportional to the amount of surface accessible β-CD sites, as outlined in Scheme 4. Moreover, the magnitude of ΔG^o for the copolymers is slightly greater than the ΔG^o of 1:1 β-CD/phth complexes (~ -25 kJ/mol).[6-7] The slight differences may be due to differences in hydration and steric effects of the annular hydroxyl groups for the β-CD copolymer materials because of cross linking effects, as suggested by Glazyrin *et al.*[46]

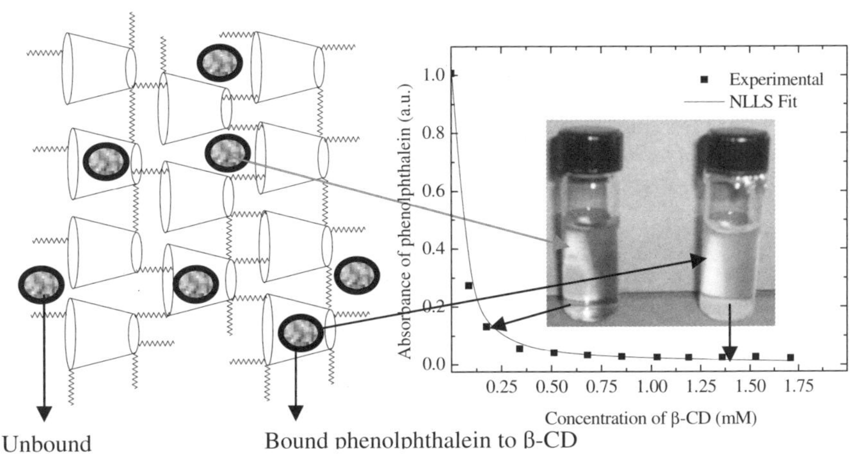

Scheme 4: A dye based decolourization method is proposed for the estimation of surface accessible inclusion binding sites for a series of β-Cyclodextrin based urethane copolymer materials.

4. Conclusions

In this research, we report the results of a dye-based study between water insoluble β-CD based copolymers and phenolphthalein in aqueous solution at pH 10.5 and 295 K. The copolymers in this study utilized relatively low

CD:linker ratios (1:1 to 1:3), and yet, the surface accessibility was remarkably different (*i.e.* 1-100%) for the five sets of copolymer sorbent materials. The Gibbs free energy change for the formation of 1:1 β-CD copolymer/phth complexes ranged from -27 to -30 kJ mol^{-1}, according to the Sips isotherm model.

This study represents the first systematic and quantitative analysis of inclusion sites in β-CD based urethane copolymers. This dye based sorption method represents a facile and relatively low cost analytical method for the study of β-CD copolymer sorbents. The formation of 1:1 β-CD/phth complexes in native β-CD and copolymer materials is highly selective towards phenolphthalein and provides useful quantitative results about the sorption properties of β-CD copolymer sorbents. Carbohydrate based copolymer materials are of increasing interest from both a basic research and an applications approach. The use of copolymer materials in sorption applications is important in chemical separations, controlled release, and environmental remediation of contaminated water. The dye based spectrophotometric technique described herein is anticipated to contribute towards the materials design of improved copolymer materials containing β-CD for applications that involve sorption based processes. We propose the use of phth as a versatile, low cost, chromophore probe for the analysis and design of novel types of β-CD based materials.

Acknowledgments

Financial assistance was provided by the Natural Sciences and Engineering Research Council and the Program of Energy Research and Development. M.H.M acknowledges the University of Saskatchewan for the award of a Graduate Teaching Fellowship and Environment Canada for the Science Horizons Program award. Mr. Joel Hardes is acknowledged for his technical assistance.

References

1. M. L. Bender, M. Komiyama, *Cyclodextrin Chemistry*; Springer-Verlag, Berlin 1978.
2. M. R. Eftink, M. L. Andy, K. Bystrom, H. D. Perlmutter, D. S. Kristol, *J. Am. Chem. Soc.* ***111***, 6765 (1989).
3. A. Buvari, J. Szejtli, L. Barcza, *J. Incl. Phenom.* ***1****(2),* 151 (1983).
4. K. Taguchi, *J. Am. Chem. Soc.*, ***108***, 2705 (1986).
5. M. E. Georgiou, C. A. Georgiou, M. A. Koupparis, *Anal. Chem.* ***67***, 114 (1995).
6. L. D. Wilson, S.R. Siddall, R.E. Verrall, *Can. J. Chem.* ***75****(7)*, 927 (1997).

7. M. H. Mohamed, L. D. Wilson, J. V. Headley, K. M. Peru, *Can. J. Chem.* ***87****(9)*, in Press (2009).
8. G. Wenz, *Angew. Chem. Int. Ed. Eng.* ***33***, 803 (1994).
9. G. Crini, S. Bertini, G. Torri, A. Naggi, D. Sforzini, C. Vecchi, L. Janus, Y. Lekchiri, M. Morcellet, *J App Poly Sci.* ***68***, 1973 (1998) .
10. M. H. Mohamed, L. D. Wilson, J. V. Headley, K. M. Peru, *IChemE: Process Safety and Environment Protection.* ***86***, 237 (2008).
11. A. Harada, A. Hashidzume, Y. Takashima, *Adv Polym Sci.* ***201***, 1 (2006).
12. G. Crini, *Prog. Polym. Sci.* ***30****(1)*, 38 (2005).
13. D. J. Wood, F. E. Hruska, W. Saenger, *J. Amer. Chem. Soc.* ***99***, 1735 (1977).
14. E. Junquera, G. Tardajos, E. Aicart, *Langmuir.* ***9****(5)*, 1213 (1993).
15. E. Saint Aman, D. A. Serve, *J. Colloid Interface Sci. 138(2)*, 365(1990).
16. U. R. Dharmawardana, S. D. Christian, E. E. Tucker, R. W. Taylor, J. F. Scamehorn, *Langmuir* ***9****(9)*, 2258 (1993).
17. W. M. Z. Wan Yunus, J. Taylor, D. M. Bloor, D. G. Hall, E. J. Wyn-Jones, *Phys. Chem.* ***96****(22)*, 8979 (1992).
18. R. I. Gelb, L. M. Schwartz, B. Cardelino, D. A. Laufer, *Anal. Biochem. 103(2)*, 362 (1980).
19. S. Fourmentin, G. G. Surpateanu, P. Blach, D. Landy, P. Decock, G. Surpateanu, *J. Incl. Phenom. and Macrocycl. Chem.* **55***(3-4)*, 263 (2006).
20. A. Buvári, L. Barcza, *Inorg. Chim. Acta.* ***33***, L179 (1979).
21. L. A. Selvidge, M. R. Eftink, *Anal. Biochem.* ***154****(2)*, 400 (1986).
22. A. Buvári, L. Barcza, M. Kajtár, *J. Chem. Soc. Perkin Trans.* **2,** 1687 (1988).
23. K. J. Sasaki, S. D. Christian, E. E. Tucker, *Fluid Phase Equilibria.* **1989**, *49*, 281.
24. J. E. Gray, S. A. MacLean, V. C. Reinsborough, *Aust. J. Chem.* ***48***, 551 (1995).
25. B. Tutaj, A. Kasprzyk, J. Czapkiewicz, *J. Incl. Phenom. Macrocycl. Chem.* ***47****(3-4)*, 133 (2003).
26. M .M. Meier, M. T. B. Luiz, P. J. Farmer, B. Szpoganicz, *J of Incl. Phenom. and Macrocycl. Chem.* ***40***, 291 (2001).
27. J. C. Harrrison, M. R. Eftink, *Biopolym.* ***21****(6)*, 1153 (2004).
28. D. Landy, M. S. Fourmentin, G. Surpateanu, *J. Incl. Phenom. Macrocycl. Chem.* ***38****(1-4)*, 187 (2000).
29. I. Suzuki, A. Yamauchi, *J. Incl. Phenom. Macrocycl. Chem.* ***54****(3-4)*, 193 (2006).
30. R. H. de Rossi, O. F. Silva, R. V. Vico, C. Gonzalez, *J. Pure Appl. Chem. 81(4)*, 755 (2009).
31. L. Janus, G. Crini, V. El-Rezzi, M. Morcellet, A. Cambiaghi, G. Torri, A. Naggi, C. Vecchi, *React. Funct. Polym. 42*, 173 (1999).
32. V. Wintgens, C. Amiel, *Langmuir 21(24)*, 11455 (2005).

33. V. Burckbuchler, V. Wintgens, C. Leborgne, S. Lecomte, N. Leygue, D. Scherman, A. Kichler, C. Amiel, *Bioconj. Chem.* ***19****(12)*, 2311 (2008).
34. I. N. Topchieva, F. A. Kalashnikov, V. V. Spiridonov, A. B. Mel'kinov, G. E. Polushina, A. V. Lezov, *Doklady Chem.* ***390****(1-3)*, 115 (2003).
35. I. Velaz, J. R. Isasi, M. Sanchez, M. Uzquenda, G. Ponchel, *J. Incl. Phenom. Macrocycl. Chem.* ***57****(1-4)*, 65 (2007).
36. R. C. de Bergamasco, G. M. Zanin, F. F. de Moraes, *J. Incl. Phenom. Macrocycl. Chem.* ***57****(1-4)*, 75 (2007).
37. E. Fontananova, G. Di Profio, E. Curcio, L. Giorno, E. Drioli, *J. Incl. Phenom. Macrocycl. Chem.* ***57****(1-4)*, 537 (2007).
38. T. Uyar, R. Havelund, Y. Nur, J. Hacaloglu, F. Besenbacher, P. Kingshott, *J. Membrane Sci. 332(1-2)*, 129 (2009).
39. R. Sips, *J. Chem. Phys. 16*, 490 (1948).
40. Y. Liu, Y-J. Lio, *Sep. and Pur. Tech.* 61, 292 (2008).
41. C. Donze A. Chatjigakis, A. W. Coleman, *J. Inclusion Phenom. Mol. Recognit. Chem.* ***13****(2)*, 155 (1992).
42. I. M. Warner, J. M. Schuette, *Spectroscopic Studies in Cyclodextrin Solutions*; I. M. Warner, L. B. McGowan, Eds.; Advances in Multidimensional Luminescence. JAI Press: Greenwich, CT. **2**, 61 (1993).
43. J. M. Schuette, I. M. Warner, *Anal. Lett.* ***27****(6)*, 1175 (1994).
44. Bertau, M.; Jorg, G. *Bioorg. Med. Chem.* ***12***, 2973 (2004).
45. B. Martel, Y. Leckchiri, A. Pollet, M. Morcellet, *Eur. Polym. J.* 31(11), 1083 (1995).
46. A. E. Glazyrin, M. K. Grachev, G. I. Kurochkina, E. E. Nifant'ev, *Russ. J. Gen. Chem.* ***74****(12)*, 1922 (2004).
47. M. Weickenmeier, G. Wenz, *Macromol. Rapid Commun.* ***17****(10)*, 731 (1996).
48. C. Yuan, Z. Jin, Z.; X. Li. *Food Chem.* ***106****(1)*, 50 (2008).